Graduate Texts in Mathematics 23

Springer
New York
Berlin
Heidelberg
Barcelona
Budapest
Hong Kong
London
Milan
Paris
Tokyo

Graduate Texts in Mathematics

continued after index

Werner Greub

Linear Algebra

Fourth Edition

With 5 Figures

Springer

Werner H. Greub
(Deceased)

Mathematics Subject Classifications: 15-01, 15A03, 15A06, 15A18, 15A21, 16-01

Library of Congress Cataloging in Publication Data

Greub, Werner Hildbert, 1925–
Linear algebra
(Graduate texts in mathematics; v. 23)
Bibliography: p. 445
1. Algebras, Linear. I. Title. II. Series.
QA184.G7313 1974 512′.5 75-19560

Printed and bound by Braun-Brumfield, Ann Arbor, MI.
Printed in the United States of America.

9 8 7 6 5 4

ISBN 0-387-90110-8 Springer-Verlag New York Berlin Heidelberg
ISBN 3-540-90110-8 Springer-Verlag Berlin Heidelberg New York

To Rolf Nevanlinna

Preface to the fourth edition

This textbook gives a detailed and comprehensive presentation of linear algebra based on an axiomatic treatment of linear spaces. For this fourth edition some new material has been added to the text, for instance, the intrinsic treatment of the classical adjoint of a linear transformation in Chapter IV, as well as the discussion of quaternions and the classification of associative division algebras in Chapter VII. Chapters XII and XIII have been substantially rewritten for the sake of clarity, but the contents remain basically the same as before. Finally, a number of problems covering new topics – e.g. complex structures, Caylay numbers and symplectic spaces – have been added.

I should like to thank Mr. M.L. Johnson who made many useful suggestions for the problems in the third edition. I am also grateful to my colleague S. Halperin who assisted in the revision of Chapters XII and XIII and to Mr. F. Gomez who helped to prepare the subject index.

Finally, I have to express my deep gratitude to my colleague J.R. Vanstone who worked closely with me in the preparation of all the revisions and additions and who generously helped with the proof reading.

Toronto, February 1975 WERNER H. GREUB

Preface to the third edition

The major change between the second and third edition is the separation of linear and multilinear algebra into two different volumes as well as the incorporation of a great deal of new material. However, the essential character of the book remains the same; in other words, the entire presentation continues to be based on an axiomatic treatment of vector spaces.

In this first volume the restriction to finite dimensional vector spaces has been eliminated except for those results which do not hold in the infinite dimensional case. The restriction of the coefficient field to the real and complex numbers has also been removed and except for chapters VII to XI, § 5 of chapter I and § 8, chapter IV we allow any coefficient field of characteristic zero. In fact, many of the theorems are valid for modules over a commutative ring. Finally, a large number of problems of different degree of difficulty has been added.

Chapter I deals with the general properties of a vector space. The topology of a real vector space of finite dimension is axiomatically characterized in an additional paragraph.

In chapter II the sections on exact sequences, direct decompositions and duality have been greatly expanded. Oriented vector spaces have been incorporated into chapter IV and so chapter V of the second edition has disappeared. Chapter V (algebras) and VI (gradations and homology) are completely new and introduce the reader to the basic concepts associated with these fields. The second volume will depend heavily on some of the material developed in these two chapters.

Chapters X (Inner product spaces) XI (Linear mappings of inner product spaces) XII (Symmetric bilinear functions) XIII (Quadrics) and XIV (Unitary spaces) of the second edition have been renumbered but remain otherwise essentially unchanged.

Chapter XII (Polynomial algebra) is again completely new and developes all the standard material about polynomials in one indeterminate. Most of this is applied in chapter XIII (Theory of a linear transformation). This last chapter is a very much expanded version of chapter XV of the second edition. Of particular importance is the generalization of the results in the second edition to vector spaces over an arbitrary coefficient field of characteristic zero. This has been accomplished without reversion to the cumbersome calculations of the first edition. Furthermore the concept of a semisimple transformation is introduced and treated in some depth.

One additional change has been made: some of the paragraphs or sections have been starred. The rest of the book can be read without reference to this material.

Last but certainly not least, I have to express my sincerest thanks to everyone who has helped in the preparation of this edition. First of all I am particularly indebted to Mr. S. HALPERIN who made a great number of valuable suggestions for improvements. Large parts of the book, in particular chapters XII and XIII are his own work. My warm thanks also go to Mr. L. YONKER, Mr. G. PEDERZOLI and Mr. J. SCHERK who did the proofreading. Furthermore I am grateful to Mrs. V. PEDERZOLI and to Miss M. PETTINGER for their assistance in the preparation of the manuscript. Finally I would like to express my thanks to professor K. BLEULER for providing an agreeable milieu in which to work and to the publishers for their patience and cooperation.

Toronto, December 1966 WERNER H. GREUB

Preface to the second edition

Besides the very obvious change from German to English, the second edition of this book contains many additions as well as a great many other changes. It might even be called a new book altogether were it not for the fact that the essential character of the book has remained the same; in other words, the entire presentation continues to be based on an axiomatic treatment of linear spaces.

In this second edition, the thorough-going restriction to linear spaces of finite dimension has been removed. Another complete change is the restriction to linear spaces with real or complex coefficients, thereby removing a number of relatively involved discussions which did not really contribute substantially to the subject. On p. 6 there is a list of those chapters in which the presentation can be transferred directly to spaces over an arbitrary coefficient field.

Chapter I deals with the general properties of a linear space. Those concepts which are only valid for finitely many dimensions are discussed in a special paragraph.

Chapter II now covers only linear transformations while the treatment of matrices has been delegated to a new chapter, chapter III. The discussion of dual spaces has been changed; dual spaces are now introduced abstractly and the connection with the space of linear functions is not established until later.

Chapters IV and V, dealing with determinants and orientation respectively, do not contain substantial changes. Brief reference should be made here to the new paragraph in chapter IV on the trace of an endomorphism – a concept which is used quite consistently throughout the book from that time on.

Special emphasis is given to tensors. The original chapter on Multilinear Algebra is now spread over four chapters: Multilinear Mappings (Ch. VI), Tensor Algebra (Ch. VII), Exterior Algebra (Ch. VIII) and Duality in Exterior Algebra (Ch. IX). The chapter on multilinear mappings consists now primarily of an introduction to the theory of the tensor-product. In chapter VII the notion of vector-valued tensors has been introduced and used to define the contraction. Furthermore, a treatment of the transformation of tensors under linear mappings has been added. In Chapter VIII the antisymmetry-operator is studied in greater detail and the concept of the skew-symmetric power is introduced. The dual product (Ch. IX) is generalized to mixed tensors. A special paragraph in this chapter covers the skew-symmetric powers of the unit tensor and shows their significance in the characteristic polynomial. The paragraph "Adjoint Tensors" provides a number of applications of the duality theory to certain tensors arising from an endomorphism of the underlying space.

There are no essential changes in Chapter X (Inner product spaces) except for the addition of a short new paragraph on normed linear spaces. In the next chapter, on linear mappings of inner product spaces, the orthogonal projections (§ 3) and the skew mappings (§ 4) are discussed in greater detail. Furthermore, a paragraph on differentiable families of automorphisms has been added here.

Chapter XII (Symmetric Bilinear Functions) contains a new paragraph dealing with Lorentz-transformations.

Whereas the discussion of quadrics in the first edition was limited to quadrics with centers, the second edition covers this topic in full.

The chapter on unitary spaces has been changed to include a more thorough-going presentation of unitary transformations of the complex plane and their relation to the algebra of quaternions.

The restriction to linear spaces with complex or real coefficients has of course greatly simplified the construction of irreducible subspaces in chapter XV. Another essential simplification of this construction was achieved by the simultaneous consideration of the dual mapping. A final paragraph with applications to Lorentz-transformation has been added to this concluding chapter.

Many other minor changes have been incorporated – not least of which are the many additional problems now accompanying each paragraph.

Last, but certainly not least, I have to express my sincerest thanks to everyone who has helped me in the preparation of this second edition. First of all, I am particularly indebted to CORNELIE J. RHEINBOLDT who assisted in the entire translating and editing work and to Dr. WERNER C. RHEINBOLDT who cooperated in this task and who also made a number of valuable suggestions for improvements, especially in the chapters on linear transformations and matrices. My warm thanks also go to Dr. H. BOLDER of the Royal Dutch/Shell Laboratory at Amsterdam for his criticism on the chapter on tensor-products and to Dr. H. H. KELLER who read the entire manuscript and offered many important suggestions. Furthermore, I am grateful to Mr. GIORGIO PEDERZOLI who helped to read the proofs of the entire work and who collected a number of new problems and to Mr. KHADJA NESAMUDDIN KHAN for his assistance in preparing the manuscript.

Finally I would like to express my thanks to the publishers for their patience and cooperation during the preparation of this edition.

Toronto, April 1963 WERNER H. GREUB

Contents

Interdependence of Chapters

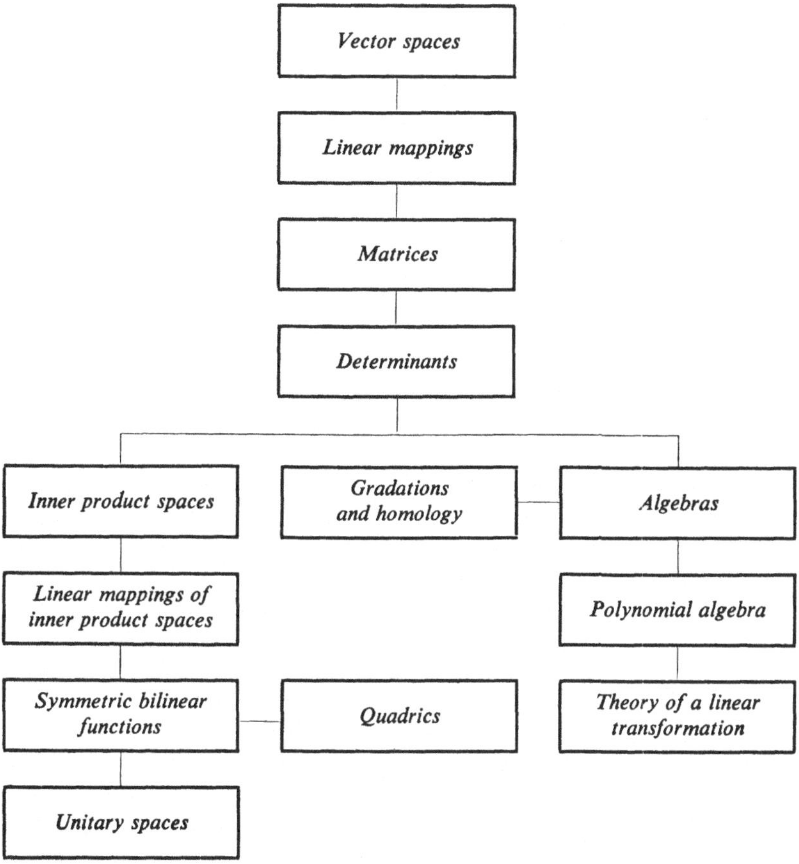

Chapter 0

Prerequisites

0.1. Sets. The reader is expected to be familiar with naive set theory up to the level of the first half of [11]. In general we shall adopt the notations and definitions of that book; however, we make two exceptions. First, the word *function* will in this book have a very restricted meaning, and what Halmos calls a function, we shall call a *mapping* or a *set mapping*. Second, we follow Bourbaki and call mappings that are one-to-one (onto, one-to-one and onto) injective (surjective, bijective).

0.2. Topology. Except for § 5 chap. I, § 8, Chap. IV and parts of chapters VII to IX we make no use at all of topology. For these parts of the book the reader should be familiar with elementary point set topology as found in the first part of [16].

0.3. Groups. A *group* is a set G, together with a binary law of composition

$$\mu: G \times G \to G$$

which satisfies the following axioms ($\mu(x, y)$ will be denoted by xy):

1. *Associativity:* $(xy)z = x(yz)$
2. *Identity:* There exists an element e, called the *identity* such that

$$xe = ex = x.$$

3. To each element $x \in G$ corresponds a second element x^{-1} such that

$$xx^{-1} = x^{-1}x = e.$$

The identity element of a group is uniquely determined and each element has a unique inverse. We also have the relation

$$(xy)^{-1} = y^{-1}x^{-1}.$$

As an example consider the set S_n of all permutations of the set $\{1 \ldots n\}$ and define the product of two permutations σ, τ by

$$(\sigma\tau)i = \sigma(\tau i) \qquad i = 1 \ldots n.$$

In this way S_n becomes a group, called the *group of permutations of n objects*. The identity element of S_n is the identity permutation.

Let G and H be two groups. Then a mapping

$$\varphi : G \to H$$

is called a *homomorphism* if

$$\varphi(x\,y) = \varphi x \, \varphi y \quad x, y \in G .$$

A homomorphism which is injective (resp. surjective, bijective) is called a monomorphism (resp. epimorphism, isomorphism). The inverse mapping of an isomorphism is clearly again an isomorphism.

A *subgroup* H of a group G is a subset H such that with any two elements $y \in H$ and $z \in H$ the product yz is contained in H and that the inverse of every element of H is again in H. Then the restriction of μ to the subset $H \times H$ makes H into a group.

A group G is called *commutative* or *abelian* if for each $x, y \in G$ $xy = yx$. In an abelian group one often writes $x+y$ instead of xy and calls $x+y$ the *sum* of x and y. Then the unit element is denoted by 0. As an example consider the set $\mathbb{Z}$ of integers and define addition in the usual way.

0.4. Factor groups of commutative groups.* Let G be a commutative group and consider a subgroup H. Then H determines an equivalence relation in G given by

$$x \sim x' \quad \text{if and only if} \quad x - x' \in H .$$

The corresponding equivalence classes are the sets $\{H + x\}$ and are called the *cosets* of H in G. Every element $x \in G$ is contained in precisely one coset $\bar{x}$. The set G/H of these cosets is called the *factor set* of G by H and the surjective mapping

$$\pi : G \to G/H$$

defined by

$$\pi x = \bar{x}, \qquad x \in \bar{x}$$

is called the *canonical projection* of G onto G/H. The set G/H can be made into a group in precisely one way such that the canonical projection becomes a homomorphism; i.e.,

$$\pi(x + y) = \pi x + \pi y . \tag{0.1}$$

To define the addition in G/H let $\bar{x} \in G/H$, $\bar{y} \in G/H$ be arbitrary and choose $x \in G$ and $y \in G$ such that

$$\pi x = \bar{x} \quad \text{and} \quad \pi y = \bar{y} .$$

*) This concept can be generalized to non-commutative groups.

Then the element $\pi(x+y)$ depends only on $\bar{x}$ and $\bar{y}$. In fact, if x', y' are two other elements satisfying $\pi x' = \bar{x}$ and $\pi y' = \bar{y}$ we have

$$x' - x \in H \quad \text{and} \quad y' - y \in H$$

whence

$$(x' + y') - (x + y) \in H$$

and so $\pi(x'+y') = \pi(x+y)$. Hence, it makes sense to define the sum $\bar{x}+\bar{y}$ by

$$\bar{x} + \bar{y} = \pi(x + y) \qquad \pi x = \bar{x}, \pi y = \bar{y}.$$

It is easy to verify that the above sum satisfies the group axioms. Relation (0.1) is an immediate consequence of the definition of the sum in G/H. Finally, since π is a surjective map, the addition in G/H is uniquely determined by (0.1).

The group G/H is called the *factor group of G with respect to the subgroup H*. Its unit element is the set H.

0.5. Fields. A *field* is a set Γ on which two binary laws of composition, called respectively addition and multiplication, are defined such that

1. Γ is a commutative group with respect to the addition.
2. The set $\Gamma - \{0\}$ is a commutative group with respect to the multiplication.
3. Addition and multiplication are connected by the *distributive law*,

$$(\alpha + \beta)\gamma = \alpha\gamma + \beta\gamma, \qquad \alpha, \beta, \gamma \in \Gamma.$$

The rational numbers $\mathbb{Q}$, the real numbers $\mathbb{R}$ and the complex numbers $\mathbb{C}$ are fields with respect to the usual operations, as will be assumed without proof.

A homomorphism $\varphi: \Gamma \to \Gamma'$ between two fields is a mapping that preserves addition and multiplication.

A subset $\Delta \subset \Gamma$ of a field which is closed under addition, multiplication and the taking of inverses is called a *subfield*. If Δ is a subfield of Γ, Γ is called an *extension field* of Δ.

Given a field Γ we define for every positive integer k the element $k\varepsilon$ (ε unit element of Γ) by

$$k\varepsilon = \underbrace{\varepsilon + \cdots + \varepsilon}_{k}$$

The field Γ is said to have *characteristic zero* if $k\varepsilon \neq 0$ for every positive integer k. If Γ has characteristic zero it follows that $k\varepsilon \neq k'\varepsilon$ whenever $k \neq k'$. Hence, a field of characteristic zero is an infinite set. Throughout this book it will be assumed without explicit mention that all fields are of characteristic zero.

For more details on groups and fields the reader is referred to [29].

0.6. Partial order. Let $\mathscr{A}$ be a set and assume that for some pairs X, Y ($X \in \mathscr{A}$, $Y \in \mathscr{A}$) a relation, denoted by $X \leqq Y$, is defined which satisfies the following conditions:

(i) $X \leqq X$ for every $X \in \mathscr{A}$ (Reflexivity)

(ii) if $X \leqq Y$ and $Y \leqq X$ then $X = Y$ (Antisymmetry)

(iii) If $X \leqq Y$ and $Y \leqq Z$, then $X \leqq Z$ (Transitivity).

Then $\leqq$ is called a *partial order* in $\mathscr{A}$.

A *homomorphism of partially ordered sets* is a map $\varphi: \mathscr{A} \to \mathscr{B}$ such that $\varphi X \leqq \varphi Y$ whenever $X \leqq Y$.

Clearly a subset of a partially ordered set is again partially ordered.

Let $\mathscr{A}$ be a partially ordered set and suppose $A \in \mathscr{A}$ is an element such that the relation $A \leqq X$ implies that $A = X$. Then A is called a *maximal element* of $\mathscr{A}$. A partial ordered set need not have a maximal element.

A partially ordered set is called *linearly ordered* or a *chain* if for every pair X, Y either $X \leqq Y$ or $Y \leqq X$.

Let $\mathscr{A}_1$ be a subset of the partially ordered set $\mathscr{A}$. Then an element $A \in \mathscr{A}$ is called an *upper bound* for $\mathscr{A}_1$ if $X \leqq A$ for every $X \in \mathscr{A}_1$.

In this book we shall assume the following axiom:

A partially ordered set in which every chain has an upper bound, contains a maximal element.

This axiom is known as Zorn's lemma, and is equivalent to the axiom of choice (cf. [11]).

0.7. Lattices. Let $\mathscr{A}$ be a partially ordered set and let $\mathscr{A}_1 \subset \mathscr{A}$ be a subset. An element $A \in \mathscr{A}$ is called a *least upper bound* (l.u.b.) for $\mathscr{A}_1$ if

1) A is an upper bound for $\mathscr{A}_1$.

2) If X is any upper bound, then $A \leqq X$. It follows from (ii) that if a l.u.b. for $\mathscr{A}_1$ exists, then it is unique.

In a similar way, lower bounds and the greatest lower bound (g.l.b.) for a subset of $\mathscr{A}$ are defined.

A partially ordered set $\mathscr{A}$ is called a *lattice*, if for any two elements X, Y the subset $\{X, Y\}$ has a l.u.b. and a g.l.b. They are denoted by $X \vee Y$ and $X \wedge Y$. It is easily checked that any finite subset $(X_1, \ldots, X_r)$ of a lattice has a l.u.b. and a g.l.b. They are denoted by $\bigvee_{i=1}^{r} X_i$ and $\bigwedge_{i=1}^{r} X_i$.

As an example of a lattice, consider the collection of subsets of a given set, X, ordered by inclusion. If U, V are any two subsets, then

$$U \wedge V = U \cap V \quad \text{and} \quad U \vee V = U \cup V.$$

Chapter I

Vector Spaces

§ 1. Vector spaces

1.1. Definition. A *vector (linear) space*, E, over the field Γ is a set of elements $x, y, \ldots$ called *vectors* with the following algebraic structure:

I. E is an additive group; that is, there is a fixed mapping $E \times E \to E$ denoted by

$$(x, y) \to x + y \tag{1.1}$$

and satisfying the following axioms:

I.1. $(x+y)+z=x+(y+z)$ (associative law)

I.2. $x+y=y+x$ (commutative law)

I.3. there exists a zero-vector 0; i.e., a vector such that $x+0=0+x=x$ for every $x \in E$.

I.4. To every vector x there is a vector $-x$ such that $x+(-x)=0$.

II. There is a fixed mapping $\Gamma \times E \to E$ denoted by

$$(\lambda, x) \to \lambda x \tag{1.2}$$

and satisfying the axioms:

II.1. $(\lambda\mu)x=\lambda(\mu x)$ (associative law)

II.2. $(\lambda+\mu)x=\lambda x+\mu x$

$\lambda(x+y)=\lambda x+\lambda y$ (distributive laws)

II.3. $1 \cdot x=x$ (1 unit element of Γ)

(The reader should note that in the left hand side of the first distributive law, + denotes the addition in Γ while in the right hand side, + denotes the addition in E. In the sequel, the name addition and the symbol + will continue to be used for both operations, but it will always be clear from the context which one is meant). Γ is called the *coefficient field* of the vector space E, and the elements of Γ are called *scalars*. Thus the mapping

(1.2) defines a multiplication of vectors by scalars, and so it is called *scalar multiplication.*

If the coefficient field Γ is the field $\mathbb{R}$ of real numbers (the field $\mathbb{C}$ of complex numbers), then E is called a real (complex) vector space. For the rest of this paragraph all vector spaces are defined over a fixed, but arbitrarily chosen field Γ of characteristic 0.

If $\{x_1, \ldots, x_n\}$ is a finite family of vectors in E, the sum $x_1 + \cdots + x_n$ will often be denoted by $\sum_{i=1}^{n} x_i$.

Now we shall establish some elementary properties of vector spaces. It follows from an easy induction argument on n that the distributive laws hold for any finite number of terms,

$$\left(\sum_{i=1}^{n} \lambda_i\right) \cdot x = \sum_{i=1}^{n} \lambda_i x$$

$$\lambda \cdot \sum_{i=1}^{n} x_i = \sum_{i=1}^{n} \lambda x_i$$

Proposition I: The equation

$$\lambda x = 0$$

holds if and only if

$$\lambda = 0 \quad \text{or} \quad x = 0 .$$

Proof: Substitution of $\mu = 0$ in the first distributive law yields

$$\lambda x = \lambda x + 0 x$$

whence $0x = 0$. Similarly, the second distributive law shows that

$$\lambda 0 = 0 .$$

Conversely, suppose that $\lambda x = 0$ and assume that $\lambda \neq 0$. Then the associative law II.1 gives that

$$1 \cdot x = (\lambda^{-1} \lambda) x = \lambda^{-1} (\lambda x) = \lambda^{-1} 0 = 0$$

and hence axiom II.3 implies that $x = 0$.

The first distributive law gives for $\mu = -\lambda$

$$\lambda x + (-\lambda) x = (\lambda - \lambda) x = 0 \cdot x = 0$$

whence

$$(-\lambda) x = -\lambda x .$$

In the same way the formula

$$\lambda(-x) = -\lambda x$$

is proved.

1.2. Examples. 1. Consider the set $\Gamma^n = \underbrace{\Gamma \times \cdots \times \Gamma}_{n}$ of n-tuples

$$x = (\xi^1, \ldots, \xi^n), \qquad \xi^i \in \Gamma$$

and define addition and scalar multiplication by

$$(\xi^1, \ldots, \xi^n) + (\eta^1, \ldots, \eta^n) = (\xi^1 + \eta^1, \ldots, \xi^n + \eta^n)$$

and

$$\lambda(\xi^1, \ldots, \xi^n) = (\lambda \xi^1, \ldots, \lambda \xi^n).$$

Then the associativity and commutativity of addition follows at once from the associativity and commutativity of addition in Γ. The zero vector is the n-tuple $(0, \ldots, 0)$ and the inverse of $(\xi^1, \ldots, \xi^n)$ is the n-tuple $(-\xi^1, \ldots, -\xi^n)$. Consequently, addition as defined above makes the set Γ^n into an additive group. The scalar multiplication satisfies II.1, II.2, and II.3, as is equally easily checked, and so these two operations make Γ^n into a vector space. This vector space is called the *n-space over* Γ. In particular, Γ is a vector space over itself in which scalar multiplication coincides with the field multiplication.

2. Let C be the set of all continuous real-valued functions, f, in the interval I : $\alpha \leqq t \leqq \beta$,

$$f : I \to \mathbb{R}.$$

If f, g are two continuous functions, then the function $f+g$ defined by

$$(f + g)(t) = f(t) + g(t)$$

is again continuous. Moreover, for any real number λ, the function λf defined by

$$(\lambda f)(t) = \lambda \cdot f(t)$$

is continuous as well. It is clear that the mappings

$$(f, g) \to f + g \quad \text{and} \quad (\lambda, f) \to \lambda \cdot f$$

satisfy the systems of axioms I. and II. and so C becomes a real vector space. The zero vector is the function 0 defined by

$$0(t) = 0$$

and the vector $-f$ is the function given by

$$(-f)(t) = -f(t).$$

Instead of the continuous functions we could equally well have considered the set of k-times differentiable functions, or the set of analytic functions.

3. Let X be an arbitrary set and E be a vector space. Consider all mappings $f: X \to E$ and define the sum of two mappings f and g as the mapping

$$(f+g)(x) = f(x) + g(x) \qquad x \in X$$

and the mapping λf by

$$(\lambda f)(x) = \lambda f(x) \qquad x \in X.$$

Under these operations the set of all mappings $f: X \to E$ becomes a vector space, which will be denoted by $(X; E)$. The zero vector of $(X; E)$ is the function f defined by $f(x) = 0$, $x \in X$.

1.3. Linear combinations. Suppose E is a vector space and $x_1, \ldots, x_r$ are vectors in E. Then a vector $x \in E$ is called a *linear combination* of the vectors x_i if it can be written in the form

$$x = \sum_{i=1}^{r} \lambda^i x_i, \quad \lambda^i \in \Gamma.$$

More generally, let $\{x_\alpha\}_{\alpha \in A}$ be any family of vectors. Then a vector $x \in E$ is called a *linear combination of the vectors* x_α if there is a family of scalars, $\{\lambda^\alpha\}_{\alpha \in A}$, only finitely many different from zero, such that

$$x = \sum_\alpha \lambda^\alpha x_\alpha$$

where the summation is extended over those α for which $\lambda^\alpha \neq 0$.

We shall simply write

$$x = \sum_{\alpha \in A} \lambda^\alpha x_\alpha$$

and it is to be understood that only finitely many λ^α are different from zero. In particular, by setting $\lambda^\alpha = 0$ for each α we obtain that the 0-vector is a linear combination of every family. It is clear from the definition that if x is a linear combination of the family $\{x_\alpha\}$ then x is a linear combination of a finite subfamily.

Suppose now that x is a linear combination of vectors x_α, $\alpha \in A$

$$x = \sum_{\alpha \in A} \lambda^\alpha x_\alpha, \quad \lambda^\alpha \in \Gamma$$

and assume further that each x_α is a linear combination of vectors $y_{\alpha\beta}$,

$\beta \in B$,

$$x_\alpha = \sum_\beta \mu_{\alpha\beta}\, y_{\alpha\beta}\,, \qquad \mu_{\alpha\beta} \in \Gamma\,.$$

Then the second distributive law yields

$$x = \sum_\alpha \lambda^\alpha x_\alpha = \sum_{\alpha,\beta} \lambda^\alpha \mu_{\alpha\beta}\, y_{\alpha\beta} = \sum_{\alpha,\beta} \varrho_{\alpha\beta}\, y_{\alpha\beta}\,, \qquad \varrho_{\alpha\beta} = \lambda^\alpha \mu_{\alpha\beta}$$

and hence x is a linear combination of the vectors $y_{\alpha\beta}$.

A subset $S \subset E$ is called a *system of generators* for E if every vector $x \in E$ is a linear combination of vectors of S. The whole space E is clearly a system of generators. Now suppose that S is a system of generators for E and that every vector of S is a linear combination of vectors of a subset $T \subset S$. Then it follows from the above discussion that T is also a system of generators for E.

1.4. Linear dependence. Let $\{x_\alpha\}_{\alpha \in A}$ be a given family of vectors. Then a non-trivial linear combination of the vectors x_α is a linear combination $\sum_\alpha \lambda^\alpha x_\alpha$ where at least one scalar λ^α is different from zero. The family $\{x_\alpha\}$ is called *linearly dependent* if zero is a non-trivial linear combination of the x_α; that is, if there exists a system of scalars λ^α such that

$$\sum_\alpha \lambda^\alpha x_\alpha = 0 \tag{1.3}$$

and at least one $\lambda^\alpha \neq 0$. It follows from the above definition that if a subfamily of the family $\{x_\alpha\}$ is linearly dependent, then so is the full family. An equation of the form (1.3) is called a *non-trivial linear relation.*

A family consisting of one vector x is linearly dependent if and only if $x = 0$. In fact, the relation

$$1 \cdot 0 = 0$$

shows that the zero vector is linearly dependent. Conversely, if the vector x is linearly dependent we have that $\lambda x = 0$ where $\lambda \neq 0$. Then Proposition I implies that $x = 0$.

It follows from the above remarks that every family containing the zero vector is linearly dependent.

Proposition II: A family of vectors $\{x_\alpha\}_{\alpha \in A}$ is linearly dependent if and only if for some $\beta \in A$, x_β is a linear combination of the vectors x_α, $\alpha \neq \beta$.

Proof: Suppose that for some $\beta \in A$,

$$x_\beta = \sum_{\beta \neq \alpha} \lambda^\alpha x_\alpha\,.$$

Then setting $\lambda^\beta = -1$ we obtain

$$\sum_\alpha \lambda^\alpha x_\alpha = 0$$

and hence the vectors x_α are linearly dependent.

Conversely, assume that

$$\sum_\alpha \lambda^\alpha x_\alpha = 0$$

and that $\lambda^\beta \neq 0$ for some $\beta \in A$. Then multiplying by $(\lambda^\beta)^{-1}$ we obtain in view of II.1 and II.2

$$0 = x_\beta + \sum_{\alpha \neq \beta} (\lambda^\beta)^{-1} \lambda^\alpha x_\alpha$$

i.e.

$$x_\beta = - \sum_{\alpha \neq \beta} (\lambda^\beta)^{-1} \lambda^\alpha x_\alpha .$$

Corollary: Two vectors x, y are linearly dependent if and only if $y = \lambda x$ (or $x = \lambda y$) for some $\lambda \in \Gamma$.

1.5. Linear independence. A family of vectors $\{x_\alpha\}_{\alpha \in A}$ is called *linearly independent* if it is not linearly dependent; i.e., the vectors x_α are linearly independent if and only if the equation

$$\sum_\alpha \lambda^\alpha x_\alpha = 0$$

implies that $\lambda^\alpha = 0$ for each $\alpha \in A$. It is clear that every subfamily of a linearly independent family of vectors is again linearly independent. If $\{x_\alpha\}_{\alpha \in A}$ is a linearly independent family, then for any two distinct indices $\alpha, \beta \in A$, $x_\alpha \neq x_\beta$, and so the map $\alpha \to x_\alpha$ is injective.

Proposition III: A family $\{x_\alpha\}_{\alpha \in A}$ of vectors is linearly independent if and only if every vector x can be written in *at most* one way as a linear combination of the x_α i.e., if and only if for each linear combination

$$x = \sum_\alpha \lambda^\alpha x_\alpha \tag{1.4}$$

the scalars λ^α are uniquely determined by x.

Proof: Suppose first that the scalars λ^α in (1.4) are uniquely determined by x. Then in particular for $x = 0$, the only scalars λ^α such that

$$\sum_\alpha \lambda^\alpha x_\alpha = 0$$

are the scalars $\lambda^\alpha = 0$. Hence, the vectors x_α are linearly independent. Conversely, suppose that the x_α are linearly independent and consider the relations

$$x=\sum_\alpha \lambda^\alpha x_\alpha, \quad x=\sum_\alpha \mu^\alpha x_\alpha .$$

Then

$$\sum_\alpha (\lambda^\alpha-\mu^\alpha)x_\alpha=0$$

whence in view of the linear independence of the x_α

$$\lambda^\alpha-\mu^\alpha=0, \qquad \alpha\in A$$

i.e., $\lambda^\alpha=\mu^\alpha$.

1.6. Basis. A family of vectors $\{x_\alpha\}_{\alpha\in A}$ in E is called a *basis* of E if it is simultaneously a system of generators and linearly independent.

In view of Proposition III and the definition of a system of generators, we have that $\{x_\alpha\}_{\alpha\in A}$ is a basis if and only if every vector $x\in E$ can be written in precisely one way as

$$x=\sum_\alpha \xi^\alpha x_\alpha, \qquad \xi^\alpha\in\Gamma .$$

The scalars ξ^α are called the *components* of x with respect to the basis $\{x_\alpha\}_{\alpha\in A}$.

As an example, consider the n-space, Γ^n, over Γ defined in example 1, sec. 1.2. It is easily verified that the vectors

$$x_i=(\underbrace{0,\dots,0}_{i-1},1,0\dots0) \qquad i=1\dots n$$

form a basis for Γ^n.

We shall prove that every non-trivial vector space has a basis. For the sake of simplicity we consider first vector spaces which admit a *finite* system of generators.

Proposition IV: (i) Every finitely generated non-trivial vector space has a finite basis

(ii) Suppose that $S=(x_1,\dots,x_m)$ is a finite system of generators of E and that the subset $R\subset S$ given by $R=(x_1,\dots,x_r)$ $(r\leqq m)$ consists of linearly independent vectors. Then there is a basis, T, of E such that $R\subset T\subset S$.

Proof: (i) Let $x_1,\dots,x_n$ be a minimal system of generators of E. Then the vectors $x_1,\dots,x_n$ are linearly independent. In fact, assume a relation

$$\sum_{\nu=1}^n \lambda^\nu x_\nu=0.$$

If $\lambda^i = 0$ for some i, it follows that

$$x_i = \sum_{\nu \neq i} \alpha_\nu x_\nu \qquad \alpha_\nu \in \Gamma \tag{1.5}$$

and so the vectors x_ν $(\nu \neq i)$ generate E. This contradicts the minimality of n.

(ii) We proceed by induction on n $(n \geqq r)$. If $n = r$ then there is nothing to prove. Assume now that the assertion is correct for $n-1$. Consider the vector space, F, generated by the vectors $x_1, \ldots, x_r, x_{r+1}, \ldots, x_{n-1}$. Then by induction, F has a basis of the form

$$x_1, \ldots, x_r, y_1, \ldots, y_s \qquad \text{where } y_j \in S \qquad (j = 1 \ldots s).$$

Now consider the vector x_n. If the vectors $x_1, \ldots, x_r, y_1, \ldots, y_s, x_n$ are linearly independent, then they form a basis of E which has the desired property. Otherwise there is a non-trivial relation

$$\sum_{\varrho=1}^{r} \alpha_\varrho x_\varrho + \sum_{\sigma=1}^{s} \beta_\sigma y_\sigma + \gamma x_n = 0.$$

Since the vectors $x_1, \ldots, x_r, y_1, \ldots, y_s$ are linearly independent, it follows that $\gamma \neq 0$. Thus

$$x_n = \sum_{\varrho=1}^{r} \lambda_\varrho x_\varrho + \sum_{\sigma=1}^{s} \mu_\sigma y_\sigma.$$

Hence the vectors $x_1, \ldots, x_n, y_1, \ldots, y_s$ generate E. Since they are linearly independent, they form a basis.

Now consider the general case.

Theorem I: Let E be a non-trivial vector space. Suppose S is a system of generators and that R is a family of linearly independent vectors in E such that $R \subset S$. Then there exists a basis, T, of E such that $R \subset T \subset S$.

Proof. Consider the collection $\mathscr{A}(R, S)$ of all subsets, X, of E such that

1) $R \subset X \subset S$
2) the vectors of X are linearly independent.

Then a partial order is defined in $\mathscr{A}(R, S)$ by inclusion (cf. sec. 0.6).

We show that every chain, $\{X_\alpha\}$, in $\mathscr{A}(R, S)$ has a maximal element A. In fact, set $A = \bigcup_\alpha X_\alpha$. We have to show that $A \in \mathscr{A}(R, S)$. Clearly, $R \subset A \subset S$. Now assume that

$$\sum_{\nu=1}^{n} \lambda^\nu x_\nu = 0 \qquad x_\nu \in A. \tag{1.6}$$

Then, for each i, $x_i \in X_\alpha$ for some α. Since $\{X_\alpha\}$ is a chain, we may assume that

$$x_i \in X_{\alpha_1} \qquad (i = 1 \dots n). \tag{1.7}$$

Since the vectors of X_{α_1} are linearly independent it follows that $\lambda^\nu = 0$ $(\nu = 1 \dots n)$ whence $A \in \mathscr{A}(R, S)$.

Now Zorn's lemma (cf. sec. 0.6) implies that there is a maximal element, T, in $\mathscr{A}(R, S)$. Then $R \subset T \subset S$ and the vectors of T are linearly independent. To show that T is a system of generators, let $x \in E$ be arbitrary. Then the vectors of $T \cup x$ are linearly dependent because otherwise it would follow that $x \cup T \in \mathscr{A}(R, S)$ which contradicts the maximality of T. Hence there is a non-trivial relation

$$\lambda x + \sum_\nu \lambda^\nu x_\nu = 0 \qquad \lambda \in \Gamma,\ \lambda^\nu \in \Gamma,\ x_\nu \in T.$$

Since the vectors of T are linearly independent, it follows that $\lambda \neq 0$ whence

$$x = \sum_\nu \alpha^\nu x_\nu .$$

This equation shows that T generates E and so it is a basis of E.

Corollary I: Every system of generators of E contains a basis. In particular, every non-trivial vector space has a basis.

Corollary II: Every family of linearly independent vectors of E can be extended to a basis.

1.7. The free vector space over a set. Let X be an arbitrary set and consider all maps $f: X \to \Gamma$ such that $f(x) \neq 0$ only for finitely many $x \in X$. Denote the set of these maps by $C(X)$. Then, if $f \in C(X)$, $g \in C(X)$ and λ, μ are scalars, $\lambda f + \mu g$ is again contained in $C(X)$. As in example 3, sec. 1.2, we make $C(X)$ into a vector space.

For any $a \in X$ denote by f_a the map given by

$$f_a(x) = \begin{cases} 1 & x = a \\ 0 & x \neq a. \end{cases}$$

Then the vectors f_a $(a \in X)$ form a basis of $C(X)$. In fact, let $f \in C(X)$ be given and let $a_1, \dots, a_n$ $(n \geqq 0)$ be the (finitely many) distinct points such that $f(a_i) \neq 0$. Then we have

$$f = \sum_{i=1}^n \alpha^i f_{a_i}$$

where

$$\alpha^i = f(a_i) \qquad (i = 1, \dots, n)$$

and so the elements f_a $(a \in X)$ generate $C(X)$. On the other hand, assume a relation

$$\sum_{j=1}^{n} \lambda^i f_{a_i} = 0.$$

Then we have for each j $(j=1 \dots n)$

$$0 = \sum_{j=1}^{n} \lambda^i f_{a_i}(a_j) = \lambda^j$$

whence $\lambda^j = 0$ $(j=1 \dots n)$. This shows that the vectors f_a $(a \in X)$ are linearly independent and hence they form a basis of $C(X)$.

Now consider the inclusion map $i_X \colon X \to C(X)$ given by

$$i_X(a) = f_a \qquad a \in X.$$

This map clearly defines a bijection between X and the basis vectors of $C(X)$. If we identify each element $a \in X$ with the corresponding map f_a, then X becomes a basis of $C(X)$. $C(X)$ is called the *free vector space over X or the vector space generated by X.*

Problems

1. Show that axiom II.3 can be replaced by the following one: The equation $\lambda x = 0$ holds only if $\lambda = 0$ or $x = 0$.

2. Given a system of linearly independent vectors $(x_1, \dots, x_p)$, prove that the system $(x_1, \dots x_i + \lambda x_j, \dots x_p)$, $i \neq j$ with arbitrary λ is again linearly independent.

3. Show that the set of all solutions of the homogeneous linear differential equation

$$\frac{d^2 y}{dt^2} + p\frac{dy}{dt} + qy = 0$$

where p and q are fixed functions of t, is a vector space.

4. Which of the following sets of functions are linearly dependent in the vector space of Example 2 sec. 1.2?

a) $f_1 = 3t;$ $\quad f_2 = t + 5;$ $\quad f_3 = 2t^2;$ $\quad f_4 = (t+1)^2$
b) $f_1 = (t+1)^2;$ $\quad f_2 = t^2 - 1;$ $\quad f_3 = 2t^2 + 2t - 3$
c) $f_1 = 1;$ $\quad f_2 = e^t;$ $\quad f_3 = e^{-t}$
d) $f_1 = t^2;$ $\quad f_2 = t;$ $\quad f_3 = 1$
e) $f_1 = 1 - t;$ $\quad f_2 = t(1-t);$ $\quad f_3 = 1 - t^2.$

5. Let E be a real linear space. Consider the set $E \times E$ of ordered pairs (x, y) with $x \in E$ and $y \in E$. Show that the set $E \times E$ becomes a complex vector space under the operations:

$$(x_1, y_1) + (x_2, y_2) = (x_1 + x_2, y_1 + y_2)$$

and

$$(\alpha + i\beta)(x, y) = (\alpha x - \beta y, \alpha y + \beta x) \qquad (\alpha, \beta \text{ real numbers}).$$

6. Which of the following sets of vectors in $\mathbb{R}^4$ are linearly independent, (a generating set, a basis)?

a) $(1, 1, 1, 1), (1, 0, 0, 0), (0, 1, 0, 0), (0, 0, 1, 0), (0, 0, 0, 1)$
b) $(1, 0, 0, 0), (2, 0, 0, 0)$
c) $(17, 39, 25, 10), (13, 12, 99, 4), (16, 1, 0, 0)$
d) $(1, \frac{1}{2}, 0, 0), (0, 0, 1, 1), (0, \frac{1}{2}, \frac{1}{2}, 1), (\frac{1}{4}, 0, 0, \frac{1}{4})$

Extend the linearly independent sets to bases.

7. Are the vectors $x_1 = (1, 0, 1)$; $x_2 = (i, 1, 0)$, $x_3 = (i, 2, 1+i)$ linearly independent in $\mathbb{C}^3$? Express $x = (1, 2, 3)$ and $y = (i, i, i)$ as linear combinations of x_1, x_2, x_3.

8. Recall that an n-tuple $(\lambda_1 \ldots \lambda_n)$ is defined by a map $f: \{1 \ldots n\} \to \Gamma$ given by

$$f(i) = \lambda_i \qquad (i = 1 \ldots n).$$

Show that the vector spaces $C\{1 \ldots n\}$ and Γ^n are equal. Show further that the basis f_i defined in sec. 1.7 coincides with the basis x_i defined in sec. 1.6.

9. Let S be any set and consider the set of maps

$$f: S \to \Gamma^n$$

such that $f(x) = 0$ for all but finitely many $x \in S$. In a manner similar to that of sec. 1.7, make this set into a vector space (denoted by $C(S, \Gamma^n)$). Construct a basis for this vector space.

10. Let $(x_\alpha)_{\alpha \in A}$ be a basis for a vector space E and consider a vector

$$a = \sum_\alpha \xi^\alpha x_\alpha .$$

Suppose that for some $\beta \in A$, $\xi^\beta \neq 0$. Show that the vectors $\{x_\alpha\}_{\alpha \neq \beta}$ together with a again form a basis for E.

11. Prove the following *exchange theorem of Steinitz:* Let $\{x_\alpha\}_{\alpha\in A}$ be a basis of E and $a_i (i=1...p)$ be a system of linearly independent vectors. Then it is possible to replace certain p of the vectors x_α by the vectors a_i such that the new system is again a basis of E. *Hint:* Use problem 10.

12. Consider the set of polynomial functions $f: \mathbb{R}\to\mathbb{R}$,

$$f(x) = \sum_{i=0}^{n} \alpha_i x^i.$$

Make this set into a vector space as in Example 3, and construct a natural basis.

§ 2. Linear mappings

In this paragraph, all vector spaces are defined over a fixed but arbitrarily chosen field Γ of characteristic zero.

1.8. Definition. Suppose that E and F are vector spaces, and let $\varphi: E\to F$ be a set mapping. Then φ will be called a *linear mapping* if

$$\varphi(x+y) = \varphi x + \varphi y \quad x, y\in E \tag{1.8}$$

and

$$\varphi(\lambda x) = \lambda\varphi x \quad \lambda\in\Gamma, x\in E \tag{1.9}$$

(Recall that condition (1.8) states that φ is a homomorphism between abelian groups). If $F=\Gamma$ then φ is called a *linear function* in E.

Conditions (1.8) and (1.9) are clearly equivalent to the condition

$$\varphi(\sum_i \lambda^i x_i) = \sum_i \lambda^i \varphi x_i$$

and so a linear mapping is a mapping which preserves linear combinations.

From (1.8) we obtain that for every linear mapping, φ,

$$\varphi 0 = \varphi(0+0) = \varphi(0) + \varphi(0)$$

whence $\varphi(0)=0$. Suppose now that

$$\sum_i \lambda^i x_i = 0 \tag{1.10}$$

is a linear relation among the vectors x_i. Then we have

$$\sum_i \lambda^i \varphi x_i = \varphi(\sum_i \lambda^i x_i) = \varphi 0 = 0$$

whence

$$\sum_i \lambda^i \varphi x_i = 0. \tag{1.11}$$

Conversely, assume that $\varphi: E \to F$ is a set map such that (1.11) holds whenever (1.10) holds. Then for any $x, y \in E$ and $\lambda \in \Gamma$ set

$$u = x + y \quad \text{and} \quad v = \lambda x.$$

Since

$$u - x - y = 0 \quad \text{and} \quad v - \lambda x = 0$$

it follows that

$$\varphi(x + y) - \varphi x - \varphi y = 0$$

and

$$\varphi(\lambda x) - \lambda \varphi x = 0$$

and hence φ is a linear mapping. This shows that linear mappings are precisely the set mappings which preserve linear relations.

In particular, it follows that if $x_1 \ldots x_r$ are linearly dependent, then so are the vectors $\varphi x_1 \ldots \varphi x_r$. If $x_1 \ldots x_r$ are linearly independent, it does not, however, follow that the vectors $\varphi x_1 \ldots \varphi x_r$ are linearly independent. In fact, the zero mapping defined by $\varphi x = 0, x \in E$, is clearly a linear mapping which maps every family of vectors into the linearly dependent set (0).

A bijective linear mapping $\varphi: E \xrightarrow{\cong} F$ is called *a linear isomorphism* and will be denoted by $\varphi: E \xrightarrow{\cong} F$. Given a linear isomorphism $\varphi: E \xrightarrow{\cong} F$ consider the set mapping $\varphi^{-1}: E \leftarrow F$. It is easy to verify that φ^{-1} again satisfies the conditions (1.8) and (1.9) and so it is a linear mapping. φ^{-1} is bijective and hence a linear isomorphism. It is called the *inverse isomorphism* of φ. Two vector spaces E and F are called *isomorphic* if there exists a linear isomorphism from E onto F.

A linear mapping $\varphi: E \to E$ is called a *linear transformation* of E. A bijective linear transformation will be called a *linear automorphism* of E.

1.9. Examples: 1. Let $E = \Gamma^n$ and define $\varphi: E \to E$ by

$$\varphi(\xi^1, \ldots, \xi^n) = (\xi^1 + \xi^2, \xi^2, \ldots, \xi^n).$$

Then φ satisfies the conditions (1.8) and (1.9) and hence it is a linear transformation of E.

2. Given a set S and a vector space E consider the vector space $(S; E)$ defined in Example 3, sec. 1.2. Let $\varphi: (S; E) \to E$ be the mapping given by

$$\varphi f = f(a) \quad f \in (S; E)$$

where $a \in S$ is a fixed element. Then φ is a linear mapping.

3. Let $\varphi: E \to E$ be the mapping defined by $\varphi x = \lambda x$, where $\lambda \in \Gamma$ is a fixed element. Then φ is a linear transformation. In particular, the *identity map* $\iota: E \to E$, $\iota x = x$, is a linear transformation.

1.10. Composition. Let $\varphi: E \to F$ and $\psi: F \to G$ be two linear mappings. Then the *composition* of φ and ψ

$$\psi \circ \varphi: E \to G$$

is defined by

$$(\psi \circ \varphi) x = \psi(\varphi x) \qquad x \in E.$$

The relation

$$\begin{aligned} \psi \circ \varphi\left(\sum_i \lambda_i x_i\right) &= \psi\left(\sum_i \lambda_i \varphi x_i\right) \\ &= \sum_i \lambda_i \psi \circ \varphi x_i \end{aligned}$$

shows that $\psi \circ \varphi$ is a linear mapping of E into G. $\psi \circ \varphi$ will often be denoted simply by $\psi\varphi$. If φ is a linear transformation in E, then we denote $\varphi \circ \varphi$ by φ^2. More generally, the linear transformation $\underbrace{\varphi \circ \ldots \circ \varphi}_{k}$ is denoted by φ^k.

We extend the definition to the case $k = 0$ by setting $\varphi^0 = \iota$. A linear transformation, φ, satisfying $\varphi^2 = \iota$ is called an *involution* in E.

1.11. Generators and basis.

Proposition I: Suppose S is a system of generators for E and $\varphi_0: S \to F$ is a set map (F a second vector space). Then φ_0 can be extended in at most one way to a linear mapping

$$\varphi: E \to F.$$

A necessary and sufficient condition for the existence of such an extension is that

$$\sum_i \lambda^i \varphi_0 x_i = 0 \tag{1.12}$$

whenever

$$\sum_i \lambda^i x_i = 0.$$

Proof: If φ is an extension of φ_0 we have for each finite set of vectors $x_i \in S$

$$\varphi \sum_i \lambda^i x_i = \sum_i \lambda^i \varphi x_i = \sum_i \lambda^i \varphi_0 x_i.$$

Since the set S generates E it follows from this relation that φ is uniquely determined by φ_0. Moreover, if

$$\sum_i \lambda^i x_i = 0 \qquad x_i \in S$$

it follows that

$$\sum_i \lambda^i \varphi_0 x_i = \sum_i \lambda^i \varphi x_i = \varphi \sum_i \lambda^i x_i = \varphi 0 = 0$$

and so condition (1.12) is necessary.

Conversely, assume that (1.12) is satisfied. Then define φ by

$$\varphi \sum_i \lambda^i x_i = \sum_i \lambda^i \varphi_0 x_i, \qquad x_i \in S. \tag{1.13}$$

To prove that φ is a well defined map assume that

$$\sum_i \lambda^i x_i = \sum_j \mu^j y_j, \qquad x_i \in S, \quad y_j \in S.$$

Then

$$\sum_i \lambda^i x_i - \sum_j \mu^j y_j = 0$$

whence in view of (1.12)

$$\sum_i \lambda^i \varphi_0 x_i - \sum_j \mu^j \varphi_0 y_j = 0$$

and so

$$\sum_i \lambda^i \varphi_0 x_i = \sum_j \mu^j \varphi_0 y_j.$$

The linearity of φ follows immediately from the definition, and it is clear that φ extends φ_0.

Proposition II: Let $\{x_\alpha\}_{\alpha \in A}$ be a basis of E and $\varphi_0: \{x_\alpha\} \to F$ be a set map. Then φ_0 can be extended in a unique way to a linear mapping $\varphi: E \to F$.

Proof: The uniqueness follows from proposition I. To prove the existence of φ consider a relation

$$\sum_\alpha \lambda^\alpha x_\alpha = 0.$$

Since the vectors x_α are linearly independent it follows that each $\lambda^\alpha = 0$, whence

$$\sum_\alpha \lambda^\alpha \varphi_0 x_\alpha = 0.$$

Now proposition *I* shows that φ_0 can be extended to a linear mapping $\varphi: E \to F$.

Corollary: Let S be a linearly independent subset of E and $\varphi_0: S \to F$ be a set map. Then φ_0 can be extended to a linear mapping $\varphi: E \to F$.

Proof: Let T be a basis of E containing S (cf. sec. 1.6). Extend φ_0 in an arbitrary way to a set map $\psi_0: T \to F$. Then ψ_0 may be extended to a linear mapping $\psi: E \to F$ and it is clear that ψ extends φ_0.

Now let $\varphi: E \to F$ be a surjective linear map, and suppose that S is a system of generators for E. Then the set

$$\varphi(S) = \{\varphi x \mid x \in S\}$$

is a system of generators for F. In fact, since φ is surjective, every vector $y \in F$ can be written as

$$y = \varphi x$$

for some $x \in E$. Since S generates E there are vectors $x_i \in S$ and scalars $\xi^i \in \Gamma$ such that

$$x = \sum_i \xi^i x_i$$

whence

$$y = \varphi x = \sum_i \xi^i \varphi x_i,$$

This shows that every vector $y \in F$ is a linear combination of vectors in $\varphi(S)$ and hence $\varphi(S)$ is a system of generators for $\varphi(S)$.

Next, suppose that $\varphi: E \to F$ is injective and that S is a linearly independent subset of E. Then $\varphi(S)$ is a linearly independent subset of F. In fact, the relation

$$\sum_i \lambda^i \varphi x_i = 0, \qquad x_i \in S$$

implies that

$$\varphi \sum_i \lambda^i x_i = 0.$$

Since φ is injective we obtain

$$\sum_i \lambda^i x_i = 0$$

whence, in view of the linear independence of the vectors x_i, $\lambda^i = 0$. Hence $\varphi(S)$ is a linearly independent set.

In particular, if $\varphi: E \to F$ is a linear isomorphism and $(x_\alpha)_{\alpha \in A}$ is a basis for E, then $(\varphi x_\alpha)_{\alpha \in A}$ is a basis for F.

Proposition III: Let $\varphi: E \to F$ be a linear mapping and $\{x_\alpha\}_{\alpha \in A}$ be a basis of E. Then φ is a linear isomorphism if and only if the vectors $y_\alpha = \varphi x_\alpha$ form a basis for F.

Proof: If φ is a linear isomorphism then the vectors y_α form a linearly independent system of generators for F. Hence they are a basis. Conversely, assume that the vectors y_α form a basis of F. Then we have for every $y\in F$

$$y=\sum_\alpha \eta^\alpha y_\alpha=\sum_\alpha \eta^\alpha \varphi x_\alpha=\varphi\sum_\alpha \eta^\alpha x_\alpha$$

and so φ is surjective.

Now assume that

$$\varphi\sum_\alpha \lambda^\alpha x_\alpha=\varphi\sum_\alpha \mu^\alpha x_\alpha .$$

Then it follows that

$$\begin{aligned}0&=\sum_\alpha \lambda^\alpha \varphi x_\alpha-\sum_\alpha \mu^\alpha \varphi x_\alpha\\&=\sum_\alpha (\lambda^\alpha-\mu^\alpha)\,y_\alpha .\end{aligned}$$

Since the vectors y_α are linearly independent, we obtain that $\lambda^\alpha=\mu^\alpha$ for each α, and so

$$\sum_\alpha \lambda^\alpha x_\alpha=\sum_\alpha \mu^\alpha x_\alpha .$$

It follows that φ is injective, and hence a linear isomorphism.

Problems

1. Consider the vector space of all real valued continuous functions defined in the interval $a\leqq t\leqq b$. Show that the mapping φ given by

$$\varphi: x(t)\to t\,x(t)$$

is linear.

2. Which of the following mappings of Γ^4 into itself are linear transformations?

a) $(\xi^1,\xi^2,\xi^3,\xi^4)\to(\xi^1\xi^2,\xi^2-\xi^1,\xi^3,\xi^4)$
b) $(\xi^1,\xi^2,\xi^3,\xi^4)\to(\lambda\xi^2,\xi^2-\xi^1,\xi^3,\xi^4)$
c) $(\xi^1,\xi^2,\xi^3,\xi^4)\to(0,\xi^3,\xi^2,\xi^1+\xi^2+\xi^3+\xi^4)$

3. Let E be a vector space over Γ, and let $f_1\dots f_r$ be linear functions in E. Show that the mapping $\varphi: E\to\Gamma^r$ given by

$$\varphi x=(f_1(x),\dots,f_r(x))$$

is linear.

4. Suppose $\varphi: E\to\Gamma^r$ is a linear map, and write

$$\varphi x=(f_1(x),\dots,f_r(x)).$$

Show that the mappings $f_i: E\to\Gamma$ are linear functions in E.

5. *The universal property of* $C(X)$. Let X be any set and consider the free vector space, $C(X)$, generated by X (cf. sec. 1.7).

(i) Show that if $f: X \to F$ is a set map from X into a vector space F then there is a unique linear map $\varphi: C(X) \to F$ such that $\varphi \circ i_X = f$ where $i_X: X \to C(X)$ is the inclusion map.

(ii) Let $\alpha: X \to Y$ be a set map. Show that there is a unique linear map $\alpha_*: C(X) \to C(Y)$ such that the diagram

$$\begin{array}{ccc} X & \xrightarrow{\alpha} & Y \\ {\scriptstyle i_X}\downarrow & & {\scriptstyle i_Y}\downarrow \\ C(X) & \xrightarrow{\alpha_*} & C(Y) \end{array}$$

commutes. If $\beta: Y \to Z$ is a second set map prove the composition formula

$$(\beta \circ \alpha)_* = \beta_* \circ \alpha_*.$$

(iii) Let E be a vector space. Forget the linear structure of E and form the space $C(E)$. Show that there is a unique linear map $\pi_E: C(E) \to E$ such that $\pi_E \circ i_E = \iota$.

(iv) Let E and F be vector spaces and let $\varphi: E \to F$ be a map between the underlying sets. Show that φ is a linear map if and only if

$$\pi_F \circ \varphi_* = \varphi \circ \pi_E.$$

(v) Denote by $N(E)$ the subspace of $C(E)$ generated by the elements of the form

$$f_{\lambda a + \mu b} - \lambda f_a - \mu f_b \qquad a, b \in E,\ \lambda, \mu \in \Gamma$$

(cf. part (iii)). Show that

$$\ker \pi_E = N(E).$$

6. Let

$$P = \sum_{\nu=0}^{n} \alpha_\nu t^\nu \qquad \alpha_\nu \in \Gamma$$

be a fixed polynomial and let f be any linear function in a vector space E. Define a function $P(f): E \to \Gamma$ by

$$P(f)x = \sum_{\nu=0}^{n} \alpha_\nu f(x)^\nu.$$

Find necessary and sufficient conditions on P that $P(f)$ be again a linear function.

§ 3. Subspaces and factor spaces

In this paragraph, all vector spaces are defined over a fixed, but arbitrarily chosen field Γ of characteristic 0.

1.12. Subspaces. Let E be a vector space over the field Γ. A non-empty subset, E_1, of E is called a *subspace* if for each $x, y \in E_1$ and every scalar $\lambda \in \Gamma$

$$x + y \in E_1 \tag{1.14}$$

and

$$\lambda x \in E_1 . \tag{1.15}$$

Equivalently, a subspace is a subset of E such that

$$\lambda x + \mu y \in E_1$$

whenever $x, y \in E_1$. In particular, the whole space E and the subset (0) consisting of the zero vector only are subspaces. Every subspace $E_1 \subset E$ contains the zero vector. In fact, if $x_1 \in E_1$ is an arbitrary vector we have that $0 = x_1 - x_1 \in E_1$. A subspace E_1 of E inherits the structure of a vector space from E.

Now consider the injective map $i: E_1 \to E$ defined by

$$i x = x , \qquad x \in E_1 .$$

In view of the definition of the linear operations in E_1 i is a linear mapping, called the *canonical injection* of E_1 into E. Since i is injective it follows from (sec. 1.11) that a family of vectors in E_1 is linearly independent (dependent) if and only if it is linearly independent (dependent) in E.

Next let S be any non-empty subset of E and denote by E_s the set of linear combinations of vectors in S. Then any linear combination of vectors in E_s is a linear combination of vectors in S (cf. sec. 1.3) and hence it belongs to E_s. Thus E_s is a subspace of E, called the *subspace generated by* S, or the *linear closure of* S.

Clearly, S is a system of generators for E_s. In particular, if the set S is linearly independent, then S is a basis of E_s. We notice that $E_s = S$ if and only if S is a subspace itself.

1.13. Intersections and sums. Let E_1 and E_2 be subspaces of E and consider the intersection $E_1 \cap E_2$ of the sets E_1 and E_2. Then $E_1 \cap E_2$ is again a subspace of E. In fact, since $0 \in E_1$ and $0 \in E_2$ we have $0 \in E_1 \cap E_2$ and so $E_1 \cap E_2$ is not empty. Moreover, it is clear that the set $E_1 \cap E_2$ satisfies again conditions (1.14) and (1.15) and so it is a subspace of E. $E_1 \cap E_2$ is called the *intersection* of the subspaces E_1 and E_2. Clearly, $E_1 \cap E_2$ is a subspace of E_1 and a subspace of E_2.

The *sum* of two subspaces E_1 and E_2 is defined as the set of all vectors of the form

$$x = x_1 + x_2 , \qquad x_1 \in E_1, x_2 \in E_2 \tag{1.16}$$

and is denoted by E_1+E_2. Again it is easy to verify that E_1+E_2 is a subspace of E. Clearly E_1+E_2 contains E_1 and E_2 as subspaces.

A vector x of E_1+E_2 can generally be written in several ways in the form (1.16). Given two such decompositions

$$x = x_1 + x_2 \quad \text{and} \quad x = x_1' + x_2'$$

it follows that

$$x_1 - x_1' = x_2' - x_2 .$$

Hence, the vector

$$z = x_1 - x_1'$$

is contained in the intersection $E_1 \cap E_2$. Conversely, let $x = x_1 + x_2$, $x_1 \in E_1$, $x_2 \in E_2$ be a decomposition of x and z be an arbitrary vector of $E_1 \cap E_2$. Then the vectors

$$x_1' = x_1 - z \in E_1 \quad \text{and} \quad x_2' = x_2 + z \in E_2$$

form again a decomposition of x. It follows from this remark that the decomposition (1.16) of a vector $x \in E_1 + E_2$ is uniquely determined if and only if $E_1 \cap E_2 = 0$. In this case $E_1 + E_2$ is called the (*internal*) *direct sum* of E_1 and E_2 and is denoted by $E_1 \oplus E_2$.

Now let S_1 and S_2 be systems of generators for E_1 and E_2. Then clearly $S_1 \cup S_2$ is a system of generators for E_1+E_2. If T_1 and T_2 are respectively bases for E_1 and E_2 and the sum is direct, $E_1 \cap E_2 = 0$, then $T_1 \cup T_2$ is a basis for $E_1 \oplus E_2$. To prove that the set $T_1 \cup T_2$ is linearly independent, suppose that

$$\sum_i \lambda^i x_i + \sum_j \mu^j y_j = 0, \quad x_i \in T_1, y_j \in T_2 .$$

Then

$$\sum_i \lambda^i x_i = -\sum_j \mu^j y_j \in E_1 \cap E_2 = 0$$

whence

$$\sum_i \lambda^i x_i = 0 \quad \text{and} \quad \sum_j \mu^j y_j = 0 .$$

Now the x_i are linearly independent, and so $\lambda^i = 0$. Similarly it follows that $\mu^j = 0$.

Suppose that

$$E = E_1 \oplus E_2 \tag{1.17}$$

is a decomposition of E as a direct sum of subspaces and let F be an arbitrary subspace of E. Then it is not in general true that

$$F = F \cap E_1 \oplus F \cap E_2 \tag{1.18}$$

as the example below will show. However, if $E_1 \subset F$, then (1.18) holds. In fact, it is clear that

$$F \cap E_1 \oplus F \cap E_2 \subset F. \tag{1.19}$$

On the other hand, if

$$y = x_1 + x_2 \qquad x_1 \in E_1, x_2 \in E_2$$

is the decomposition of any vector $y \in F$, then

$$x_1 \in E_1 = F \cap E_1, \quad x_2 = y - x_1 \in F \cap E_2.$$

It follows that

$$F \subset F \cap E_1 \oplus F \cap E_2. \tag{1.20}$$

The relations (1.19) and (1.20) imply (1.18).

Example: Let E be a vector space with a basis e_1, e_2. Define E_1, E_2 and F as the subspaces generated by e_1, e_2 and $e_1 + e_2$ respectively. Then

$$E = E_1 \oplus E_2$$

while on the other hand

$$F \cap E_1 = F \cap E_2 = 0.$$

Hence

$$F \neq F \cap E_1 \oplus F \cap E_2.$$

1.14. Arbitrary families of subspaces. Next consider an arbitrary family of subspaces $E_\alpha \subset E$, $\alpha \in A$. Then the intersection $\bigcap_\alpha E_\alpha$ is again a subspace of E. The *sum* $\sum_\alpha E_\alpha$ is defined as the set of all vectors which can be written as finite sums,

$$x = \sum_\alpha x_\alpha, \qquad x_\alpha \in E_\alpha \tag{1.21}$$

and is a subspace of E as well. If for every $\alpha \in A$

$$E_\alpha \cap \sum_{\beta \neq \alpha} E_\beta = 0$$

then each vector of the sum $\sum_\alpha E_\alpha$ can be *uniquely* represented in the form (1.21). In this case the space $\sum_\alpha E_\alpha$ is called the (*internal*) *direct sum* of the subspaces E_α, and is denoted by $\sum_\alpha E_\alpha$.

If S_α is a system of generators for E_α, then the set $\bigcup_\alpha S_\alpha$ is a system of generators for $\sum_\alpha E_\alpha$. If the sum of the E_α is direct and T_α is a basis of E_α, then $\bigcup_\alpha T_\alpha$ is a basis for $\sum_\alpha E_\alpha$.

Example: Let $\{x_\alpha\}_{\alpha\in A}$ be a basis of E and E_α be the subspace generated by x_α. Then

$$E = \sum_\alpha E_\alpha .$$

Suppose

$$E = \sum_\alpha E_\alpha \tag{1.22}$$

is a direct sum of subspaces. Then we have the canonical injections $i_\alpha : E_\alpha \to E$. We define the *canonical projections* $\pi_\alpha : E \to E_\alpha$ determined by

$$\pi_\alpha x = x_\alpha$$

where

$$x = \sum_\alpha x_\alpha \qquad x_\alpha \in E_\alpha .$$

It is clear that the π_α are surjective linear mappings. Moreover, it is easily verified that the following relations hold:

$$\pi_\alpha \circ i_\beta = \begin{cases} \iota & \beta = \alpha \\ 0 & \beta \neq \alpha \end{cases}$$

$$\sum_\alpha i_\alpha \pi_\alpha x = x \qquad x \in E .$$

1.15. Complementary subspaces. An important property of vector spaces is given in the

Proposition I: If E_1 is a subspace of E, then there exists a second subspace E_2 such that

$$E = E_1 \oplus E_2 .$$

E_2 is called a *complementary* subspace for E_1 in E.

Proof: We may assume that $E_1 \neq E$ and $E_1 \neq (0)$ since the proposition is trivial in these cases. Let $\{x_\alpha\}$ be a basis of E_1 and extend it with vectors y_β to form a basis of E (cf. Corollary II to Theorem I, sec. 1.6). Let E_2 be the subspace of E generated by the vectors y_β. Then

$$E = E_1 \oplus E_2 .$$

In fact, since $\{x_\alpha\} \cup \{y_\beta\}$ is a system of generators for E, we have that

$$E = E_1 + E_2 . \tag{1.23}$$

On the other hand, if $x \in E_1 \cap E_2$, then we may write

$$x = \sum_\alpha \lambda^\alpha x_\alpha \quad \text{and} \quad x = \sum_\beta \mu^\beta y_\beta$$

whence

$$\sum_\alpha \lambda^\alpha x_\alpha - \sum_\beta \mu^\beta y_\beta = 0.$$

Now since the set $\{x_\alpha\} \cup \{y_\beta\}$ is linearly independent, we obtain

$$\lambda^\alpha = 0 \quad \text{and} \quad \mu^\beta = 0$$

whence $x=0$. It follows that $E_1 \cap E_2 = 0$ and so the decomposition (1.23) is direct.

As an immediate consequence of the proposition we have

Corollary I. Let E_1 be a subspace of E and $\varphi_1 : E_1 \to F$ a linear mapping (F a second vector space). Then φ_1 may be extended (in several ways) to a linear map $\varphi : E \to F$.

Proof: Let E_2 be a complementary subspace for E_1 in E,

$$E = E_1 \oplus E_2 \tag{1.24}$$

and define φ by

$$\varphi x = \varphi_1 y$$

where

$$x = y + z$$

is the decomposition of x determined by (1.24). Then

$$\begin{aligned} \varphi \sum_i \lambda^i x_i &= \varphi\left(\sum_i \lambda^i y_i + \sum_i \lambda^i z_i\right) \qquad x_i = y_i + z_i \\ &= \varphi_1 \sum_i \lambda^i y_i \\ &= \sum_i \lambda^i \varphi_1 y_i \\ &= \sum_i \lambda^i \varphi x_i \end{aligned}$$

and so φ is linear. It is trivial that φ extends φ_1.

As a special example we have:

Corollary II: Let E_1 be a subspace of E. Then there exists a surjective linear map

$$\varphi : E \to E_1$$

such that

$$\varphi x = x \qquad x \in E_1.$$

Proof: Simply extend the identity map $\iota: E_1 \to E_1$ to a linear map $\varphi: E \to E_1$.

1.16. Factor spaces. Suppose E_1 is a subspace of the vector space E. Two vectors $x \in E$ and $x' \in E$ are called *equivalent* mod E_1 if $x' - x \in E_1$. It is easy to verify that this relation is reflexive, symmetric and transitive and hence is indeed an equivalence relation. (The equivalence classes are the cosets of the additive subgroup E_1 in E (cf. sec. 0.4)). Let E/E_1 denote the set of the equivalence classes so obtained and let

$$\pi: E \to E/E_1$$

be the set mapping given by

$$\pi x = \bar{x}, \qquad x \in E$$

where $\bar{x}$ is the equivalence class containing x. Clearly π is a surjective map.

Proposition II: There exists precisely one linear structure in E/E_1 such that π is a linear mapping.

Proof: Assume that E/E_1 is made into a vector space such that π is a linear mapping. Then the equations

$$\pi(x + y) = \pi x + \pi y$$

and

$$\pi(\lambda x) = \lambda \pi x$$

show that the linear operations in E/E_1 are uniquely determined by the linear operations in E.

It remains to be shown that a linear structure can be defined in E/E_1 such that π becomes a linear mapping. Let $\bar{x}$ and $\bar{y}$ be two arbitrary elements of E/E_1 and choose vectors $x \in E$ and $y \in E$ such that

$$\pi x = \bar{x}, \quad \pi y = \bar{y}.$$

Then the class $\pi(x+y)$ depends only on $\bar{x}$ and $\bar{y}$. Assume for instance that $x' \in E$ is another vector such that $\pi x' = \bar{x}$.
Then $\pi x' = \pi x$ and hence we may write

$$x' = x + z, \qquad z \in E_1.$$

It follows that

$$x' + y = (x + y) + z$$

whence

$$\pi(x' + y) = \pi(x + y).$$

We now define the sum of the elements $\bar{x} \in E/E_1$ and $\bar{y} \in E/E_1$ by

$$\bar{x} + \bar{y} = \pi(x + y) \quad \text{where} \quad \bar{x} = \pi x \quad \text{and} \quad \bar{y} = \pi y. \tag{1.25}$$

It is easy to verify that E/E_1 becomes an abelian group under this operation and that the class $\bar{0} = E_1$ is the zero-element.

Now let $\bar{x} \in E/E_1$ be an arbitrary element and $\lambda \in \Gamma$ be a scalar. Choose $x \in E$ such that $\pi x = \bar{x}$. Then a similar argument shows that the class $\pi(\lambda x)$ depends only on $\bar{x}$ (and not on the choice of the vector x). We now define the scalar multiplication in E/E_1 by

$$\lambda \cdot \bar{x} = \pi(\lambda x) \quad \text{where} \quad \bar{x} = \pi x. \tag{1.26}$$

Again it is easy to verify that the multiplication satisfies axioms II.1–II.3 and so E/E_1 is made into a vector space. It follows immediately from (1.25) and (1.26) that

$$\begin{aligned} \pi(x + y) &= \pi x + \pi y \qquad & x, y \in E \\ \pi(\lambda x) &= \lambda \pi x \qquad & \lambda \in \Gamma \end{aligned}$$

i.e., π is a linear mapping.

The vector space E/E_1 obtained in this way is called the *factor space* of E with respect to the subspace E_1. The linear mapping π is called the *canonical projection* of E onto E_1. If $E_1 = E$, then the factor space reduces to the vector $\bar{0}$. On the other hand, if $E_1 = 0$, two vectors $x \in E$ and $y \in E$ are equivalent mod E_1 if and only if $y = x$. Thus the elements of $E/(0)$ are the singleton sets $\{x\}$ where x is any element of E, and π is the linear isomorphism $x \to \{x\}$. Consequently we identify E and $E/(0)$.

1.17. Linear dependence mod a subspace. Let E_1 be a subspace of E, and suppose that (x_α) is a family of vectors in E. Then the x_α will be called *linearly dependent* mod E_1 if there are scalars λ^α, not all zero, such that

$$\sum_\alpha \lambda^\alpha x_\alpha \in E_1 .$$

If the x_α are not linearly dependent mod E_1 they will be called *linearly independent* mod E_1.

Now consider the canonical projection

$$\pi: E \to E/E_1 .$$

It follows immediately from the definition that the vectors x_α are linearly dependent (independent) mod E_1 if and only if the vectors πx_α are linearly dependent (independent) in E/E_1.

1.18. Basis of a factor space. Suppose that $\{y_\alpha\}\cup\{z_\beta\}$ is a basis of E such that the vectors y_α form a basis of E_1. Then the vectors πz_β form a basis of E/E_1. To prove this let E_2 be the subspace of E generated by the vectors z_β. Then

$$E = E_1 \oplus E_2 . \tag{1.27}$$

Now consider the linear mapping $\varphi: E_2 \to E/E_1$ defined by

$$\varphi z = \pi z \qquad z \in E_2 .$$

Then φ is surjective. In fact, let $\bar{x} \in E/E_1$ be an arbitrary vector. Since $\pi: E \to E/E_1$ is surjective we can write

$$\bar{x} = \pi x , \qquad x \in E .$$

In view of (1.27) the vector x can be decomposed in the form

$$x = y + z \qquad y \in E_1, z \in E_2 . \tag{1.28}$$

Equation (1.28) yields

$$\bar{x} = \pi x = \pi y + \pi z = \pi z = \varphi z$$

and so φ is surjective.

To show that φ is injective assume that

$$\varphi z = \varphi z' \qquad z, z' \in E_2 .$$

Then

$$\pi(z' - z) = \varphi(z' - z) = 0$$

and hence $z' - z \in E_1$. On the other hand we have that $z' - z \in E_2$ and thus

$$z' - z \in E_1 \cap E_2 = 0 .$$

It follows that $\varphi: E_2 \to E/E_1$ is a linear isomorphism and now Proposition III of sec. 1.11 shows that the vectors πz_β form a basis of E/E_1.

Problems

1. Let (ξ^1, ξ^2, ξ^3) be an arbitrary vector in Γ^3. Which of the following subsets are subspaces?

a) all vectors with $\xi^1 = \xi^2 = \xi^3$
b) all vectors with $\xi^3 = 0$
c) all vectors with $\xi^1 = \xi^2 - \xi^3$
d) all vectors with $\xi^1 = 1$

2. Find the subspaces F_a, F_b, F_c, F_d generated by the sets of problem 1, and construct bases for these subspaces.

3. Construct bases for the factor spaces determined by the subspaces of problem 2.

4. Find complementary spaces for the subspaces of problem 2, and construct bases for these complementary spaces. Show that there exists more than one complementary space for each given subspace.

5. Show that

a) $\Gamma^3 = F_a + F_b$

b) $\Gamma^3 = F_b + F_c$

c) $\Gamma^3 = F_a + F_c$

Find the intersections $F_a \cap F_b$, $F_b \cap F_c$, $F_a \cap F_c$ and decide in which cases the sums above are direct.

6. Let S be an arbitrary subset of E and let E_S be its linear closure. Show that E_S is the intersection of all subspaces of E containing S.

7. Assume a direct composition $E = E_1 \oplus E_2$. Show that in each class of E with respect to E_1 (i.e. in each coset $\bar{x} \in E/E_1$) there is exactly one vector of E_2.

8. Let E be a plane and let E_1 be a straight line through the origin. What is the geometrical meaning of the equivalence classes (cf. sec. 1.16) with respect to E_1. Give a geometrical interpretation of the fact that $x \sim x'$ and $y \sim y'$ implies that $x + y \sim x' + y'$.

9. Suppose S is a set of linearly independent vectors in E, and suppose T is a basis of E. Prove that there is a subset of T which, together with S, is again a basis of E.

10. Let ω be an involution in E. Show that the sets E_+ and E_- defined by

$$E_+ = \{x \in E;\ \omega x = x\}, \quad E_- = \{x \in E;\ \omega x = -x\}$$

are subspaces of E and that

$$E = E_+ \oplus E_- .$$

11. Let E_1, E_2 be subspaces of E. Show that $E_1 + E_2$ is the linear closure of $E_1 \cup E_2$. Prove that

$$E_1 + E_2 = E_1 \cup E_2$$

if and only if

$$E_1 \supset E_2 \quad \text{or} \quad E_2 \supset E_1 .$$

12. Find subspaces E_1, E_2, E_3 of Γ^3 such that

i) $E_i \cap E_j = 0 \quad (i \neq j)$

ii) $E_1 + E_2 + E_3 = \Gamma^3$

iii) the sum in ii) is not direct.

§ 4. Dimension

In this paragraph all vector spaces are defined over a fixed, but arbitrarily chosen field Γ of characteristic 0.

1.19. Finitely generated vector spaces. Suppose E is a finitely generated vector space, and consider a surjective linear mapping $\varphi: E \to F$. Then F is finitely generated as well. In fact, if $x_1 \ldots x_n$ is a system of generators for E, then the vectors $\varphi x_1, \ldots, \varphi x_n$ generate F. In particular, the factor space of a finitely generated space with respect to any subspace is finitely generated.

Now consider a subspace E_1 of E. In view of Cor. II to Proposition I, sec. 1.15 there exists a surjective linear mapping $\varphi: E \to E_1$. It follows that E_1 is finitely generated.

1.20. Dimension. Recall that every system of generators of a non-trivial vector space contains a basis. It follows that a finitely generated non-trivial vector space has a finite basis. In the following it will be shown that in this case every basis of E consists of the same number of vectors. This number will be called the *dimension of* E and will be denoted by $\dim E$. E will be called a finite-dimensional vector space. We extend the definition to the case $E=(0)$ by assigning the dimension 0 to the space (0). If E does not have finite dimension it will be called an *infinite-dimensional vector space.*

Proposition I: Suppose a vector space has a basis of n vectors. Then every family of $(n+1)$ vectors is linearly dependent. Consequently, n is the maximum number of linearly independent vectors in E and hence every basis of E consists of n vectors.

Proof: We proceed by induction on n. Consider first the case $n=1$ and let a be a basis vector of E. Then if $x \neq 0$ and $y \neq 0$ are two arbitrary vectors we have that

$$x = \lambda a, \quad \lambda \neq 0 \quad \text{and} \quad y = \mu a, \quad \mu \neq 0$$

whence

$$\mu x - \lambda y = 0.$$

Thus the vectors x and y are linearly dependent.

Now assume by induction that the proposition holds for every vector space having a basis of $r \leq n-1$ vectors.

Let E be a vector space, and let a_μ $(\mu = 1 \ldots n)$ be a basis of E and $x_1 \ldots x_{n+1}$ a family of $n+1$ vectors. We may assume that $x_{n+1} \neq 0$ because otherwise it would follow immediately that the vectors $x_1 \ldots x_{n+1}$ were linearly dependent.

Consider the factor space $E_1 = E/(x_{n+1})$ and the canonical projection

$$\pi: E \to E/(x_{n+1})$$

where (x_{n+1}) denotes the subspace generated by x_{n+1}. Since the system $\bar{a}_1, \ldots, \bar{a}_n$ generates E_1 it contains a basis of E_1 (cf. Cor. I to Theorem I, sec. 1.6). On the other hand the equation

$$x_{n+1} = \sum_{\nu=1}^{n} \lambda^\nu a_\nu$$

implies that

$$\sum_{\nu=1}^{n} \lambda^\nu \bar{a}_\nu = 0$$

and so the vectors $(\bar{a}_1, \ldots, \bar{a}_n)$ are linearly dependent. It follows that E_1 has a basis consisting of less than n vectors. Hence, by the induction hypothesis, the vectors $\bar{x}_1 \ldots \bar{x}_n$ are linearly dependent. Consequently, there exists a non-trivial relation

$$\sum_{\nu=1}^{n} \xi^\nu \bar{x}_\nu = 0$$

and so

$$\sum_{\nu=1}^{n} \xi^\nu x_\nu = \xi^{n+1} x_{n+1}.$$

This formula shows that the vectors $x_1 \ldots x_{n+1}$ are linearly dependent and closes the induction.

Example: Since the space Γ^n (cf. Example 1, sec. 1.2) has a basis of n vectors it follows that

$$\dim \Gamma^n = n.$$

Proposition II: Two finite dimensional vector spaces E and F are isomorphic if and only if they have the same dimension.

Proof: Let $\varphi: E \to F$ be an isomorphism. Then it follows from Proposition III, sec. 1.11 that φ maps a basis of E injectively onto a basis of F and so $\dim E = \dim F$. Conversely, assume that $\dim E = \dim F = n$ and let x_μ and y_μ $(\mu = 1 \ldots n)$ be bases of E and F respectively. According to Proposition II, sec. 1.11 there exists a linear mapping $\varphi: E \to F$ such that $\varphi x_\mu = y_\mu$ $(\mu = 1 \ldots n)$. Then φ maps the basis x_μ onto the basis y_μ and hence it is a linear isomorphism by Proposition III, sec. 1.11.

1.21. Subspaces and factor spaces. Let E_1 be a subspace of the n-dimensional vector space E. Then E_1 is finitely generated and so it has finite dimension m. Let $x_1 \dots x_m$ be a basis of E_1. Then the vectors $x_1 \dots x_m$ are linearly independent in E and so Cor. II to Theorem I, sec. 1.6 implies that the vectors x_i may be extended to a basis of E. Hence

$$\dim E_1 \leqq \dim E. \tag{1.29}$$

If equality holds, then the vectors $x_1 \dots x_m$ form a basis of E and it follows that $E_1 = E$.

Now it will be shown that

$$\dim E = \dim E_1 + \dim E/E_1. \tag{1.30}$$

If $E_1 = (0)$ or $E_1 = E$ (1.30) is trivial and so we may assume that E_1 is a proper non-trivial subspace of E,

$$0 < \dim E_1 < \dim E.$$

Let $x_1 \dots x_r$ be a basis of E_1 and extend it to a basis $x_1 \dots x_r \dots x_n$ of E. Then the vectors $\bar{x}_{r+1} \dots \bar{x}_n$ form a basis of E/E_1 (cf. sec. 1.18) and so (1.30) follows.

Finally, suppose that E is a direct sum of two subspaces E_1 and E_2,

$$E = E_1 \oplus E_2.$$

Then

$$\dim E = \dim E_1 + \dim E_2. \tag{1.31}$$

In fact, if $x_1 \dots x_p$ is a basis of E_1 and $x_{p+1} \dots x_{p+q}$ is a basis of E_2, then $x_1 \dots x_{p+q}$ is a basis of E, whence (1.31). More generally, if E is the direct sum of several subspaces,

$$E = \sum_{i=1}^{r} E_i$$

then

$$\dim E = \sum_{i=1}^{r} \dim E_i.$$

Formula (1.31) can also be generalized in the following way. Let E_1 and E_2 be arbitrary subspaces of E. Then

$$\dim(E_1 + E_2) + \dim(E_1 \cap E_2) = \dim E_1 + \dim E_2. \tag{1.32}$$

In fact, let $z_1 \dots z_r$ be a basis of $E_1 \cap E_2$ and extend it to a basis $z_1 \dots z_r$,

$x_{r+1} \dots x_p$ of E_1 and to a basis $z_1 \dots z_r$, $y_{r+1} \dots y_q$ of E_2. Then the vectors

$$z_1 \dots z_r, \quad x_{r+1}, \dots x_p, \quad y_{r+1} \dots y_q \tag{1.33}$$

form a basis of $E_1 + E_2$. Clearly, the vectors (1.33) generate $E_1 + E_2$.

To show that they are linearly independent, we comment first that the vectors x_i are linearly independent $\operatorname{mod}(E_1 \cap E_2)$. In fact, the relation

$$\sum_i \lambda^i x_i \in E_1 \cap E_2$$

implies that

$$\sum_i \lambda^i x_i = \sum_k \mu^k z_k$$

whence $\lambda^i = 0$ and $\mu^k = 0$. Now assume a relation

$$\sum_k \zeta^k z_k + \sum_i \xi^i x_i + \sum_j \eta^j y_j = 0 .$$

Then

$$\sum_i \xi^i x_i = -\sum_j \eta^j y_j - \sum_k \zeta^k z_k \in E_2$$

whence

$$\sum_i \xi^i x_i \in E_1 \cap E_2 .$$

Since the vectors x_i are linearly independent $\operatorname{mod}(E_1 \cap E_2)$ it follows that $\xi^i = 0$. In the same way it is shown that $\eta^j = 0$. Now it follows that $\zeta^k = 0$ and so the vectors (1.33) are linearly independent. Hence, they form a basis of $E_1 + E_2$ and we obtain that

$$\begin{aligned} \dim(E_1 + E_2) &= r + (p - r) + (q - r) \\ &= p + q - r \\ &= \dim E_1 + \dim E_2 - \dim(E_1 \cap E_2). \end{aligned}$$

Problems

1. Let (x_1, x_2) be a basis of a 2-dimensional vector space. Show that the vectors

$$\bar{x}_1 = x_1 + x_2 , \quad \bar{x}_2 = x_1 - x_2$$

again form a basis. Let (ξ^1, ξ^2) and $(\bar{\xi}^1, \bar{\xi}^2)$ be the components of a vector x relative to the bases (x_1, x_2) and $(\bar{x}_1, \bar{x}_2)$ respectively. Express the components $(\bar{\xi}^1, \bar{\xi}^2)$ in terms of the components (ξ^1, ξ^2).

2. Consider an n-dimensional complex vector space E. Since the multiplication with real coefficients in particular is defined in E, this space may

also be considered as a *real* vector space. Let $(z_1 \dots z_n)$ be a basis of E. Prove that the vectors $z_1 \dots z_n$, $iz_1 \dots iz_n$ form a basis of E if E is considered as a real vector space.

3. Let E be an n-dimensional real vector space and C the complex linear space as constructed in § 1, Problem 5. If $x_\nu (\nu = 1 \dots n)$ is a basis of E, prove that the vectors $(x_\nu, 0)(\nu = 1 \dots n)$ form a basis of C.

4. Consider the space Γ^n of n-tuples of scalars $\lambda \in \Gamma$. Choose as basis the vectors:

$$\begin{aligned} e_1 &= (1, 1, \dots, 1, 1) \\ e_2 &= (0, 1, \dots, 1, 1) \\ &\vdots \\ e_n &= (0, 0, \dots, 0, 1). \end{aligned}$$

Compute the components $\eta^1, \eta^2, \dots, \eta^n$ of the vector $x = (\xi^1, \xi^2, \dots, \xi^n)$ relative to the above basis. For which basis in Γ^n is the connection between the components of x and the scalars $\xi^1, \xi^2, \dots, \xi^n$ particularly simple?

5. In Γ^4 consider the subspace T of all vectors $(\xi^1, \xi^2, \xi^3, \xi^4)$ satisfying $\xi^1 + 2\xi^2 = \xi^3 + 2\xi^4$. Show that the vectors: $x_1 = (1, 0, 1, 0)$ and $x_2 = (0, 1, 0, 1)$ are linearly independent and lie in T; then extend this set of two vectors to a basis of T.

6. Let $\alpha_1, \alpha_2, \alpha_3$ be fixed real numbers. Show that all vectors $(\eta^1, \eta^2, \eta^3, \eta^4)$ in $\mathbb{R}^4$ obeying $\eta^4 = \alpha_1\eta^1 + \alpha_2\eta^2 + \alpha_3\eta^3$ form a subspace V. Show that V is generated by

$$x_1 = (1, 0, 0, \alpha_1); \; x_2 = (0, 1, 0, \alpha_2); \; x_3 = (0, 0, 1, \alpha_3).$$

Verify that x_1, x_2, x_3 form a basis of the subspace V.

7. In the space P of all polynomials of degree $\leqq n-1$ consider the two bases p_ν and q_ν defined by

$$\begin{aligned} p_\nu(t) &= t^\nu \\ q_\nu(t) &= (t-a)^\nu \quad (a, \text{constant}; \; \nu = 0, \dots, n-1). \end{aligned}$$

Express the vectors q_ν explicitly in terms of the vectors p_ν.

8. A subspace E_1 of a vector space E is said to have *co-dimension* n if the factor space E/E_1 has dimension n. Let E_1 and F_1 be subspaces of finite codimension, and let E_2, F_2 be complementary subspaces,

$$E_1 \oplus E_2 = E, \quad F_1 \oplus F_2 = E.$$

Show that

$$\dim E_2 = \operatorname{codim} E_1, \quad \dim F_2 = \operatorname{codim} F_1.$$

Prove that $E_1 \cap F_1$ has finite codimension, and that

$$\operatorname{codim}(E_1 \cap F_1) \leqq \dim E_2 + \dim F_2 .$$

9. Under the hypothesis of problem 8, construct a decomposition $E = H_1 \oplus H_2$ such that H_1 has finite codimension and

i) $H_1 \subset E_1 \cap F_1$
ii) $H_2 \supset E_2 + F_2$.

Show that

$$H_2 = E_2 \oplus (E_1 \cap H_2)$$

and

$$H_2 = F_2 \oplus (F_1 \cap H_2).$$

10. Let $(x_\alpha)_{\alpha \in A}$ and $(y_\beta)_{\beta \in B}$ be two bases for a vector space E. Establish a $1-1$ correspondence between the sets A and B.

11. Let E be an n-dimensional real vector space and E_1 be an $(n-1)$-dimensional subspace. Denote by E^1 the set of all vectors $x \in E$ which are not contained in E_1. Define an equivalence relation in E^1 as follows: Two vectors $x \in E^1$ and $y \in E^1$ are equivalent, if the straight segment

$$x(t) = (1-t)x + ty \qquad 0 \leqq t \leqq 1$$

is disjoint to E_1. Prove that there are precisely two equivalence classes.

12. Show that a vector space is not the union of finitely many proper subspaces.

13. Let E be an n-dimensional vector space. Let F_i $(i = 1 \dots k)$ be subspaces such that

$$\dim F_i \leqq r \qquad (i = 1 \dots k)$$

where $r < n$ is a given integer. Show that there is a subspace $F \subset E$ of dimension $n - r$ such that $F \cap F_i = 0$ $(i = 1 \dots k)$. *Hint:* Use problem 12.

§ 5. The topology of a real finite-dimensional vector space

1.22. Real topological vector spaces. Let E be a real vector space in which a topology is defined. Then E is called a *topological vector space* if the linear operations

$$E \times E \to E \quad \text{and} \quad \mathbb{R} \times E \to E \quad \text{defined by}$$
$$(x, y) \to x + y$$

and

$$(\lambda, x) \to \lambda x$$

are continuous.

Example: Consider the space $\mathbb{R}^n$. Since the set $\mathbb{R}^n$ is the Cartesian product of n copies of $\mathbb{R}$, a topology is induced in $\mathbb{R}^n$ by the topology in $\mathbb{R}$. It is easy to verify that the linear operations are continuous with respect to this topology and so $\mathbb{R}^n$ is a topological vector space. A second example is given in problem 6.

In the following it will be shown that a real vector space of finite dimension carries a natural topology.

Proposition: Let E be an n-dimensional vector space over $\mathbb{R}$. Then there exists precisely one topology in E satisfying the conditions

T_1: E is a topological vector space

T_2: Every linear function in E is continuous.

Proof: To prove the existence of such a topology let e_ν ($\nu = 1, \ldots, n$) be a fixed basis of E and consider the linear isomorphism $\varphi: \mathbb{R}^n \to E$ given by

$$(\xi^1, \ldots, \xi^n) \to \sum_\nu \xi^\nu e_\nu .$$

Then define the open sets in E by $\varphi(U)$ where U is an open set in $\mathbb{R}^n$. Clearly φ becomes a homeomorphism and the linear operations in E are continuous in this topology. Now let f be a linear function in E. Then we have for every $x_0 \in E$, $x \in E$

$$f(x) - f(x_0) = f(x - x_0) = \sum_\nu (\xi^\nu - \xi_0^\nu) f(e_\nu) .$$

Given an arbitrary positive number $\varepsilon > 0$ consider the neighbourhood, φU, of x_0 defined by

$$|\xi^\nu - \xi_0^\nu| < \delta \qquad \nu = 1, \ldots, n$$

where $\delta > 0$ is a number such that

$$\delta \cdot \sum_\nu |f(e_\nu)| < \varepsilon .$$

Then if $x \in \varphi U$ we have that

$$|f(x) - f(x_0)| < \delta \sum_\nu |f(e_\nu)| < \varepsilon$$

which proves the continuity of f at $x = x_0$.

It remains to be shown that the topology of E is uniquely determined by T_1 and T_2. In fact, suppose that an arbitrary topology is defined in E which satisfies T_1 and T_2.

Let e_ν ($\nu = 1, \ldots, n$) be a basis of E and define mappings $\varphi: \mathbb{R}^n \to E$ and $\psi: E \to \mathbb{R}^n$ by

$$\varphi(\xi^1, \ldots, \xi^n) = \sum_\nu \xi^\nu e_\nu$$

and

$$\psi x = (\xi^1(x), \ldots, \xi^n(x))$$

where

$$x = \sum_\nu \xi^\nu(x) e_\nu$$

T_1 implies that φ is continuous. On the other hand, the functions $x \to \xi^\nu(x)$ are linear and hence it follows from T_2 that ψ is continuous. Since

$$\psi \circ \varphi = \iota_{\mathbb{R}^n} \quad \text{and} \quad \varphi \circ \psi = \iota_E$$

we obtain that φ is a homeomorphism of $\mathbb{R}^n$ onto E. Hence the topology of E is uniquely determined by T_1 and T_2.

Corollary: The topology of E constructed above is independent of the basis e_ν.

Let F be a second finite-dimensional real vector space and let $\varphi: E \to F$ be a linear mapping. Then φ is continuous. In fact, if y_μ ($\mu = 1, \ldots, m$) is a basis of F we can write

$$\varphi x = \sum_\mu \eta^\mu(x) y_\mu$$

where the η^μ are linear functions in E. Now the continuity of φ follows from T_1 and T_2.

1.23. Complex topological vector spaces. The reader should verify that the results of sec. 1.22 carry over word for word in the case of complex spaces (if $\xi \in \mathbb{C}$, we write $|\xi| = \sqrt{\xi\bar{\xi}}$).

Problems

1. Let f be a real valued continuous function in the real n-dimensional linear space E such that

$$f(x + y) = f(x) + f(y) \qquad x, y \in E.$$

Prove that f is linear.

2. Let $\varphi: E_1 \to E_2$ be a surjective linear mapping of finite dimensional real vector spaces. Show that φ is open (the image of an open set in E_1 under φ is open in E_2).

3. Let $\pi: E \to E/F$ be the canonical projection, where E is a real finite dimensional vector space, and F is a subspace. Then the topology in E determines a topology in E/F (a subset $U \subset E/F$ is open if and only if $\pi^{-1}U$ is open in E).

a) Prove that this topology coincides with the natural topology in the vector space E/F.

b) Prove that the subspace topology of F coincides with the natural topology of F.

4. Show that every subspace of a finite dimensional real vector space is a closed set.

5. Construct a topology for finite dimensional real vector spaces that satisfies T_1 but not T_2, and a topology that satisfies T_2 but not T_1.

6. Let E be a real vector space. Then every finite dimensional subspace of E carries a natural topology. Let E_1 be any finite dimensional subspace of E, and let $U_1 \subset E_1$ be an open set. Moreover let E_2 be a complementary subspace in E, $E = E_1 \oplus E_2$. Then U_1 and E_2 determine a set O given by

$$O = \{x + y;\ x \in U_1, y \in E_2\}. \tag{1.34}$$

Suppose that

$$O' = \{x + y;\ x \in U'_1,\ y \in E'_2\}$$

is a second set of this form. Prove that $O \cap O'$ is again a set of this form. *Hint:* Use problems 8 and 9, § 4.

Conclude that the sets $O \subset E$ of the form (1.34) form a basis for a topology in E.

7. Prove that the topology defined in problem 6 satisfies T_1 and T_2.

8. Prove that the topology of problem 7 is regular. Show that E is not metrizable if it has infinite dimension.

9. Let $a_1, \ldots, a_n$ and $b_1, \ldots, b_n$ be bases of an n-dimensional vector space E. Fix some p $(1 \leq p \leq n)$. Show that for a suitable permutation σ, the vectors

$$a_1, \ldots, a_p, \qquad b_{\sigma[1]}, \ldots, b_{\sigma[n]}$$

and

$$b_{\sigma[1]}, \ldots, b_{\sigma[p]}, \qquad a_{p+1}, \ldots, a_n$$

are both bases of E.

Chapter II

Linear Mappings

In this chapter all vector spaces are defined over a fixed but arbitrarily chosen field, Γ, of characteristic 0.

§ 1. Basic properties

2.1. Kernel and image space. Suppose E, F are vector spaces and let $\varphi: E\to F$ be a linear mapping. Then the *kernel* of φ, denoted by $\ker\varphi$, is the subset of vectors $x\in E$ such that $\varphi x=0$. It follows from (1.8) and (1.9) that $\ker\varphi$ is a subspace of E.

The mapping φ is injective if and only if

$$\ker\varphi = (0). \tag{2.1}$$

In fact, if φ is injective there is at most one vector $x\in E$ such that $\varphi x=0$. But $\varphi 0=0$ and so it follows that $\ker\varphi=(0)$. Conversely, assume that (2.1) holds. Then if

$$\varphi x_1 = \varphi x_2$$

for two vectors $x_1, x_2\in E$ we have

$$\varphi(x_1 - x_2) = 0$$

whence $x_1 - x_2\in\ker\varphi$. It follows that $x_1-x_2=0$ and so $x_1=x_2$. Hence φ is injective.

The *image space* of φ, denoted by $\operatorname{Im}\varphi$, is the set of vectors $y\in F$ of the form $y=\varphi x$ for some $x\in E$. $\operatorname{Im}\varphi$ is a subspace of F. It is clear that φ is surjective if and only if $\operatorname{Im}\varphi=F$.

Example: Let E_1 be a subspace of E and consider the canonical projection

$$\pi: E\to E/E_1 .$$

Then

$$\ker\pi = E_1 \quad\text{and}\quad \operatorname{Im}\pi = E/E_1 .$$

2.2. The restriction of a linear mapping. Suppose $\varphi: E \to F$ is a linear mapping and let $E_1 \subset E$, $F_1 \subset F$ be subspaces such that

$$\varphi x \in F_1 \quad \text{for} \quad x \in E_1 .$$

Then the linear mapping

$$\varphi_1 : E_1 \to F_1$$

defined by

$$\varphi_1 x = \varphi x \qquad x \in E_1$$

is called the *restriction* of φ to E_1, F_1. It satisfies the relation

$$\varphi \circ i_E = i_F \circ \varphi_1$$

where $i_E: E_1 \to E$ and $i_F: F_1 \to F$ are the canonical injections. Equivalently, the diagram

$$\begin{array}{ccc} E & \xrightarrow{\varphi} & F \\ {\scriptstyle i_E}\uparrow & & \uparrow{\scriptstyle i_F} \\ E_1 & \xrightarrow{\varphi_1} & F_1 \end{array}$$

is commutative.

2.3. The induced mapping in the factor spaces. Let $\varphi: E \to F$ be a linear mapping and $\varphi_1 : E_1 \to F_1$ be its restriction to subspaces $E_1 \subset E$ and $F_1 \subset F$. Then there exists precisely one linear mapping

$$\bar{\varphi}: E/E_1 \to F/F_1$$

such that

$$\bar{\varphi} \circ \pi_E = \pi_F \circ \varphi \tag{2.2}$$

where

$$\pi_E : E \to E/E_1 \quad \text{and} \quad \pi_F : F \to F/F_1$$

are the canonical projections.

Since π_E is surjective, the mapping $\bar{\varphi}$ is uniquely determined by (2.2) if it exists. To define $\bar{\varphi}$ we notice first that

$$\pi_F \varphi x_1 = \pi_F \varphi x_2 \tag{2.3}$$

whenever

$$\pi_E x_1 = \pi_E x_2 . \tag{2.4}$$

In fact, (2.4) implies that

$$x_1 - x_2 \in \ker \pi_E = E_1 .$$

But by the hypothesis

$$\varphi x_1 - \varphi x_2 = \varphi(x_1 - x_2) \in F_1 = \ker \pi_F$$

and so

$$\pi_F \varphi x_1 = \pi_F \varphi x_2 .$$

It follows from (2.3) and (2.4) that there is a set map $\bar{\varphi}: E/E_1 \to E/F_1$ satisfying (2.2). To prove that $\bar{\varphi}$ is linear let $\bar{x} \in E/E_1$ and $\bar{y} \in E/E_1$ be arbitrary and choose vectors $x \in E$ and $y \in E$ such that $\pi_E x = \bar{x}$ and $\pi_E y = \bar{y}$. Then it follows from (2.2) that

$$\begin{aligned} \bar{\varphi}(\lambda \bar{x} + \mu \bar{y}) &= \bar{\varphi}\, \pi_E(\lambda x + \mu y) = \pi_F\, \varphi(\lambda x + \mu y) \\ &= \lambda\, \pi_F\, \varphi\, x + \mu\, \pi_F\, \varphi\, y = \lambda\, \bar{\varphi}\, \bar{x} + \mu\, \bar{\varphi}\, \bar{y} \end{aligned}$$

and hence $\bar{\varphi}$ is a linear mapping.

The reader should notice that the relation (2.2) is equivalent to the requirement that the diagram

$$\begin{array}{ccc} E & \xrightarrow{\varphi} & F \\ {\scriptstyle \pi_E}\downarrow & & \downarrow{\scriptstyle \pi_F} \\ E/E_1 & \xrightarrow{\bar{\varphi}} & F/F_1 \end{array}$$

be commutative. Setting $\pi_E x = \bar{x}$, $x \in E$ and $\pi_F y = \bar{y}$, $y \in F$ we can rewrite (2.2) in the form

$$\bar{\varphi}\, \bar{x} = \overline{\varphi x}\,.$$

2.4. The factoring of a linear mapping. Let $\varphi: E \to F$ be a linear mapping and consider the subspaces $E_1 = \ker \varphi$ and $F_1 = (0)$. Since $\varphi x = 0$, $x \in E_1$ a linear mapping

$$\bar{\varphi}: E/\ker \varphi \to F$$

is induced by φ (cf. sec. 2.3) such that

$$\bar{\varphi} \circ \pi = \varphi \tag{2.5}$$

where π denotes the canonical projection

$$\pi: E \to E/\ker \varphi\,.$$

The mapping $\bar{\varphi}$ is injective. In fact, if $\bar{\varphi}\pi x = 0$ we have that $\varphi x = 0$. Hence $x \in \ker \varphi$ and so $\pi x = 0$. It follows that $\bar{\varphi}$ is injective. In particular, the restriction of $\bar{\varphi}$ to $E/\ker \varphi$, $\operatorname{Im} \varphi$ (also denoted by $\bar{\varphi}$) is a linear isomorphism

$$\bar{\varphi}: E/\ker \varphi \xrightarrow{\cong} \operatorname{Im} \varphi\,.$$

Formula (2.5) shows that every linear mapping $\varphi: E \to F$ can be written as the composition of a surjective and an injective linear mapping,

$$\begin{array}{ccc} E & \xrightarrow{\varphi} & F \\ {\scriptstyle \pi}\downarrow & \nearrow{\scriptstyle \bar{\varphi}} & \\ E/\ker \varphi & & \end{array}$$

As an application it will now be shown that for any two subspaces $E_1 \subset E$ and $E_2 \subset E$ there is a natural isomorphism

$$E_1/(E_1 \cap E_2) \xrightarrow{\cong} (E_1 + E_2)/E_2 . \tag{2.6}$$

Consider the canonical projection

$$\pi: E_1 + E_2 \to (E_1 + E_2)/E_2$$

and let φ be the restriction of π to $E_1, (E_1+E_2)/E_2$. Then φ is surjective. In fact, if

$$x = x_1 + x_2, \qquad x_1 \in E_1, x_2 \in E_2$$

is any vector of $E_1 + E_2$ we have

$$\pi x = \pi(x_1 + x_2) = \pi x_1 = \varphi x_1 .$$

Since

$$\ker \varphi = \ker \pi \cap E_1 = E_2 \cap E_1$$

it follows that φ induces a linear isomorphism

$$\bar{\varphi}: E_1/(E_1 \cap E_2) \xrightarrow{\cong} (E_1 + E_2)/E_2 .$$

Now consider the special case that

$$E = E_1 \oplus E_2 .$$

Then $E_1 \cap E_2 = 0$ and hence the relation (2.6) reduces to

$$E_1 \xrightarrow{\cong} E/E_2 .$$

As a second example, let $f_i (i=1 \ldots r)$ be r linear functions in E and define a subspace $F \subset E$ by

$$F = \bigcap_{i=1}^{r} \ker f_i .$$

Now consider the linear mapping $\varphi: E \to \Gamma^r$ defined by

$$\varphi x = (f_1(x), \ldots, f_r(x)) .$$

Then clearly

$$\ker \varphi = \bigcap_{i=1}^{r} \ker f_i = F$$

and so φ determines a linear isomorphism

$$\bar{\varphi}: E/F \xrightarrow{\cong} \operatorname{Im} \varphi \subset \Gamma^r .$$

It follows that Im φ, and hence E/F, has dimension $\leqq r$,

$$\dim E/F \leqq r.$$

Proposition I: Suppose $\varphi: E \to F$ and $\psi: E \to G$ are linear mappings such that

$$\ker \varphi \subset \ker \psi.$$

Then ψ can be factored over φ; that is, there exists a linear mapping $\chi: F \to G$ such that

$$\chi \circ \varphi = \psi.$$

Proof: Since ψ maps $\ker \varphi$ into 0 it induces a linear mapping $\bar{\psi}: E/\ker \varphi \to G$ such that

$$\bar{\psi} \circ \pi = \psi$$

where

$$\pi: E \to E/\ker \varphi$$

is the canonical projection. Let

$$\bar{\varphi}: E/\ker \varphi \xrightarrow{\cong} \operatorname{Im} \varphi$$

be the linear isomorphism determined by φ, and define a linear mapping $\bar{\psi}_1: \operatorname{Im} \varphi \to G$ by

$$\bar{\psi}_1 = \bar{\psi} \circ \bar{\varphi}^{-1}.$$

Finally, let $\chi: F \to G$ be any linear mapping which extends $\bar{\psi}_1$. Then we have that

$$\bar{\varphi}^{-1} \circ \varphi = \bar{\varphi}^{-1} \circ \bar{\varphi} \circ \pi = \pi$$

whence

$$\chi \circ \varphi = \bar{\psi}_1 \circ \varphi = \bar{\psi} \circ \bar{\varphi}^{-1} \circ \varphi = \bar{\psi} \circ \pi = \psi.$$

Our result is expressed in the commutative diagram

$$\begin{array}{ccc} E & \xrightarrow{\varphi} & F \\ \psi \downarrow & \swarrow \chi & \\ G & & \end{array}$$

2.5. Exact sequences. Exact sequences provide a sophisticated method for describing elementary properties of linear mappings.

A sequence of linear mappings

$$F \xrightarrow{\varphi} E \xrightarrow{\psi} G \tag{2.7}$$

is called *exact at E* if

$$\operatorname{Im}\varphi = \ker\psi .$$

We exhibit the following special cases:

1. $F=0$. Then the exact sequence (2.7) reads

$$0 \xrightarrow{\varphi} E \xrightarrow{\psi} G . \tag{2.8}$$

Since $\operatorname{Im}\varphi=0$ it follows that $\ker\psi=0$ i.e., ψ is injective. Conversely, suppose ψ: $E\to G$ is injective. Then $\ker\psi=0$, and so the sequence (2.8) is exact at E.

2. $G=0$. Then the exact sequence (2.7) has the form

$$F \xrightarrow{\varphi} E \xrightarrow{\psi} 0 . \tag{2.9}$$

Since ψ is the zero mapping it follows that

$$\operatorname{Im}\varphi = \ker\psi = E$$

and so φ is surjective. Conversely, if the linear mapping $\varphi: F\to E$ is surjective, then the sequence (2.9) is exact.

A *short exact sequence* is a sequence of the form

$$0 \to F \xrightarrow{\varphi} E \xrightarrow{\psi} G \to 0 \quad {}^{*)} \tag{2.10}$$

which is exact at F, E and G. As an example consider the sequence

$$0 \to E_1 \xrightarrow{i} E \xrightarrow{\pi} E/E_1 \to 0 \tag{2.11}$$

where E_1 is a subspace of E and i, π denote the canonical injection and projection respectively. Then

$$\operatorname{Im} i = E_1 = \ker\pi$$

and so (2.11) is exact at E. Moreover, since i and π are respectively injective and surjective, it follows that (2.11) is exact at E_1 and E/E_1 and so (2.11) is a short exact sequence.

The example above is essentially the only example of a short exact sequence.

---

*) It is clear that the first and the last mapping in the above diagram are the zero mappings.

In fact, suppose

$$0 \to F \xrightarrow{\varphi} E \xrightarrow{\psi} G \to 0$$

is a short exact sequence. Let

$$E_1 = \operatorname{Im}\varphi = \ker\psi$$

and consider the exact sequence

$$0 \to E_1 \xrightarrow{i} E \xrightarrow{\pi} E/E_1 \to 0.$$

Since the mapping $\varphi: F \to E$ is injective its restriction φ_1 to F, E_1 is a linear isomorphism, $\varphi_1: F \xrightarrow{\cong} E_1$. On the other hand, ψ induces a linear isomorphism

$$\bar{\psi}: E/E_1 \xrightarrow{\cong} G.$$

Now it follows easily from the definitions that the diagram

$$\begin{array}{ccccccc} 0 \to & F & \to & E & \to & G & \to 0 \\ & {\scriptstyle\varphi_1}\downarrow\cong & & {\scriptstyle\iota}\downarrow\cong & & {\scriptstyle\bar{\psi}^{-1}}\downarrow\cong & \\ 0 \to & E_1 & \to & E & \to & E/E_1 & \to 0 \end{array} \tag{2.12}$$

is commutative.

2.6. Homomorphisms of exact sequences. A commutative diagram of the form

$$\begin{array}{ccccccc} 0 \to & F_1 & \xrightarrow{\varphi_1} & E_1 & \xrightarrow{\psi_1} & G_1 & \to 0 \\ & \downarrow{\scriptstyle\varrho} & & \downarrow{\scriptstyle\sigma} & & \downarrow{\scriptstyle\tau} & \\ 0 \to & F_2 & \xrightarrow{\varphi_2} & E_2 & \xrightarrow{\psi_2} & G_2 & \to 0 \end{array} \tag{2.13}$$

where both horizontal sequences are short exact sequences, and ϱ, σ, τ are linear mappings, is called a *homomorphism* of exact sequences. If ϱ, σ, τ are linear isomorphisms, then (2.13) is called an *isomorphism* between the two short exact sequences. In particular, (2.12) is an isomorphism of short exact sequences.

2.7. Split short exact sequences. Suppose that

$$0 \to F \xrightarrow{\varphi} E \xrightarrow{\psi} G \to 0 \tag{2.10}$$

is a short exact sequence, and assume that $\chi: E \leftarrow G$ is a linear mapping such that

$$\psi \circ \chi = \iota.$$

Then χ is said to *split* the sequence (2.10) and the sequence

$$0 \to F \xrightarrow{\varphi} E \underset{\chi}{\overset{\psi}{\rightleftarrows}} G \to 0$$

is called a *split short exact sequence.*

Proposition II: Every short exact sequence can be split.

Proof: Given a short exact sequence, (2.10) let E_1 be a complementary subspace of $\ker\psi$ in E,

$$E = E_1 \oplus \ker\psi$$

and consider the restriction, ψ_1, of ψ to E_1, G. Since $\ker\psi_1 = 0$, ψ_1 is a linear isomorphism, $\psi_1 : E_1 \xrightarrow{\cong} G$. Then the mapping $\chi : E_1 \leftarrow G$ defined by $\chi = \psi_1^{-1}$ satisfies the relation

$$\psi\chi z = \psi\psi_1^{-1} z = \psi_1\psi_1^{-1} z = \iota z \qquad z \in G$$

and hence χ splits the sequence.

2.8. Stable subspaces. Consider now the case $F = E$; i.e., let φ be a linear transformation of the vector space E. Then a subspace $E_1 \subset E$ will be called *stable under* φ if

$$\varphi x \in E_1 \quad \text{for} \quad x \in E_1 .$$

It is easy to verify that the subspaces $\ker\varphi$ and $\operatorname{Im}\varphi$ are stable. If E_1 is a stable subspace, the restriction, φ_1, of φ to E_1, E_1 will be called the *restriction of* φ *to* E_1. Clearly, φ_1 is a linear transformation of E_1. We also have that the induced map

$$\bar{\varphi} : E/E_1 \to E/E_1$$

is a linear transformation of E/E_1.

Problems

1. Let C be the space of continuous functions $f : \mathbb{R} \to \mathbb{R}$ and define the mapping $\varphi : C \to C$ by

$$\varphi : f(t) \to \int_0^t f(s)\, ds .$$

Prove that $\operatorname{Im}\varphi$ consists of all continuously differentiable functions while the kernel of φ is 0. Conclude that φ is injective but not bijective.

2. Find the image spaces and kernels of the following linear transformations of Γ^4:

a) $\psi(\xi^1, \xi^2, \xi^3, \xi^4) = (\xi^1 - \xi^2, \xi^1 + \xi^2, \xi^3, \xi^4)$
b) $\psi(\xi^1, \xi^2, \xi^3, \xi^4) = (\xi^1, \xi^1, \xi^1, \xi^2)$
c) $\psi(\xi^1, \xi^2, \xi^3, \xi^4) = (\xi^4, \xi^1 + \xi^2, \xi^1 + \xi^3, \xi^4)$.

3. Find the image spaces and kernels of the following linear mappings of Γ^4 into Γ^5:

a) $\varphi(\xi^1,\xi^2,\xi^3,\xi^4)=(5\xi^1-\xi^2,\xi^1+\xi^2,\xi^3,\xi^4,\xi^1)$
b) $\varphi(\xi^1,\xi^2,\xi^3,\xi^4)=(\xi^1+\xi^2+7\xi^3+\xi^4,2\xi^3+\xi^4,\xi^1,\xi^2,\xi^1-\xi^2)$
c) $\varphi(\xi^1,\xi^2,\xi^3,\xi^4)=(\xi^4-\xi^2+\xi^3+\xi^1,\xi^3-\xi^2,17\xi^1+13\xi^2,16\xi^1+ + 5\xi^4,\xi^2-\xi^3)$

4. Construct bases for the factor spaces $\Gamma^4/\ker\psi$ and $\Gamma^4/\ker\varphi$ of problems 2 and 3. Determine the action of the induced mappings on these bases and verify that the induced mappings are injective.

5. Prove that if $\varphi:E\to F$ and $\psi:E\to G$ are linear mappings, then the relation

$$\ker\varphi\subset\ker\psi$$

is necessary for the existence of a linear mapping $\chi:F\to G$ such that $\psi=\chi\circ\varphi$.

6. Consider the pairs (ψ,φ) in parts a, b, c of problems 2 and 3. Decide in each case if ψ can be factored over φ, or if φ can be factored over ψ, or if both factorings are possible. Whenever ψ can be factored over φ (or conversely) construct an explicit factoring map.

7. a) Use formula (2.6) to obtain an elegant proof of formula (1.32).
b) Establish a linear isomorphism

$$(E/F)/(E_1/F)\xrightarrow{\cong}E/E_1$$

where $F\subset E_1\subset E$.

8. Consider the short exact sequence

$$0\to E_1\xrightarrow{i}E\xrightarrow{\pi}E/E_1\to 0.$$

Show that the relation $\chi\rightleftarrows\operatorname{Im}\chi$ defines a $1-1$ correspondence between linear mappings $\chi:E\leftarrow E/E_1$ which split the sequence, and complementary subspaces of E_1 in E.

9. Show that a short exact sequence $0\to F\xrightarrow{\varphi}E\xrightarrow{\psi}G\to 0$ is split if and only if there exists a linear mapping $\omega:F\leftarrow E$ such that $\omega\circ\varphi=\iota$.
In the process establish a $1-1$ correspondence between the split short exact sequences of the form

$$0\to F\xrightarrow{\varphi}E\underset{\chi}{\overset{\psi}{\rightleftarrows}}G\to 0$$

and of the form

$$0\to F\underset{\omega}{\overset{\varphi}{\rightleftarrows}}E\xrightarrow{\psi}G\to 0$$

such that the diagram

$$0 \leftarrow F \overset{\omega}{\leftarrow} E \overset{\chi}{\leftarrow} G \leftarrow 0$$

is again a short exact sequence.

10. Assume a commutative diagram of linear maps

$$\begin{array}{ccccccccc} E_1 & \xrightarrow{\alpha_1} & E_2 & \xrightarrow{\alpha_2} & E_3 & \xrightarrow{\alpha_3} & E_4 & \xrightarrow{\alpha_4} & E_5 \\ \downarrow{\scriptstyle \varphi_1} & & \downarrow{\scriptstyle \varphi_2} & & \downarrow{\scriptstyle \varphi_3} & & \downarrow{\scriptstyle \varphi_4} & & \downarrow{\scriptstyle \varphi_5} \\ F_1 & \xrightarrow{\beta_1} & F_2 & \xrightarrow{\beta_2} & F_3 & \xrightarrow{\beta_3} & F_4 & \xrightarrow{\beta_4} & F_5 \end{array}$$

where both horizontal sequences are exact.

i) Show that if φ_4 is injective and φ_1 is surjective, then

$$\ker \varphi_3 = \alpha_2(\ker \varphi_2).$$

ii) Show that if φ_2 is surjective and φ_5 is injective, then

$$\operatorname{Im} \varphi_3 = \beta_3^{-1}(\operatorname{Im} \varphi_4).$$

iii) Conclude that if the maps $\varphi_1, \varphi_2, \varphi_4, \varphi_5$ are linear isomorphisms, then so is φ_3 (5-lemma).

11. Consider a system of linear mappings

$$\begin{array}{ccccccc} & & 0 & & 0 & & 0 \\ & & \downarrow & & \downarrow & & \downarrow \\ 0 & \to & E_{00} & \xrightarrow{\varphi_{00}} & E_{01} & \xrightarrow{\varphi_{01}} & E_{02} \to \\ & & \psi_{00}\downarrow & & \psi_{01}\downarrow & & \psi_{02}\downarrow \\ 0 & \to & E_{10} & \xrightarrow{\varphi_{10}} & E_{11} & \xrightarrow{\varphi_{11}} & E_{12} \to \\ & & \psi_{10}\downarrow & & \psi_{11}\downarrow & & \psi_{12}\downarrow \\ 0 & \to & E_{20} & \xrightarrow{\varphi_{20}} & E_{21} & \xrightarrow{\varphi_{21}} & E_{22} \to \\ & & \downarrow & & \downarrow & & \downarrow \end{array}$$

where all the horizontal and the vertical sequences are exact at each E_{ij}. Assume that the diagram is commutative. Define spaces H_{ij} $(i \geqq 1, j \geqq 1)$ by

$$H_{ij} = (\ker \varphi_{ij} \cap \ker \psi_{ij})/\operatorname{Im}(\psi_{i-1\,j} \circ \varphi_{i-1\,j-1}).$$

Construct a linear isomorphism between $H_{i,j+1}$ and $H_{i+1,j}$.

12. Given an exact sequence

$$E \xrightarrow{\varphi} F \xrightarrow{\psi} G \xrightarrow{\chi} H$$

prove that φ is surjective if and only if χ is injective.

13. Prove the *hexagonal lemma*: Given a diagram of linear maps

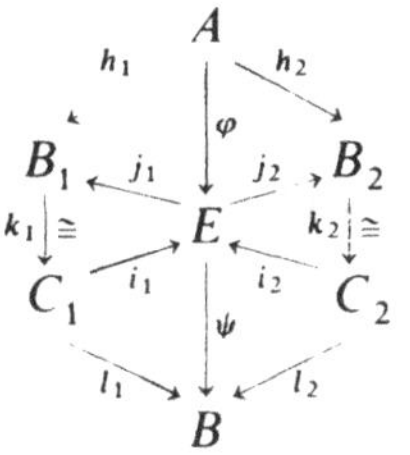

in which all triangles are commutative, k_1 and k_2 are isomorphisms and the diagonals $j_2 \circ i_1$ and $j_1 \circ i_2$ are exact at E, show that

$$l_1 \circ k_1 \circ h_1 + l_2 \circ k_2 \circ h_2 = \psi \circ \varphi .$$

§ 2. Operations with linear mappings

2.9. The space $L(E; F)$. Let E and F be vector spaces and consider the set $L(E; F)$ of linear mappings $\varphi: E \to F$. If φ and ψ are two such mappings $\varphi + \psi$ and $\lambda\varphi$ of E into F are defined by

$$(\varphi + \psi)x = \varphi x + \psi x$$

and

$$(\lambda\varphi)x = \lambda\varphi x \qquad x \in E .$$

It is easy to verify that $\varphi + \psi$ and $\lambda\varphi$ are again linear mappings, and so the set $L(E; F)$ becomes a linear space, called the *space of linear mappings of E into F*. The zero vector of $L(E; F)$ is the linear mapping 0 defined by $0\,x = 0$, $x \in E$.

In the case that $F = \Gamma$ (φ and ψ are linear *functions*) $L(E; \Gamma)$ is denoted simply by $L(E)$.

2.10. Composition. Recall (sec. 1.10) that if $\varphi: E \to F$ and $\psi: F \to G$ are linear mappings then the mapping $\psi \circ \varphi: E \to G$ defined by

$$(\psi \circ \varphi)x = \psi(\varphi x)$$

is again linear. If H is a fourth linear space and $\chi: G \to H$ is a linear mapping, we have for each $x \in E$

$$[\chi \circ (\psi \circ \varphi)]\,x = \chi(\psi \circ \varphi)x = \chi[\psi(\varphi x)] = (\chi \circ \psi)\varphi x = [(\chi \circ \psi) \circ \varphi]\,x$$

whence

$$\chi \circ (\psi \circ \varphi) = (\chi \circ \psi) \circ \varphi . \tag{2.14}$$

Consequently, we can simply write $\chi \circ \psi \circ \varphi$.

If $\varphi: E \to F$ is a linear mapping and ι_E and ι_F are the identity mappings of E and F we have clearly

$$\varphi \circ \iota_E = \varphi \quad \text{and} \quad \iota_F \circ \varphi = \varphi. \tag{2.15}$$

Moreover, if φ is a linear isomorphism and φ^{-1} is the inverse isomorphism we have the relations

$$\varphi^{-1} \circ \varphi = \iota_E \quad \text{and} \quad \varphi \circ \varphi^{-1} = \iota_F. \tag{2.16}$$

Finally, if $\varphi_i: E \to F$ and $\psi_i: F \to G$ are linear mappings, then it is easily checked that

$$\Big(\sum_i \lambda^i \psi_i\Big) \circ \varphi = \sum_i \lambda^i (\psi_i \circ \varphi)$$

and (2.17)

$$\psi \circ \Big(\sum_i \lambda^i \varphi_i\Big) = \sum_i \lambda^i (\psi \circ \varphi_i).$$

2.11. Left and right inverses. Let $\varphi: E \to F$ and $\psi: E \leftarrow F$ be linear mappings. Then ψ is called a *right inverse* of φ if $\varphi \circ \psi = \iota_F$.
ψ is called a *left inverse* of φ if $\psi \circ \varphi = \iota_E$.

Proposition I: A linear mapping $\varphi: E \to F$ is surjective if and only if it has a right inverse. It is injective if and only if it has a left inverse.

Proof: Suppose φ has a right inverse, ψ. Then we have for every $y \in F$

$$y = \varphi \psi y$$

and so $y \in \operatorname{Im} \varphi$; i.e., φ is surjective. Conversely, if φ is surjective, let E_1 be a complementary subspace of $\ker \varphi$ in E,

$$E = E_1 \oplus \ker \varphi.$$

Then the restriction φ_1 of φ to E_1, F is a linear isomorphism. Define the linear mapping $\psi: E \leftarrow F$ by $\psi = i_1 \varphi_1^{-1}$, where $i_1: E_1 \to E$ is the canonical injection. Then

$$\varphi \psi y = \varphi_1 \varphi_1^{-1} y = y, \qquad y \in F$$

i.e., $\varphi \circ \psi = \iota_F$.

For the proof of the second part of the proposition assume that φ has a left inverse. Then if $x \in \ker \varphi$ we have that

$$x = \psi \varphi x = \psi 0 = 0$$

whence $\ker \varphi = 0$. Consequently φ is injective.

Conversely, if φ is injective, consider the restriction φ_1 of φ to E, $\operatorname{Im}\varphi$. Then φ_1 is a linear isomorphism. Let $\pi: F \to \operatorname{Im}\varphi$ be a linear mapping such that

$$\pi y = y \quad \text{for} \quad y \in \operatorname{Im}\varphi$$

(cf. Cor. II, Proposition I, sec. 1.15) and define $\psi: E \leftarrow F$ by

$$\psi = \varphi_1^{-1} \circ \pi .$$

Then we have that

$$\psi \varphi x = \varphi_1^{-1} \pi \varphi x = \varphi_1^{-1} \varphi x = \varphi_1^{-1} \varphi_1 x = x$$

whence $\psi \circ \varphi = \iota_E$. Hence φ has a left inverse. This completes the proof.

Corollary: A linear isomorphism $\varphi: E \to F$ has a uniquely determined right (left) inverse, namely, φ^{-1}.

Proof: Relation (2.16) shows that φ^{-1} is a left (and right) inverse to φ. Now let ψ be any left inverse of φ,

$$\psi \circ \varphi = \iota_E .$$

Then multiplying by φ^{-1} from the right we obtain

$$\psi \circ \varphi \circ \varphi^{-1} = \varphi^{-1}$$

whence $\psi = \varphi^{-1}$. In the same way it is shown that the only right inverse of φ is φ^{-1}.

2.12. Linear automorphisms. Consider the set $GL(E)$ of all linear automorphisms of E. Clearly, $GL(E)$ is closed under the composition $(\varphi, \psi) \to \psi \circ \varphi$ and it satisfies the following conditions:

i) $\chi \circ (\psi \circ \varphi) = (\chi \circ \psi) \circ \varphi$ (associative law)

ii) there exists an element ι (the identity map) such that $\varphi \circ \iota = \iota \circ \varphi = \varphi$ for every $\varphi \in GL(E)$

iii) to every $\varphi \in GL(E)$ there is an element $\varphi^{-1} \in GL(E)$ such that $\varphi^{-1} \circ \varphi = \varphi \circ \varphi^{-1} = \iota$.

In other words, the linear automorphisms of E form a *group*.

Problems

1. Show that if E, F are vector spaces, then the inclusions

$$L(E; F) \subset C(E; F) \subset (E; F)$$

are proper ($(E; F)$ is defined in Example 3, sec. 1.2 and $C(E; F)$ is defined in problem 9, § 1, chap. I). Under which conditions do any of these spaces have finite dimension?

2. Suppose

$$\varphi_1, \psi_1, \chi_1 : E \to F \quad \text{and} \quad \varphi_2, \psi_2, \chi_2 : F \to G$$

are linear mappings. Assume that φ_1, φ_2 are injective, ψ_1, ψ_2 are surjective and χ_1, χ_2 are bijective. Prove that

a) $\varphi_2 \circ \varphi_1$ is injective

b) $\psi_2 \circ \psi_1$ is surjective

c) $\chi_2 \circ \chi_1$ is bijective

3. Let $\varphi: E \to F$ be a linear mapping. a) Consider the space $M^l(\varphi)$ of linear mappings $\psi: E \leftarrow F$ such that $\psi \circ \varphi = 0$. Prove that if φ is surjective then $M^l(\varphi) = 0$.

b) Consider the space $M^r(\varphi)$ of linear mappings $\psi: E \leftarrow F$ such that $\varphi \circ \psi = 0$. Prove that if φ is injective then $M^r(\varphi) = 0$.

4. Suppose that $\varphi: E \to F$ is injective and let $M^l(\varphi)$ be the subspace defined in problem 3. Show that the set of left inverses of φ is a coset in the factor space $L(F; E)/M^l(\varphi)$, and conclude that the left inverse of φ is uniquely determined if and only if φ is surjective. Establish a similar result for surjective linear mappings.

5. Show that the space $M^l(\varphi)$ of problem 3 is the set of linear mappings $\psi: E \leftarrow F$ such that $\operatorname{Im} \varphi \subset \ker \psi$. Construct a natural linear isomorphism between $M^l(\varphi)$ and $L(F/\operatorname{Im} \varphi; E)$.

Construct a natural linear isomorphism between $M^r(\varphi)$ (cf. problem 3) and $L(F; \ker \varphi)$.

6. Assume that $\varphi: E \to E$ is a linear transformation such that $\varphi \circ \psi = \psi \circ \varphi$ for every linear transformation ψ. Prove that $\varphi = \lambda \iota$ where λ is a scalar. *Hint:* Show first that, for every vector $x \in E$ there is a scalar $\lambda(x)$ such that $\varphi x = \lambda(x) x$. Then prove that $\lambda(x)$ does not depend on x.

7. Prove that the group $GL(E)$ is not commutative for $\dim E > 1$. If $\dim E = 1$, show that $GL(E)$ is isomorphic to the multiplicative group of the field Γ.

8. Let E be a vector space and S be a set of linear transformations of E. A subspace $F \subset E$ is called *stable* with respect to S if F is stable under every $\varphi \in S$. The space E is called *irreducible* with respect to S if the only stable subspaces are $F = 0$ and $F = E$.

Prove *Schur's Lemma:* Let E and F be vector spaces and $\alpha: E \to F$ be a linear mapping. Assume that S_E and S_F are two sets of linear transformations of E and F with respect to which E and F are irreducible and such that

$$\alpha S_E = S_F \alpha$$

i.e. to every transformation $\varphi \in S_E$ there exists a transformation $\psi \in S_F$ such that $\alpha \circ \varphi = \psi \circ \alpha$ and conversely. Prove that $\alpha = 0$ or α is a linear isomorphism of E onto F.

§ 3. Linear isomorphisms

2.13. It is customary to state simply that a linear isomorphism preserves all linear properties. We shall attempt to make this statement more precise, by listing without proof (the proofs being all trivial) some of the important properties which are preserved under an isomorphism $\varphi: E \xrightarrow{\cong} F$.

Property I: The image under φ of a generating set (linearly independent set, basis) in E is a generating set (linearly independent set, basis) in F.

Property II: If E_1 is any subspace in E, and E/E_1 is the corresponding factor space, then φ determines linear isomorphisms

$$E_1 \xrightarrow{\cong} \varphi E_1$$

and

$$E/E_1 \xrightarrow{\cong} \varphi E/\varphi E_1 .$$

Property III: If G is a third vector space, then the mappings

$$\psi \to \psi \circ \varphi^{-1} \qquad \psi \in L(E; G)$$

and

$$\psi \to \varphi \circ \psi \qquad \psi \in L(G; E)$$

are linear isomorphisms

$$L(E; G) \xrightarrow{\cong} L(F; G)$$

and

$$L(G; E) \xrightarrow{\cong} L(G; F) .$$

2.14. Identification: Suppose $\varphi: E \to F$ is an injective linear mapping. Then φ determines a linear isomorphism

$$\varphi_1 : E \xrightarrow{\cong} \operatorname{Im} \varphi .$$

It may be convenient not to distinguish between E and $\operatorname{Im}\varphi$, but to regard them as the *same* vector space. This is called *identification*, and while in some sense it is sloppy mathematics, it leads to a great deal of economy of formulae and a much clearer presentation. Of course we shall only identify spaces whenever there is no possibility of confusion.

§ 4. Direct sum of vector spaces

2.15. Definition. Let E and F be two vector spaces and consider the set $E \times F$ of all ordered pairs (x, y), $x \in E$, $y \in F$. It is easy to verify that the set $E \times F$ becomes a vector space under the operations

$$(x_1, y_1) + (x_2, y_2) = (x_1 + x_2, y_1 + y_2)$$

and

$$\lambda(x, y) = (\lambda x, \lambda y)$$

This vector space is called the *(external) direct sum* of E and F and is denoted by $E \oplus F$. If $(x_\alpha)_{\alpha \in A}$ and $(y_\beta)_{\beta \in B}$ are bases of E and F respectively then the pairs $(x_\alpha, 0)$ and $(0, y_\beta)$ form a basis of $E \oplus F$. In particular, if E and F are finite dimensional we have that

$$\dim(E \oplus F) = \dim E + \dim F.$$

2.16. The canonical injections and projections. Consider the linear mappings

$$i_1: E \to E \oplus F \quad i_2: F \to E \oplus F$$

defined by

$$i_1 x = (x, 0) \quad i_2 y = (0, y)$$

and the linear mappings

$$\pi_1: E \oplus F \to E \quad \pi_2: E \oplus F \to F$$

given by

$$\pi_1(x, y) = x \quad \pi_2(x, y) = y.$$

It follows immediately from the definitions that

$$\pi_1 \circ i_1 = \iota_E \quad \pi_2 \circ i_2 = \iota_F \tag{2.18}$$

$$\pi_1 \circ i_2 = 0 \quad \pi_2 \circ i_1 = 0 \tag{2.19}$$

and

$$i_1 \circ \pi_1 + i_2 \circ \pi_2 = \iota_{E \oplus F}. \tag{2.20}$$

The relations (2.18) imply that the mappings $i_\lambda (\lambda = 1, 2)$ are injective and the mappings $\pi_\lambda (\lambda = 1,2)$ are surjective. The mappings i_λ are called respectively the *canonical injections* and π_λ the *canonical projections* associated with the external direct sum $E \oplus F$. Since i_1 and i_2 are injective we can identify E with $\text{Im}\, i_1$ and F with $\text{Im}\, i_2$. Then E and F become subspaces of $E \oplus F$, and $E \oplus F$ is the internal direct sum of E and F.

The reader will have noticed that we have used the same symbol to denote the external and the internal direct sums of two subspaces of a vector space. However, it will always be clear from the context whether the internal or the external direct sum is meant. (If we perform the identification, then the distinction vanishes). In the discussion of direct sums of families of subspaces (see sec. 2.17) we adopt different notations.

If $F = E$ we define an injective mapping $\Delta : E \to E \oplus E$ by

$$\Delta x = (x, x).$$

Δ is called the *diagonal mapping*. In terms of i_1 and i_2 the diagonal mapping can be written as

$$\Delta = i_1 + i_2.$$

Relations (2.18) and (2.19) imply that

$$\pi_1 \circ \Delta = \pi_2 \circ \Delta = \iota_E.$$

The following proposition shows that the direct sum of two vector spaces is characterized by its canonical injections and projections up to an isomorphism.

Proposition I: Let E, F, G be three vector spaces and suppose that a system of linear mappings

$$\begin{aligned} \varphi_1 &: E \to G, & \psi_1 &: G \to E \\ \varphi_2 &: F \to G, & \psi_2 &: G \to F \end{aligned}$$

is given subject to the conditions

$$\begin{aligned} \psi_1 \circ \varphi_1 &= \iota_E & \psi_2 \circ \varphi_2 &= \iota_F \\ \psi_1 \circ \varphi_2 &= 0 & \psi_2 \circ \varphi_1 &= 0 \end{aligned}$$

and

$$\varphi_1 \circ \psi_1 + \varphi_2 \circ \psi_2 = \iota_G.$$

Then there exists a linear isomorphism $\tau: E\oplus F\xrightarrow{\cong} G$ such that

$$\begin{aligned}\varphi_1 &= \tau\circ i_1 \quad \psi_1 = \pi_1\circ\tau^{-1}\\ \text{and}\qquad\qquad & \\ \varphi_2 &= \tau\circ i_2 \quad \psi_2 = \pi_2\circ\tau^{-1}.\end{aligned} \tag{2.21}$$

The φ_i, ψ_i are called (as before) canonical injections and projections.

Proof: Define linear mappings

$$\sigma: G\to E\oplus F \quad\text{and}\quad \tau: E\oplus F\to G$$

by

$$\sigma z = (\psi_1 z, \psi_2 z), \qquad z\in G$$

and

$$\tau(x,y) = \varphi_1 x + \varphi_2 y, \qquad x\in E,\, y\in F.$$

Then for every vector $z\in G$

$$\tau\sigma z = \varphi_1\psi_1 z + \varphi_2\psi_2 z = z$$

and for every vector $(x,y)\in E\oplus F$

$$\sigma\tau(x,y) = (\psi_1\varphi_1 x + \psi_1\varphi_2 y, \psi_2\varphi_1 x + \psi_2\varphi_2 y) = (x,y).$$

These relations show that τ and σ are inverse isomorphisms. Formulae (2.21) are immediate consequences of the definition of τ.

Example: Let E be a real vector space. Then $E\oplus E$ can be made into a complex vector space as follows:

$$(\alpha + i\beta)(x,y) = (\alpha x - \beta y, \alpha y + \beta x) \qquad \alpha,\beta\in\mathbb{R}.$$

The complex vector space so obtained is called the *complexification* of E and is denoted by $E_{\mathbb{C}}$.

Every vector $(x,y)\in E_{\mathbb{C}}$ can be uniquely represented in the form

$$(x,y) = (x,0) + (0,y) = (x,0) + i(y,0).$$

Now identify the (real) subspace $E\oplus 0$ of $E_{\mathbb{C}}$ with E under the inclusion map $i_1: E\to E\oplus E$. Then the equation above reads

$$(x,y) = x + iy \qquad x,y\in E.$$

If E has finite dimension n, and if $x_1,\dots,x_n$ is a basis of E, then the vectors $x_1,\dots,x_n$ form a basis of the complex space $E_{\mathbb{C}}$ as is easily verified. Thus

$$\dim_{\mathbb{C}} E_{\mathbb{C}} = \dim_{\mathbb{R}} E.$$

2.17. Direct sum of an arbitrary family of vector spaces. Let $\{E_\alpha\}_{\alpha\in A}$ be an arbitrary family of vector spaces. To define the direct sum of the family E_α consider all mappings

$$x: A \to \bigcup_\alpha E_\alpha \tag{2.22}$$

such that

i) $x(\alpha) \in E_\alpha$, $\alpha \in A$

ii) all but finitely many $x(\alpha)$ are zero.

We denote $x(\alpha)$ by x_α. Then the mapping (2.22) can be written as

$$x: \alpha \to x_\alpha .$$

The sum of two mappings x and y is defined by

$$(x + y)(\alpha) = x_\alpha + y_\alpha$$

and the mapping λx is given by

$$(\lambda x)(\alpha) = \lambda x_\alpha .$$

Under these operations the set of all mappings (2.22) is made into a vector space. This vector space is called the *(external) direct sum* of the vector spaces E_α and will be denoted by $\bigoplus_\alpha E_\alpha$. The zero vector of $\bigoplus_\alpha E_\alpha$ is the mapping x given by

$$x(\alpha) = 0_\alpha \qquad (0_\alpha \text{ zero vector of } E_\alpha).$$

For every fixed $\varrho \in A$ we define the canonical injection $i_\varrho: E_\varrho \to \bigoplus_\alpha E_\alpha$ by

$$i_\varrho x: \alpha \to \begin{cases} 0_\alpha & \varrho \neq \alpha \\ x & \varrho = \alpha \end{cases} \qquad x \in E_\varrho \tag{2.23}$$

and the canonical projection $\pi_\varrho: \bigoplus_\alpha E_\alpha \to E_\varrho$ by

$$\pi_\varrho x = x_\varrho \qquad x \in \bigoplus_\alpha E_\alpha \tag{2.24}$$

It follows from (2.23) and (2.24) that

$$\pi_\varrho \circ i_\sigma = \delta_{\varrho\sigma}\, \iota \tag{2.25}$$

and

$$\sum_\varrho i_\varrho \pi_\varrho x = x \quad x \in \bigoplus_\alpha E_\alpha . \tag{2.26}$$

By 'abus de langage' we shall write (2.26) simply as

$$\sum_\varrho i_\varrho \pi_\varrho = \iota .$$

Proposition II: Suppose that a decomposition of a vector space E as a direct sum of a family of subspaces E_α is given. Then E is isomorphic to the external direct sum of the vector spaces E_α.

Proof: Let $\bigoplus_\alpha E_\alpha = \tilde{E}$. Then a linear mapping $\sigma: E \to \tilde{E}$ is defined by

$$\sigma x = \sum_\alpha i_\alpha x_\alpha \quad \text{where} \quad x = \sum_\alpha x_\alpha, x_\alpha \in E_\alpha.$$

Conversely, a linear mapping $\tau: \tilde{E} \to E$ is given by

$$\tau \tilde{x} = \sum_\alpha \pi_\alpha \tilde{x}.$$

Relations (2.25) and (2.26) imply that

$$\tau \circ \sigma = \iota \quad \text{and} \quad \sigma \circ \tau = \iota$$

and hence σ is an isomorphism of E onto $\tilde{E}$ and τ is the inverse isomorphism.

2.18. Direct sum of linear mappings. Suppose $\varphi_1: E_1 \to F_1$ and $\varphi_2: E_2 \to F_2$ are linear mappings. Then a linear mapping $\varphi_1 \oplus \varphi_2: E_1 \oplus E_2 \to F_1 \oplus F_2$ is defined by

$$(\varphi_1 \oplus \varphi_2)(x_1, x_2) = (\varphi_1 x_1, \varphi_2 x_2).$$

It follows immediately from the definition that

$$\operatorname{Im}(\varphi_1 \oplus \varphi_2) = \operatorname{Im}\varphi_1 \oplus \operatorname{Im}\varphi_2$$

and

$$\ker(\varphi_1 \oplus \varphi_2) = \ker\varphi_1 \oplus \ker\varphi_2.$$

Now suppose E_1, E_2 are subspaces of E and F_1, F_2 are subspaces of F such that

$$E = E_1 \oplus E_2 \quad \text{and} \quad F = F_1 \oplus F_2. \tag{2.27}$$

If $\varphi_i: E_i \to F_i$ are linear maps then $\varphi_1 \oplus \varphi_2$ is again a linear map, defined by

$$(\varphi_1 \oplus \varphi_2)(x_1 + x_2) = \varphi_1 x_1 + \varphi_2 x_2$$

where $x = x_1 + x_2$ is the decomposition of any vector $x \in E$ determined by (2.27). $\varphi_1 \oplus \varphi_2$ may be characterized as the unique linear map of E into F which extends φ_1 and φ_2.

2.19. Projection operators. A linear transformation $\varphi: E \to E$ is called a *projection operator* in E, if $\varphi^2 = \varphi$. If φ is a projection operator in E, then

$$E = \ker\varphi \oplus \operatorname{Im}\varphi. \tag{2.28}$$

Moreover,

$$\varphi = \iota_{\operatorname{Im}\varphi} \oplus 0_{\ker\varphi}\,. \tag{2.29}$$

To prove (2.28) let $x \in E$ be an arbitrary vector. Writing

$$x = y + \varphi x \quad (\text{i.e. } y = x - \varphi x)$$

we obtain that

$$\varphi y = \varphi x - \varphi^2 x = 0$$

whence $y \in \ker\varphi$. It follows that

$$E = \ker\varphi + \operatorname{Im}\varphi\,. \tag{2.30}$$

To show that the decomposition (2.30) is direct let $z = \varphi x$ be an arbitrary vector of $\ker\varphi \cap \operatorname{Im}\varphi$. Then we have that

$$0 = \varphi z = \varphi^2 x = \varphi x = z$$

and thus $\ker\varphi \cap \operatorname{Im}\varphi = 0$.

To prove (2.29) we observe that the subspaces $\operatorname{Im}\varphi$ and $\ker\varphi$ are stable under φ (cf. sec. 2.8) and that the induced transformations are the identity and the zero mapping respectively.

Conversely, if a direct decomposition

$$E = E_1 \oplus E_2$$

is given, then the linear mapping

$$\varphi = \iota_{E_1} \oplus 0_{E_2}$$

is clearly a projection operator in E.

Proposition III: Let $\varrho_i\,(i = 1 \dots r)$ be projection operators in E such that

$$\varrho_i \circ \varrho_j = 0\,, \qquad i \neq j \tag{2.31}$$

and

$$\sum_i \varrho_i = \iota\,.$$

Then

$$E = \bigoplus_{i=1}^{r} \operatorname{Im}\varrho_i\,.$$

Proof: Let $x \in E$ be arbitrary. Then the relation

$$x = \sum_i \varrho_i x \in \sum_{i=1}^{r} \operatorname{Im}\varrho_i$$

shows that

$$E = \sum_{i=1}^{r} \operatorname{Im} \varrho_i . \tag{2.32}$$

To prove that the sum (2.32) is direct suppose that

$$x \in \operatorname{Im} \varrho_i \cap \sum_{j \neq i} \operatorname{Im} \varrho_j .$$

Then $x = \varrho_i y$ (some $y \in E$), so that

$$\varrho_i x = \varrho_i^2 y = \varrho_i y = x . \tag{2.33}$$

On the other hand, we have that for some vectors $y_j \in E$,

$$x = \sum_{j \neq i} \varrho_j y_j$$

whence, in view of (2.31),

$$\varrho_i x = \sum_{j \neq i} \varrho_i \varrho_j y_j = 0. \tag{2.34}$$

Relations (2.33) and (2.34) yield $x = 0$ and hence the decomposition (2.32) is direct.

Suppose now that

$$E = \sum_{\nu} E_{\nu}$$

is a decomposition of E as a direct sum of subspaces E_{ν}. Let $\pi_{\nu}: E \to E_{\nu}$ and $i_{\nu}: E_{\nu} \to E$ denote the canonical projections and injections, and consider the linear mappings $\varrho_{\nu}: E \to E$ defined by

$$\varrho_{\nu} = i_{\nu} \pi_{\nu} .$$

Then the ϱ_{ν} are projection operators satisfying (2.31) as follows from (2.25) and (2.26). Moreover, $\operatorname{Im} \varrho_{\nu} = E_{\nu}$ and so the decomposition of E determined by the ϱ_{ν} agrees with the original decomposition.

Problems

1. Assume a decomposition

$$E = E_1 + E_2 .$$

Consider the external direct sum $E_1 \oplus E_2$ and define a linear mapping $\varphi: E_1 \oplus E_2 \to E$ by

$$\varphi(x_1, x_2) = x_1 + x_2 \qquad x_1 \in E_1, x_2 \in E_2 .$$

Prove that the kernel of φ is the subspace of $E_1 \ominus E_2$ consisting of the pairs

$(x, -x)$ where $x \in E_1 \cap E_2$. Show that φ is a linear isomorphism if and only if the decomposition $E = E_1 + E_2$ is direct.

2. Given two vector spaces E and F, consider subspaces $E_1 \subset E$, $F_1 \subset F$ and the canonical projections

$$\pi_E: E \to E/E_1, \qquad \pi_F: F \to F/F_1.$$

Define a mapping

$$\varphi: E \oplus F \to E/E_1 \oplus F/F_1$$

by

$$\varphi(x, y) = (\pi_E x, \pi_F y).$$

Show that φ induces a linear isomorphism

$$\bar{\varphi}: (E \oplus F)/(E_1 \oplus F_1) \to E/E_1 \oplus F/F_1.$$

3. Let $E = E_1 \oplus E_2$ and $F = F_1 \oplus F_2$ be decompositions of E and F as direct sums of subspaces. Show that the external direct sum, G, of E and F can be written as $G = G_1 \oplus G_2$ where G_1 and G_2 are subspaces of G and G_i is the external direct sum of E_i and $F_i (i = 1, 2)$.

4. Prove that from every projection operator π in E an involution ω is obtained by $\omega = 2\pi - \iota$ and that every involution can be written in this form.

5. Let $\pi_i (i = 1 \dots r)$ be projection operators in E such that

$$\operatorname{Im} \pi_i = F \qquad (i = 1 \dots r)$$

where F is a fixed subspace of E. Let $\lambda^i (i = 1 \dots r)$ be scalars. Show that

a) If $\sum_i \lambda^i \neq 0$ then $\operatorname{Im} \sum_i \lambda^i \pi_i = F$

b) $\sum \lambda^i \pi_i$ is a non-zero projection operator in E if and only if $\sum_i \lambda^i = 1$.

6. Let E be a vector space with a countable basis. Construct a linear isomorphism between E and $E \oplus E$.

7. Let X be any set and let $C(X)$ be the free vector space generated by X (cf. sec. 1.7). Show that if Y is a second set, then $C(X \uplus Y) \cong C(X) \oplus C(Y)$.

§ 5. Dual vector spaces

2.20. Bilinear functions. Let E and F be vector spaces. Then a mapping $\Phi: E \times F \to \Gamma$ satisfying

$$\Phi(\lambda x_1 + \mu x_2, y) = \lambda \Phi(x_1, y) + \mu \Phi(x_2, y) \quad x_1, x_2 \in E, y \in F \tag{2.35}$$

and

$$\Phi(x, \lambda y_1 + \mu y_2) = \lambda \Phi(x, y_1) + \mu \Phi(x, y_2) \quad x \in E, y_1, y_2 \in F \tag{2.36}$$

is called a *bilinear function* in $E \times F$. If Φ is a bilinear function in $E \times F$ and $E_1 \subset E$, $F_1 \subset F$ are subspaces, then Φ induces a bilinear function Φ_1 in $E_1 \times F_1$ defined by

$$\Phi_1(x, y) = \Phi(x, y) \qquad x \in E_1, y \in F_1$$

Φ_1 is called the *restriction of* Φ to $E_1 \times F_1$.

Conversely, every bilinear function Φ_1 in $E_1 \times F_1$ may be extended (in several ways) to a bilinear function in $E \times F$. In fact, let

$$\varrho : E \to E_1, \quad \sigma : F \to F_1$$

be surjective linear mappings such that ϱ and σ reduce to the identity in E_1 and F_1 respectively (cf. Cor. II, Proposition I, sec. 1.15). Define Φ by

$$\Phi(x, y) = \Phi_1(\varrho x, \sigma y).$$

Then Φ is a bilinear function in $E \times F$ and for $x_1 \in E_1$, $y_1 \in F_1$ we have that

$$\Phi(x_1, y_1) = \Phi_1(\varrho x_1, \sigma y_1) = \Phi_1(x_1, y_1)$$

Thus Φ extends Φ_1.

Now let

$$E = \sum_\alpha E_\alpha \quad \text{and} \quad F = \sum_\beta F_\beta \tag{2.37}$$

be decompositions of E and F as direct sums of subspaces. Then every system of bilinear functions

$$\Phi_{\alpha\beta} : E_\alpha \times F_\beta \to \Gamma$$

can be extended in precisely one way to a bilinear function Φ in $E \times F$. The function Φ is given by

$$\Phi(x, y) = \sum_{\alpha, \beta} \Phi_{\alpha\beta}(\pi_\alpha x, \pi_\beta y)$$

where $\pi_\alpha : E \to E_\alpha$ and $\pi_\beta : F \to F_\beta$ denote the canonical projections associated with the decompositions (2.37).

2.21. Nullspaces. A bilinear function Φ in $E \times F$ determines two subspaces $N_E \subset E$ and $N_F \subset F$ defined by

$$N_E = \{x \mid \Phi(x, y) = 0\} \quad \text{for every} \quad y \in F$$

and

$$N_F = \{y \mid \Phi(x, y) = 0\} \quad \text{for every} \quad x \in E.$$

It follows immediately from (2.35) and (2.36) that N_E and N_F are subspaces of E and F. They are called the *nullspaces* of Φ. If $N_E = 0$ and $N_F = 0$ then the bilinear function Φ is called *non-degenerate*.

Given an arbitrary bilinear function Φ consider the canonical projections

$$\pi_E: E \to E/N_E, \quad \pi_F: F \to F/N_F.$$

Then Φ induces a non-degenerate bilinear function $\bar{\Phi}$ in $E/N_E \times F/N_F$ such that

$$\bar{\Phi}(\pi_E x, \pi_F y) = \Phi(x, y).$$

To show that $\bar{\Phi}$ is well defined, suppose that $x' \in E$ and $y' \in F$ are two other vectors such that $\pi_E x = \pi_E x'$ and $\pi_F y = \pi_F y'$. Then $x' - x \in N_E$ and $y' - y \in N_F$ and hence we can write $x' = x + u$, $u \in N_E$ and $y' = y + v$, $v \in N_F$. It follows that

$$\begin{aligned}\Phi(x', y') &= \Phi(x + u, y + v) \\ &= \Phi(x, y) + \Phi(x, v) + \Phi(u, y) + \Phi(u, v) \\ &= \Phi(x, y).\end{aligned}$$

Clearly $\bar{\Phi}$ is bilinear. It remains to be shown that $\bar{\Phi}$ is non-degenerate. In fact, assume that

$$\bar{\Phi}(\pi_E x, \pi_F y) = 0 \tag{2.38}$$

for a fixed $\pi_E x$ and every $\pi_F y$. Then $\Phi(x, y) = 0$ for every $y \in F$. It follows that $x \in N_E$ whence $\pi_E x = 0$. Similarly, if (2.38) holds for a fixed $\pi_F y$ and every $\pi_E x$, then $\pi_F y = 0$. Hence $\bar{\Phi}$ is non-degenerate.

A non-degenerate bilinear function Φ in $E \times F$ will often be denoted by $\langle\,,\rangle$. Then we write

$$\Phi(x, y) = \langle x, y\rangle \qquad x \in E, y \in F.$$

2.22. Dual spaces. Suppose E^*, E is a pair of vector spaces, and assume that a fixed non-degenerate bilinear function, $\langle\,,\rangle$, in $E^* \times E$ is defined. Then E and E^* will be called *dual* with respect to the bilinear function $\langle\,,\rangle$. The scalar $\langle x^*, x\rangle$, is called the *scalar product* of x^* and x, and the bilinear function $\langle\,,\rangle$ is called a *scalar product between* E^* and E.

Examples. 1 Let $E = E^* = \Gamma$ and define a mapping $\langle\,,\rangle$ by

$$\langle\lambda, \mu\rangle = \lambda\mu \qquad \lambda, \mu \in \Gamma.$$

Clearly $\langle\,,\rangle$ is a non-degenerate bilinear function, and hence Γ can be regarded as a self-dual space.

2. Let $E = E^* = \Gamma^n$ and consider the bilinear mapping $\langle\,,\rangle$ defined by

$$\langle x^*, x\rangle = \sum_{i=1}^{n} \xi^i \xi_i$$

where

$$x^* = (\xi^1, \dots, \xi^n) \quad \text{and} \quad x = (\xi_1, \dots, \xi_n).$$

It is easy to verify that the bilinear mapping $\langle , \rangle$ is non-degenerate and hence Γ^n is dual to itself.

3. Let E be any vector space and $E^* = L(E)$ the space of linear functions in E. Define a bilinear mapping $\langle , \rangle$ by

$$\langle f, x \rangle = f(x), \qquad f \in L(E), x \in E.$$

Since $f(x) = 0$ for each $x \in E$ if and only if $f = 0$, it follows that $N_{L(E)} = 0$.

On the other hand, let $a \in E$ be a non-zero vector and E_1 be the one-dimensional subspace of E generated by a. Then a linear function g is defined in E_1 by

$$g(x) = \lambda \quad \text{where} \quad x = \lambda a.$$

In view of sec. 1.15, g can be extended to a linear function f in E. Then

$$\langle f, a \rangle = f(a) = g(a) = 1 \neq 0.$$

It follows that $N_E = 0$ and hence the bilinear function $\langle , \rangle$ is non-degenerate.

This example is of particular importance because of the following

Proposition I: Let E^*, E be a pair of vector spaces which are dual with respect to a scalar product $\langle , \rangle$. Then an injective linear map $\Phi: E^* \to L(E)$ is defined by

$$\Phi(x^*)(x) = \langle x^*, x \rangle \qquad x^* \in E^*, \ x \in E. \tag{2.39}$$

Proof: Fix a vector $a^* \in E^*$. Then a linear function, f_{a^*}, is defined in E by

$$f_{a^*}(x) = \langle a^*, x \rangle \qquad x \in E. \tag{2.40}$$

Since $\langle , \rangle$ is bilinear, f_{a^*} depends linearly on a^*. Now define Φ by setting

$$\Phi(a^*) = f_{a^*}. \tag{2.41}$$

To show that Φ is injective, assume that $\Phi(a^*) = 0$ for some $a^* \in E^*$. Then $\langle a^*, x \rangle = 0$ for every $x \in E$. Since $\langle , \rangle$ is non-degenerate, it follows that $a^* = 0$.

Note: It will be shown in sec. 2.33 that Φ is surjective (and hence a linear isomorphism) if E has finite dimension.

2.23. Orthogonal complements. Two vectors $x^* \in E^*$ and $x \in E$ are called *orthogonal* if $\langle x^*, x\rangle = 0$. Now let E_1 be a subspace of E. Then the vectors of E^* which are orthogonal to E_1 form a subspace $E_1^\perp$ of E^*. $E_1^\perp$ is called the *orthogonal complement* of E_1. In the same way every subspace $E_1^* \subset E^*$ determines an orthogonal complement $(E_1^*)^\perp \subset E$. The fact that the bilinear function $\langle\,,\rangle$ is non-degenerate can now be expressed by the relations

$$E^\perp = 0 \quad \text{and} \quad (E^*)^\perp = 0 .$$

It follows immediately from the definition that for every subspace $E_1 \subset E$

$$E_1 \subset (E_1^\perp)^\perp \tag{2.42}$$

Suppose next that E^*, E are a pair of dual spaces and that F is a subspace of E. Then a scalar product is induced in the pair $E^*/F^\perp$, F by

$$\langle \bar{x}^*, y\rangle = \langle x^*, y\rangle , \qquad \bar{x}^* \in E^*/F^\perp,\ y \in F ,$$

where x^* is a representative of the class $\bar{x}^*$. In fact, let Φ be the restriction of the scalar product $\langle\,,\rangle$ to $E^* \times F$. Then the nullspaces of Φ are given by

$$N_{E^*} = F^\perp \quad \text{and} \quad N_F = 0 .$$

Now our result follows immediately from sec. 2.21.

More generally, suppose $F \subset E$ and $H^* \subset E^*$ are any subspaces. Then a scalar product in the pair $H^*/H^* \cap F^\perp$, $F/F \cap (H^*)^\perp$, is determined by

$$\langle \bar{x}^*, \bar{x}\rangle = \langle x^*, x\rangle$$

as a similar argument will show.

2.24. Dual mappings. Suppose that E, E^* and F, F^* are two pairs of dual spaces and $\varphi: E \to F$, $\varphi^*: E^* \leftarrow F^*$ are linear mappings. The mappings φ and φ^* are called *dual* if

$$\langle y^*, \varphi x\rangle = \langle \varphi^* y^*, x\rangle \qquad y^* \in F^*, x \in E .$$

To a given linear mapping $\varphi: E \to F$ there exists at most one dual mapping. If φ_1^* and φ_2^* are dual to φ we have that

$$\langle y^*, \varphi x\rangle = \langle \varphi_1^* y^*, x\rangle \quad \text{and} \quad \langle y^*, \varphi x\rangle = \langle \varphi_2^* y^*, x\rangle$$

whence

$$\langle \varphi_1^* y^* - \varphi_2^* y^*, x\rangle = 0 \quad x \in E, y^* \in F .$$

This implies, in view of the duality of E and E^*, that $\varphi_1^* y^* = \varphi_2^* y^*$ whence $\varphi_1^* = \varphi_2^*$.

As an example of dual mappings consider the dual pairs E^*, E and $E^*/E_1^\perp, E_1$ where E_1 is a subspace of E (cf. sec. 2.23) and let π be the canonical projection of E^* onto $E^*/E_1^\perp$,

$$\pi: E^*/E_1^\perp \leftarrow E^* .$$

Then the canonical injection

$$i: E_1 \to E$$

is dual to π. In fact, if $x \in E_1$, and $y^* \in E^*$ are arbitrary, we have

$$\langle y^*, i\,x\rangle = \langle y^*, x\rangle = \langle \bar{y}^*, x\rangle = \langle \pi\, y^*, x\rangle$$

and thus

$$\pi = i^* .$$

2.25. Operations with dual mappings. Assume that E^*, E and F^*, F are two pairs of dual vector spaces. Assume further that $\varphi: E \to F$ and $\psi: E \to F$ are linear mappings and that there exist dual mappings φ^*: $E^* \leftarrow F^*$ and $\psi^*: E^* \leftarrow F^*$. Then there are mappings dual to $\varphi + \psi$ and $\lambda\varphi$ and these dual mappings are given by

$$(\varphi + \psi)^* = \varphi^* + \psi^* \tag{2.43}$$

and

$$(\lambda\,\varphi)^* = \lambda\,\varphi^* . \tag{2.44}$$

(2.43) follows from the relation

$$\begin{aligned}\langle(\varphi^* + \psi^*)\,y^*, x\rangle &= \langle\varphi^*\,y^*, x\rangle + \langle\psi^*\,y^*, x\rangle \\ &= \langle y^*, \varphi\;x\rangle + \langle y^*, \psi\;x\rangle = \langle y^*, (\varphi + \psi)x\rangle\end{aligned}$$

and (2.44) is proved in the same way. Now let G, G^* be a third pair of dual spaces and let $\chi: F \to G$, $\chi^*: F^* \leftarrow G^*$ be a pair of dual mappings. Then the dual mapping of $\chi \circ \varphi$ exists, and is given by

$$(\chi \circ \varphi)^* = \varphi^* \circ \chi^* .$$

In fact, if $z^* \in G^*$ and $x \in E$ are arbitrary vectors we have that

$$\langle\varphi^*\,\chi^*\,z^*, x\rangle = \langle\chi^*\,z^*, \varphi\,x\rangle = \langle z^*, \chi\,\varphi\,x\rangle .$$

For the identity map we have clearly

$$\iota_{E^*} = (\iota_E)^* .$$

Now assume that $\varphi: E \to F$ has a left inverse $\varphi_1: F \to E$,

$$\varphi_1 \circ \varphi = \iota_E \tag{2.45}$$

and that the dual mappings $\varphi^*: E^* \leftarrow F^*$ and $\varphi_1^*: F^* \leftarrow E^*$ exist. Then we obtain from (2.45) that

$$\varphi^* \circ \varphi_1^* = (\varphi_1 \circ \varphi)^* = (\iota_E)^* = \iota_{E^*} . \tag{2.46}$$

In view of sec. 2.11 the relations (2.45) and (2.46) are equivalent to

$$\varphi \text{ injective,} \quad \varphi_1 \text{ surjective}$$

and

$$\varphi_1^* \text{ injective,} \quad \varphi^* \text{ surjective.}$$

In particular, if φ and φ_1 are inverse linear isomorphisms, then so are φ^* and φ_1^*.

2.26. Kernel and image space. Let $\varphi: E \to F$ and $\varphi^*: E^* \leftarrow F^*$ be a pair of dual mappings. In this section we shall study the relations between the subspaces

$$\ker \varphi \subset E, \quad \operatorname{Im} \varphi \subset F$$

and

$$\ker \varphi^* \subset F^*, \quad \operatorname{Im} \varphi^* \subset E^* .$$

First we establish the formulae

$$\ker \varphi^* = (\operatorname{Im} \varphi)^{\perp} \tag{2.47}$$

$$\ker \varphi = (\operatorname{Im} \varphi^*)^{\perp} . \tag{2.48}$$

In fact, for any two vectors $y^* \in \ker \varphi^*$, $\varphi x \in \operatorname{Im} \varphi$ we have

$$\langle y^*, \varphi x \rangle = \langle \varphi^* y^*, x \rangle = 0$$

and hence the subspaces $\ker \varphi^*$ and $\operatorname{Im} \varphi$ are orthogonal, $\ker \varphi^* \subset (\operatorname{Im} \varphi)^{\perp}$. Now let $y^* \in (\operatorname{Im} \varphi)^{\perp}$ be any vector. Then for every $x \in E$

$$\langle \varphi^* y^*, x \rangle = \langle y^*, \varphi x \rangle = 0 .$$

It follows that $\varphi^* y^* = 0$, whence $y^* \in \ker \varphi^*$. This completes the proof of (2.47). (2.48) is proved by the same argument.

Now assume that φ is surjective. Then $\operatorname{Im}\varphi=F$ and hence formula (2.47) implies that $\ker\varphi^*=0$; i.e., φ^* is injective. If φ is injective we obtain from (2.48) that $(\operatorname{Im}\varphi^*)^\perp=0$. However, this does not imply that $\operatorname{Im}\varphi^*=E^*$ and so we can not conclude that the dual of an injective mapping is surjective (cf. problem 9).

2.27. Relations between the induced mappings. Again let $\varphi: E\to F$ and $\varphi^*: E^*\leftarrow F^*$ be a pair of dual mappings. Then it follows from (2.48) and from the discussion in sec. 2.23 that a scalar product is induced in the pair $\operatorname{Im}\varphi^*$, $E/\ker\varphi$, by

$$\langle x^*,\bar{x}\rangle=\langle x^*,x\rangle \quad x^*\in\operatorname{Im}\varphi^*,\bar{x}\in E/\ker\varphi\,.$$

In particular, if φ is injective, then the restriction of the scalar product in E^*, E to $\operatorname{Im}\varphi^*$, E is non-degenerate.

The same argument as above shows that the vector spaces $F^*/\ker\varphi^*$ and $\operatorname{Im}\varphi$ are dual with respect to the bilinear functions given by

$$\langle\bar{x}^*,x\rangle=\langle x^*,x\rangle \qquad \bar{x}^*\in F^*/\ker\varphi^*, x\in\operatorname{Im}\varphi\,.$$

Now consider the surjective linear mapping

$$\varphi_1: E\to\operatorname{Im}\varphi$$

induced by φ and the injective linear mapping

$$\bar{\varphi}^*: E^*\leftarrow F^*/\ker\varphi^*$$

induced by φ^*. The mappings φ_1 and $\bar{\varphi}^*$ are dual. In fact, if $\bar{x}^*\in F^*/\ker\varphi^*$ and $x\in E$ are arbitrary vectors, we have that

$$\begin{aligned}\langle\bar{\varphi}^*\bar{x}^*,x\rangle&=\langle\varphi^*x^*,x\rangle\\&=\langle x^*,\varphi x\rangle\\&=\langle\bar{x}^*,\varphi_1x\rangle\,.\end{aligned}$$

In the same way it follows that the surjective mapping

$$\varphi_1^*:\operatorname{Im}\varphi^*\leftarrow F^*$$

induced by φ^* and the injective mapping

$$\varphi: E/\ker\varphi\to F$$

induced by φ are dual. Finally, the induced isomorphisms

$$E/\ker\varphi\xrightarrow{\cong}\operatorname{Im}\varphi$$

and

$$\operatorname{Im}\varphi^* \overset{\cong}{\leftarrow} F^*/\ker\varphi^*$$

are dual as well.

2.28. The space of linear functions as a dual space. Let E be a vector space and $L(E)$ be the space of linear functions in E. Then the spaces E, $L(E)$ are dual with respect to the scalar product defined in sec. 2.22. For these spaces we have three important results, which are not valid for arbitrary pairs of dual spaces.

Proposition II: Let F, F^* be arbitrary dual spaces and $\varphi: E \to F$ be a linear mapping. Then a dual mapping $\varphi^*: L(E) \leftarrow F^*$ exists, and is given by

$$(\varphi^* y^*)(x) = \langle y^*, \varphi x\rangle \qquad \begin{matrix} y^* \in F^* \\ x \in E \end{matrix} \tag{2.49}$$

Proof: It is easy to verify that the correspondence $y^* \to \varphi^* y^*$ defined by (2.49) determines a linear mapping. Moreover, the relation

$$\langle \varphi^* y^*, x\rangle = (\varphi^* y^*)(x) = \langle y^*, \varphi x\rangle$$

shows that φ^* is dual to φ. If $F^* = L(F)$ as well, (2.49) can be written in the form

$$\varphi^* f = f \circ \varphi, \qquad f \in L(F). \tag{2.50}$$

Proposition III: Suppose $\varphi: E \to F$ is a linear mapping, and consider the dual mapping

$$\varphi^*: L(E) \leftarrow L(F).$$

Then

$$\operatorname{Im}\varphi^* = (\ker\varphi)^\perp. \tag{2.51}$$

Proof: From (2.42) and (2.48) we obtain that

$$\operatorname{Im}\varphi^* \subset (\operatorname{Im}\varphi^*)^{\perp\perp} = (\ker\varphi)^\perp. \tag{2.52}$$

On the other hand, suppose that $f \in (\ker\varphi)^\perp$. Then

$$\ker f \supset \ker\varphi$$

and hence (cf. sec. 2.4) there exists a linear function g in F such that

$$g \circ \varphi = f.$$

Now (2.50) yields

$$\varphi^* g = g \circ \varphi = f$$

and so $f \in \operatorname{Im}\varphi^*$. Thus $\operatorname{Im}\varphi^* \supset (\ker\varphi)^\perp$ which together with (2.52) proves (2.51).

Corollary I: If φ is injective, then φ^* is surjective.

Proof: If $\ker\varphi = 0$ formula (2.51) yields

$$\operatorname{Im}\varphi^* = (\ker\varphi)^\perp = (0)^\perp = L(E)$$

and so φ^* is surjective.

Corollary II: $(\ker\varphi)^{\perp\perp} = \ker\varphi$

Proof: Proposition III together with the relation (2.48) yields

$$(\ker\varphi)^{\perp\perp} = (\operatorname{Im}\varphi^*)^\perp = \ker\varphi .$$

Proposition IV: If $E_1 \subset E$ is any subspace, then

$$E_1^{\perp\perp} = E_1 . \tag{2.53}$$

Proof: Consider the canonical projection $\pi: E \to E/E_1$. Then $\ker\pi = E_1$. Now the result follows immediately from corollary II.

Corollary I: If $\varphi: E \to F$ is a linear mapping and $\varphi^*: L(E) \leftarrow L(F)$ is the dual mapping, then

$$(\ker\varphi^*)^\perp = \operatorname{Im}\varphi .$$

Proof: It follows from (2.47) and (2.53) that

$$(\ker\varphi^*)^\perp = (\operatorname{Im}\varphi)^{\perp\perp} = \operatorname{Im}\varphi .$$

Corollary II: The bilinear function

$$\langle x^*, \bar{x}\rangle = \langle x^*, x\rangle \qquad x^* \in E_1^\perp, \bar{x} \in E/E_1$$

defines a scalar product in the pair $E_1^\perp$, E/E_1.

2.29. Dual exact sequences. As an application suppose the sequence

$$F \xrightarrow{\varphi} E \xrightarrow{\psi} G$$

is exact at E. Then the dual sequence

$$L(F) \xleftarrow{\varphi^*} L(E) \xleftarrow{\psi^*} L(G)$$

is exact at $L(E)$. In fact, it follows from (2.47) and (2.51) that

$$\ker\varphi^* = (\operatorname{Im}\varphi)^\perp = (\ker\psi)^\perp = \operatorname{Im}\psi^* .$$

In particular, if

$$0 \to F \xrightarrow{\varphi} E \xrightarrow{\psi} G \to 0$$

is a short exact sequence, then the dual sequence

$$0 \leftarrow L(F) \overset{\varphi^*}{\leftarrow} L(E) \overset{\psi^*}{\leftarrow} L(G) \leftarrow 0$$

is again a short exact sequence.

2.30. Direct decompositions. *Proposition V*: Suppose

$$E = E_1 \oplus E_2 \tag{2.54}$$

is a decomposition of E as a direct sum of subspaces. Then

$$L(E) = E_1^\perp \oplus E_2^\perp$$

and the pairs $E_1^\perp, E_2$ and $E_2^\perp, E_1$ are dual with respect to the induced scalar products. Moreover, the induced injections

$$E_1^\perp \to L(E_2), \quad E_2^\perp \to L(E_1)$$

are surjective, and hence

$$L(E) = L(E_1) \oplus L(E_2).$$

Finally, $(E_1^\perp)^{\perp\perp} = E_1^\perp$ and $(E_2^\perp)^{\perp\perp} = E_2^\perp$.

Proof: Let $\pi_1 : E \to E_1$ and $\pi_2 : E \to E_2$ be the canonical projections associated with the direct decomposition (2.54). Let $f \in L(E)$ be any linear function, and define functions f_1, f_2 by

$$f_1(x) = f(\pi_2 x) \quad \text{and} \quad f_2(x) = f(\pi_1 x).$$

It follows that $f_i \in E_i^\perp$ $(i = 1, 2)$ and

$$f = f_1 + f_2.$$

Consequently,

$$L(E) = E_1^\perp + E_2^\perp. \tag{2.55}$$

To show that the decomposition (2.55) is direct, assume that $f \in E_1^\perp \cap E_2^\perp$. Then

$$f(x) = 0 \qquad x \in E_1, x \in E_2$$

and hence $f(x) = 0$ for every $x \in E$. Thus $f = 0$, and so the decomposition (2.55) is direct. The rest of the proposition is trivial.

Corollary: If $E = E_1 \oplus \cdots \oplus E_r$ is a decomposition of E as a direct sum of r subspaces, then

$$L(E) = F_1^\perp \oplus \cdots \oplus F_r^\perp$$

where

$$F_i = \sum_{j \neq i} E_j.$$

Moreover, the restriction of the scalar product to E_i, $F_i^{\perp}$ is again non-degenerate, and

$$F_i^{\perp} \cong L(E_i).$$

Proposition V has the following converse:

Proposition VI: Let $E_1 \subset E$ be any subspace, and let $E_1^* \subset L(E)$ be a subspace dual to E_1 such that

$$(E_1^*)^{\perp\perp} = E_1^*.$$

Then

$$E = E_1 \oplus (E_1^*)^{\perp} \tag{2.56}$$

and

$$L(E) = E_1^* \oplus E_1^{\perp}. \tag{2.57}$$

Proof: We have (cf. problem 2 below)

$$(E_1 + E_1^{*\perp})^{\perp} = E_1^{\perp} \cap (E_1^{*\perp})^{\perp} = E_1^{\perp} \cap E_1^* = 0$$

whence

$$E = 0^{\perp} = (E_1 + E_1^{*\perp})^{\perp\perp} = E_1 + E_1^{*\perp}. \tag{2.58}$$

On the other hand, since E_1 and E_1^* are dual, it follows that

$$E_1 \cap E_1^{*\perp} = 0$$

which together with (2.58) proves (2.56). (2.57) follows from Proposition V and (2.56).

Problems

1. Given two pairs of dual spaces E^*, E and F^*, F prove that the spaces $E^* \oplus F^*$ and $E \oplus F$ are dual with respect to the bilinear function

$$\langle (x^*, y^*), (x, y) \rangle = \langle x^*, x \rangle + \langle y^*, y \rangle.$$

2. Consider two subspaces E_1 and E_2 of E. Establish the relation

$$(E_1 + E_2)^{\perp} = E_1^{\perp} \cap E_2^{\perp}.$$

3. Given a vector space E consider the mapping $\lambda: E \to L(L(E))$ defined by

$$\lambda_a(f) = f(a) \qquad a \in E, \; f \in L(E).$$

Prove that λ is injective. Show that λ is surjective if and only if $\dim E < \infty$.

4. Suppose $\pi: E \to E$ and $\pi^*: E^* \leftarrow E^*$ are dual mappings. Assume that π is a projection operator in E. Prove that π^* is a projection operator in E^* and that

$$\operatorname{Im}\pi^* = (\ker\pi)^\perp, \qquad \operatorname{Im}\pi = (\ker\pi^*)^\perp.$$

Conclude that the subspaces Im π, Im π^* and ker π, ker π^* are dual pairs.

5. Suppose E, E^* is a pair of dual spaces such that every linear function $f: E \to \Gamma$ induces a dual mapping $f^*: E^* \leftarrow \Gamma$. Show that the natural injection $E^* \to L(E)$ is surjective.

6. Suppose that E is an infinite dimensional vector space. Show that there exists a dual space E^* such that the natural injection $E^* \to L(E)$ is *not* surjective.

7. Consider the vector space E of sequences

$$(\lambda_0, \lambda_1 \ldots) \quad \lambda_i \in \Gamma$$

and the subspace F consisting of those sequences for which only finitely many λ_i are different from zero (addition and scalar multiplication being defined as in the case of Γ^n). Show that the mapping $E \times F \to \Gamma$ given by

$$(\lambda_0, \lambda_1 \ldots), \quad (\mu_0, \mu_1 \ldots) \to \sum_i \lambda_i \mu_i$$

defines a scalar product in E and F. Show further that the induced injection $E \to L(F)$ is surjective.

8. Let S be any set. Construct a scalar product between $(S; \Gamma)$ and $C(S)$ (cf. Example 3, sec. 1.2 and sec. 1.7) which determines a linear isomorphism $(S; \Gamma) \xrightarrow{\cong} L(C(S))$.

Hint: See problem 7.

9. Let E be any vector space of infinite dimension. Show that there is a dual space E^* and a second pair of dual spaces F, F^* such that there exist dual mappings

$$\varphi: E \to F, \quad \varphi^*: E^* \leftarrow F^*$$

where φ is injective but φ^* is *not* surjective.

Prove that E, Im φ^* is again a dual pair of spaces.

10. Let $\varphi: E \to F$ be a linear mapping with restriction $\varphi_1: E_1 \to F_1$. Suppose that $\varphi^*: E^* \leftarrow F^*$ is a dual mapping. Prove that φ^* can be restricted to the pair $(F_1^\perp, E_1^\perp)$. Show that the induced mapping

$$\bar{\varphi}^*: E/E_1^\perp \leftarrow F/F_1^\perp$$

is dual to φ_1 with respect to the induced scalar product.

11. Show that if F is a vector space with a countable basis then there is no vector space E such that $F \cong L(E)$.

12. Let E be an infinite dimensional vector space. Construct a linear automorphism of $L(E)$ which is not the dual of an automorphism of E.

13. Let $\Phi: E \times F \to \Gamma$ be a bilinear map such that the linear maps $\varphi: E \to L(F)$ and $\psi: F \to L(E)$ given by $\varphi_a(y) = \Phi(a, y)$ and $\psi_b(x) = \Phi(x, b)$ are isomorphisms. Show that the canonical map λ defined in problem 3 is given by $\lambda = (\psi^{-1})^* \circ \varphi$. Conclude that E and F must have the same finite dimension.

§ 6. Finite dimensional vector spaces

2.31. The space $L(E; F)$. Let E and F be vector spaces of dimension n and m respectively. Then the space $L(E; F)$ has dimension nm,

$$\dim L(E; F) = \dim E \cdot \dim F. \tag{2.59}$$

To prove (2.59) let x_ν $(\nu = 1, \dots, n)$ be a basis of E and y_μ $(\mu = 1, \dots, m)$ be a basis of F. Consider the linear mappings $\varphi_\sigma^\lambda: E \to F$ defined by

$$\varphi_\sigma^\lambda x_\nu = \delta_\nu^\lambda y_\sigma \quad {}^{*)} \qquad \begin{matrix} \lambda, \nu = 1, \dots, n \\ \sigma = 1, \dots, m \end{matrix}$$

Now let $\varphi: E \to F$ be any linear mapping, and define scalars α_ν^μ by

$$\varphi x_\nu = \sum_{\mu=1}^{m} \alpha_\nu^\mu y_\mu.$$

Then, for $\lambda = 1, \dots, n$,

$$(\varphi - \sum_{\mu,\nu} \alpha_\nu^\mu \varphi_\mu^\nu) x_\lambda = \sum_{\varrho} \alpha_\lambda^\varrho y_\varrho - \sum_{\mu,\nu} \alpha_\nu^\mu \delta_\lambda^\nu y_\mu = \sum_{\varrho} \alpha_\lambda^\varrho y_\varrho - \sum_{\mu} \alpha_\lambda^\mu y_\mu = 0,$$

whence

$$\varphi = \sum_{\nu,\mu} \alpha_\nu^\mu \varphi_\nu^\mu.$$

It follows that the mappings φ_ν^μ generate the space $L(E; F)$. A similar argument shows that the mappings φ_ν^μ are linearly independent and hence they form a basis of $L(E; F)$. This basis is called the basis *induced* by the bases of E and F. Since the basis φ_ν^μ consists of nm vectors, formula (2.59) follows.

---

*) δ_ν^λ is the *Kronecker symbol* defined by $\delta_\nu^\lambda = \begin{cases} 1 & \lambda = \nu \\ 0 & \lambda \neq \nu \end{cases}$.

2.32. The space $L(E)$. Now consider the case that $F=\Gamma$ and choose in Γ the basis consisting of the unit element. Then the basis of $L(E)$ induced by the basis x_ν $(\nu=1, \ldots, n)$ consists of n linear functions f^μ given by

$$f^\mu(x_\nu) = \delta_\nu^\mu . \tag{2.60}$$

The basis f^μ of $L(E)$ is called the *dual* of the basis x_ν of E. In particular, we have

$$\dim L(E) = \dim E .$$

Since the functions f^μ form a basis of $L(E)$ every linear function f on E can be uniquely written in the form

$$f = \sum_\mu \lambda_\mu f^\mu ,$$

where the scalars λ_μ are given by

$$\lambda_\mu = f(x_\mu) \qquad \mu = 1, \ldots, n .$$

This formula shows that the components of f with respect to the basis f^μ are obtained by evaluating the function f on the basis x_μ.

2.33. Dual spaces. We shall now prove the assertion quoted at the end of sec. 2.22.

Proposition I: Let E, E^* be a pair of dual spaces and assume that E has finite dimension. Then the injection $\Phi: E^* \to L(E)$ defined by formula (2.39) sec. 2.22 is surjective and hence a linear isomorphism. In particular, E^* has finite dimension and

$$\dim E^* = \dim E . \tag{2.61}$$

Proof: Since Φ is injective and $\dim L(E) = \dim E$ it follows that

$$\dim E^* \leq \dim E .$$

Hence E^* has finite dimension. In view of the symmetry between E and E^* we also have that

$$\dim E \leq \dim E^*$$

whence (2.61). On the other hand, $\dim L(E) = \dim E$ and hence Φ is surjective.

Corollary I: Let E, E^* be a pair of dual finite dimensional spaces. Then the results of sec. 2.28, 2.29 and 2.30 hold.

Proof: Each result needs to be verified independently, but the proofs can all be obtained by using the linear isomorphism $E^* \xrightarrow{\cong} L(E)$. The actual verifications are left to the reader.

Corollary II: Let E_1^* and E_2^* be any two vector spaces dual to E. Then there exists a unique linear isomorphism $\varphi: E_1^* \xrightarrow{\cong} E_2^*$ such that

$$\langle \varphi x^*, x\rangle = \langle x^*, x\rangle \qquad x^* \in E_1^*, x \in E.$$

Two bases x_ν and $x^{*\nu}$ $(\nu = 1 \dots n)$ of E and E^* are called *dual* if

$$\langle x^{*\nu}, x_\mu\rangle = \delta_\mu^\nu. \tag{2.62}$$

Given a basis x_ν $(\nu = 1 \dots n)$ of E there exists precisely one dual basis of E^*. It is clear that the vectors $x^{*\nu}$ are uniquely determined by (2.62). To prove the existence of the dual basis let f^ν be the basis of $L(E)$ defined in sec. 2.32 and set

$$x^{*\nu} = \Phi^{-1} f^\nu \qquad \nu = 1 \dots n$$

where Φ is the linear isomorphism of E^* onto $L(E)$. Then we have

$$\langle x^{*\nu}, x_\mu\rangle = \Phi(x^{*\nu})(x_\mu) = f^\nu(x_\mu) = \delta_\mu^\nu.$$

Given a pair of dual bases x_ν, $x^{*\nu}$ $(\nu = 1 \dots n)$ consider two vectors

$$x^* = \sum_\nu \xi_\nu x^{*\nu} \quad \text{and} \quad x = \sum_\nu \xi^\nu x_\nu.$$

It follows from (2.62) that

$$\langle x^*, x\rangle = \sum_\nu \xi_\nu \xi^\nu.$$

Replacing x^* by $x^{*\lambda}$ in this relation we obtain the formula

$$\xi^\lambda = \langle x^{*\lambda}, x\rangle$$

which shows that the components of a vector $x \in E$ with respect to a basis of E are the scalar products of x with the dual basis vectors.

Proposition II: Let F be a subspace of E and consider the orthogonal complement $F^\perp$. Then

$$\dim F + \dim F^\perp = \dim E. \tag{2.63}$$

Proof: Consider the factor space $E^*/F^\perp$. In view of sec. 2.23, $E^*/F^\perp$ is dual to F which implies (2.63).

Proposition III: Let E, E^* be a pair of dual vector spaces and consider a bilinear function $\Phi: E^* \times E \to \Gamma$. Then there exists precisely one linear transformation $\varphi: E \to E$ such that

$$\Phi(x^*, x) = \langle x^*, \varphi x\rangle \qquad x^* \in E^*, x \in E.$$

Proof: Let $x \in E$ be a fixed vector and consider the linear function f_x in E^* defined by

$$f_x(x^*) = \Phi(x^*, x).$$

In view of proposition I there is precisely one vector $\varphi x \in E$ such that

$$f_x(x^*) = \langle x^*, \varphi x \rangle.$$

The two above equations yield

$$\langle x^*, \varphi x \rangle = \Phi(x^*, x) \qquad x^* \in E^*, x \in E$$

and so a mapping $\varphi: E \to E$ is defined. The linearity of φ follows immediately from the bilinearity of Φ. Suppose now that φ_1 and φ_2 are two linear transformations of E such that

$$\Phi(x^*, x) = \langle x^*, \varphi_1 x \rangle \quad \text{and} \quad \Phi(x^*, x) = \langle x^*, \varphi_2 x \rangle$$

Then we have that

$$\langle x^*, \varphi_1 x - \varphi_2 x \rangle = 0$$

whence $\varphi_1 = \varphi_2$.
Proposition III establishes a canonical linear isomorphism between the spaces $B(E^*, E)$ and $L(E; E)$,

$$B(E^*, E) \cong L(E; E).$$

Here $B(E^*, E)$ is the space of bilinear functions $\Phi: E^* \times E \to \Gamma$ with addition and scalar multiplication defined by

$$(\Phi_1 + \Phi_2)(x^*, x) = \Phi_1(x^*, x) + \Phi_2(x^*, x)$$

and

$$(\lambda \Phi)(x^*, x) = \lambda \cdot \Phi(x^*, x).$$

2.34. The rank of a linear mapping. Let $\varphi: E \to F$ be a linear mapping of finite dimensional vector spaces. Then $\ker \varphi \subset E$ and $\operatorname{Im} \varphi \subset F$ have finite dimension as well. We define the *rank* of φ as the dimension of $\operatorname{Im} \varphi$

$$r(\varphi) = \dim \operatorname{Im} \varphi.$$

In view of the induced linear isomorphism

$$E/\ker \varphi \xrightarrow{\cong} \operatorname{Im} \varphi$$

we have at once

$$r(\varphi) + \dim \ker \varphi = \dim E. \tag{2.64}$$

φ is called *regular* if it is injective. (2.64) implies that φ is regular if and only if $r(\varphi) = \dim E$.

In the special case dim $E = \dim F$ (and hence in particular in the case of a linear transformation) we have that φ is regular if and only if it is surjective and hence an isomorphism.

2.35. Dual mappings. Let E^*, E and F^*, F be dual pairs and $\varphi: E \to F$ be a linear mapping. Since E^* is canonically isomorphic to the space $L(E)$ there exists a dual mapping $\varphi^*: E^* \leftarrow F^*$. Hence we have the relations (cf. sec. 2.28)

$$\operatorname{Im} \varphi = (\ker \varphi^*)^{\perp}$$

and

$$\operatorname{Im} \varphi^* = (\ker \varphi)^{\perp}.$$

The first relation states that the equation

$$\varphi x = y$$

has a solution for a given $y \in F$ if and only if y satisfies the condition

$$\langle x^*, y \rangle = 0 \quad \text{for every} \quad x^* \in \ker \varphi^*.$$

The second relation implies that dual mappings have the same rank. In fact, from (2.63) and (2.64) we have

$$\dim \operatorname{Im} \varphi^* = \dim (\ker \varphi)^{\perp} = \dim E - \dim \ker \varphi = \dim \operatorname{Im} \varphi$$

whence

$$r(\varphi^*) = r(\varphi). \tag{2.65}$$

Problems

(In problems 1–10 it will be assumed that all vector spaces have finite dimension).

1. Let E, E^* be a pair of dual vector spaces, and let E_1, E_2 be subspaces of E. Prove that

$$(E_1 \cap E_2)^{\perp} = E_1^{\perp} + E_2^{\perp}.$$

Hint: Use problem 2, § 5.

2. Given subspaces $U \subset E$ and $V^* \subset E^*$ prove that

$$\dim (U^{\perp} \cap V^*) + \dim U = \dim (U \cap V^{*\perp}) + \dim V^*.$$

3. Let E, E^* be a pair of non-trivial dual vector spaces and let $\varphi: E \to E^*$ be a linear mapping such that $\varphi \circ \tau = (\tau^*)^{-1} \circ \varphi$ for every linear automorphism τ of E. Prove that $\varphi = 0$. Conclude that there exists no linear mapping $\varphi: E \to E^*$ which transforms every basis of E into its dual basis.

4. Given a pair of dual bases $x^{*\nu}$, x_ν $(\nu=1\ldots n)$ of E, E^* show that the bases $(x^{*1}+\sum_{\nu=2}^{n}\lambda_\nu x^{*\nu}, x^{*2}, \ldots, x^{*n})$ and $(x_1, x_2-\lambda_2 x_1, \ldots, x_n-\lambda_n x_1)$ are again dual.

5. Let E, F, G be three vector spaces. Given two linear mappings $\varphi: E\to F$ and $\psi: F\to G$ prove that

$$r(\psi\circ\varphi)\leqq r(\varphi) \quad \text{and} \quad r(\psi\circ\varphi)\leqq r(\psi).$$

6. Let E be a vector space of dimension n and consider a system of n linear transformations $\sigma_i: E\to E$, $\sigma_i \neq 0$, such that

$$\sigma_i\circ\sigma_j=\sigma_i\delta_{ij} \qquad (i,j=1\ldots n).$$

a) Show that every σ_i has rank 1

b) If σ_i' $(i=1\ldots n)$ is a second system with the same property, prove that there exists a linear automorphism τ of E such that

$$\sigma_i'=\tau^{-1}\circ\sigma_i\circ\tau.$$

7. Given two linear mappings $\varphi: E\to F$ and $\psi: E\to F$ prove that

$$|r(\varphi)-r(\psi)|\leqq r(\varphi+\psi)\leqq r(\varphi)+r(\psi).$$

8. Show that the dimensions of the spaces $M^l(\varphi)$, $M^r(\varphi)$ in problem 3, § 2 are given by

$$\dim M^l(\varphi)=(\dim F-r(\varphi))\cdot\dim E$$
$$\dim M^r(\varphi)=\dim\ker\varphi\cdot\dim F.$$

9. Show that the mapping $\Phi: \varphi\to\varphi^*$ defines a linear isomorphism,

$$\Phi: L(E;F)\xrightarrow{\cong} L(F^*;E^*).$$

10. Prove that

$$\Phi M^l(\varphi)=M^r(\varphi^*)$$

and

$$\Phi M^r(\varphi)=M^l(\varphi^*)$$

where the notation is defined in problems 8 and 9. Hence obtain the formula

$$r(\varphi)=r(\varphi^*).$$

11. Let $\varphi: E\to F$ be a linear mapping (E, F possibly of infinite dimension). Prove that Im φ has finite dimension if and only if ker φ has finite codimension (recall that the codimension of a subspace is the dimension

of the corresponding factor space), and that in this case

$$\operatorname{codim}\ker\varphi = \dim\operatorname{Im}\varphi\,.$$

12. Let E and F be vector spaces of finite dimension and consider a bilinear function Φ in $E\times F$. Prove that

$$\dim E - \dim N_E = \dim F - \dim N_F$$

where N_E and N_F denote the null spaces of Φ.

Conclude that if $\dim E=\dim F$, then $N_E=0$ if and only if $N_F=0$.

Chapter III

Matrices

In this chapter all vector spaces will be defined over a fixed, but arbitrarily chosen field Γ of characteristic 0.

§ 1. Matrices and systems of linear equations

3.1. Definition. A rectangular array

$$A = \begin{pmatrix} \alpha_1^1 \dots \alpha_1^m \\ \vdots \quad\quad \vdots \\ \alpha_n^1 \dots \alpha_n^m \end{pmatrix} \tag{3.1}$$

of nm scalars α_ν^μ is called a *matrix* of n rows and m columns or, in brief, an $n \times m$-matrix. The scalars α_ν^μ are called the *entries* or the *elements* of the matrix A. The rows

$$a_\nu = (\alpha_\nu^1 \dots \alpha_\nu^m) \qquad (\nu = 1 \dots n)$$

can be considered as vectors of the space Γ^m and therefore are called the *row-vectors* of A. Similarly, the columns

$$b^\mu = (b_1^\mu \dots b_n^\mu) \qquad (\mu = 1 \dots m)$$

considered as vectors of the space Γ^n, are called the *column-vectors* of A.

Interchanging rows and columns we obtain from A the *transposed matrix*

$$A^* = \begin{pmatrix} \alpha_1^1 \dots \alpha_n^1 \\ \vdots \quad\quad \vdots \\ \alpha_1^m \dots \alpha_n^m \end{pmatrix}. \tag{3.2}$$

In the following, matrices will rarely be written down explicitly as in (3.1) but rather be abbreviated in the form $A = (\alpha_\nu^\mu)$. This notation has the disadvantage of not identifying which index counts the rows and which the columns. It has to be mentioned in this connection that it would be very undesirable – as we shall see – to agree once and for all to always let

6*

the subscript count the rows, etc. If the above abbreviation is used, it will be stated explicitly which index indicates the rows.

3.2. The matrix of a linear mapping. Consider two linear spaces E and F of dimensions n and m and a linear mapping $\varphi: E \to F$. With the aid of bases $x_\nu (\nu = 1 \dots n)$ and $y_\mu (\mu = 1 \dots m)$ in E and in F respectively, every vector φx_ν can be written as a linear combination of the vectors y_μ $(\mu = 1 \dots m)$,

$$\varphi x_\nu = \sum_{\mu=1}^{m} \alpha_\nu^\mu y_\mu \qquad (\nu = 1 \dots n). \tag{3.3}$$

In this way, the mapping φ determines an $n \times m$-matrix (α_ν^μ), where ν counts the rows and μ counts the columns. This matrix will be denoted by $M(\varphi, x_\nu, y_\mu)$ or simply by $M(\varphi)$ if no ambiguity is possible.

Conversely, every $n \times m$-matrix (α_ν^μ) determines a linear mapping $\varphi: E \to F$ by the equations (3.3). Thus, the operator

$$M: \varphi \to M(\varphi)$$

defines a one-to-one correspondence between all linear mappings $\varphi: E \to F$ and all $n \times m$-matrices.

3.3. The matrix of the dual mapping. Let E^* and F^* be dual spaces of E and F, respectively, and $\varphi: E \to F$, $\varphi^*: E^* \leftarrow F^*$ a pair of dual mappings. Consider two pairs of dual bases $x^{*\nu}, x_\nu (\nu = 1 \dots n)$ and $y^{*\mu}, y_\mu (\mu = 1 \dots m)$ of E^*, E and F^*, F, respectively. We shall show that the two corresponding matrices $M(\varphi)$ and $M(\varphi^*)$ (relative to these bases) are transposed, i.e., that

$$M(\varphi^*) = M(\varphi)^*. \tag{3.4}$$

The matrices $M(\varphi)$ and $M(\varphi^*)$ are defined by the representations

$$\varphi x_\nu = \sum_\mu \alpha_\nu^\mu y_\mu \quad \text{and} \quad \varphi^* y^{*\mu} = \sum_\nu \alpha_\nu^{*\mu} x^{*\nu}.$$

Note here that the subscript ν indicates in the first formula the rows of the matrix α_ν^μ and in the second the columns of the matrix α_ν^μ. Substituting $x = x_\nu$ and $y = y^{*\mu}$ in the relation

$$\langle y^*, \varphi x \rangle = \langle \varphi^* y^*, x \rangle \tag{3.5}$$

we obtain

$$\langle y^{*\mu}, \varphi x_\nu \rangle = \langle \varphi^* y^{*\mu}, x_\nu \rangle. \tag{3.6}$$

Now

$$\langle y^{*\mu}, \varphi x_\nu \rangle = \sum_\kappa \alpha_\nu^\kappa \langle y^{*\mu}, y_\kappa \rangle = \alpha_\nu^\mu \tag{3.7}$$

and

$$\langle\varphi^* y^{*\mu}, x_\nu\rangle = \sum_\lambda \alpha_\lambda^{*\mu}\langle x^{*\lambda}, x_\nu\rangle = \alpha_\nu^{*\mu}. \tag{3.8}$$

The relations (3.6), (3.7) and (3.8) then yield

$$\alpha_\nu^{*\mu} = \alpha_\nu^\mu.$$

Observing – as stated before – that the subscript ν indicates rows of (α_ν^μ) and columns of $(\alpha_\nu^{*\mu})$ we obtain the desired equation (3.4).

3.4. Rank of a matrix. Consider an $n\times m$-matrix A. Denote by r_1 and by r_2 the maximal number of linearly independent row-vectors and column-vectors, respectively. It will be shown that $r_1 = r_2$. To prove this let E and F be two linear spaces of dimensions n and m. Choose a basis $x_\nu\ (\nu = 1 \ldots n)$ and $y_\mu\ (\mu = 1 \ldots m)$ in E and in F and define the linear mapping $\varphi: E \to F$ by

$$\varphi x_\nu = \sum_\mu \alpha_\nu^\mu y_\mu.$$

Besides φ, consider the isomorphism

$$\beta: F \to \Gamma^m$$

defined by

$$\beta: y \to (\eta^1 \ldots \eta^m),$$

where

$$y = \sum_\mu \eta^\mu y_\mu.$$

Then $\beta\circ\varphi$ is a linear mapping of E into Γ^m. From the definition of β it follows that $\beta\circ\varphi$ maps x_ν into the ν-th row-vector,

$$\beta\varphi x_\nu = a_\nu.$$

Consequently, the rank of $\beta\circ\varphi$ is equal to the maximal number r_1 of linearly independent row-vectors. Since β is a linear isomorphism, $\beta\circ\varphi$ has the same rank as φ and hence r_1 is equal to the rank r of φ.

Replacing φ by φ^* we see that the maximal number r_2 of linearly independent column-vectors is equal to the rank of φ^*. But φ^* has the same rank as φ (cf. sec. 2.35) and thus $r_1 = r_2 = r$. The number r is called the *rank* of the matrix A.

3.5. Systems of linear equations. Matrices play an important role in the discussion of systems of linear equations in a field. Such a system

$$\sum_\nu \alpha_\nu^\mu \xi^\nu = \eta^\mu \qquad (\mu = 1 \ldots m) \tag{3.9}$$

of m equations with n unknowns is called *inhomogeneous* if at least one η^μ is different from zero. Otherwise it is called *homogeneous.*

From the results of Chapter II it is easy to obtain theorems about the existence and uniqueness of solutions of the system (3.9). Let E and F be two linear spaces of dimensions n and m. Choose a basis $x_\nu\,(\nu=1\ldots n)$ of E as well as a basis $y_\mu\,(\mu=1\ldots m)$ of F and define the linear mapping $\varphi: E\to F$ by

$$\varphi\, x_\nu = \sum_\mu \alpha_\nu^\mu\, y_\mu\,.$$

Consider two vectors

$$x = \sum_\nu \xi^\nu x_\nu \tag{3.10}$$

and

$$y = \sum_\mu \eta^\mu\, y_\mu\,. \tag{3.11}$$

Then

$$\varphi\, x = \sum_\nu \xi^\nu \varphi\, x_\nu = \sum_{\nu,\,\mu} \alpha_\nu^\mu\, \xi^\nu\, y_\mu\,. \tag{3.12}$$

Comparing the representations (3.9) and (3.12) we see that the system (3.9) is equivalent to the vector-equation

$$\varphi\, x = y\,.$$

Consequently, the system (3.9) has a solution if and only if the vector y is contained in the image-space Im φ. Moreover, this solution is uniquely determined if and only if the kernel of φ consists only of the zero-vector.

3.6. The homogeneous system. Consider the homogeneous system

$$\sum_\nu \alpha_\nu^\mu\, \xi^\nu = 0 \qquad (\mu = 1\ldots m)\,. \tag{3.13}$$

From the foregoing discussion it is immediately clear that $(\xi^1\ldots\xi^n)$ is a solution of this system if and only if the vector x defined by (3.10) is contained in the kernel ker φ of the linear mapping φ. In sec. 2.34 we have shown that the dimension of ker φ equals $n-r$ where r denotes the rank of φ.

Since the rank of φ is equal to the rank of the matrix (α_ν^μ), we therefore obtain the following theorem:

A homogeneous system of m equations with n unknowns whose coefficient-matrix is of rank r has exactly $n-r$ linearly independent solutions. In the special case that the number m of equations is less than the number n of unknowns we have $n-r \geqq n-m \geqq 1$. Hence the theorem asserts that the system (3.13) always has non-trivial solutions if m is less than n.

3.7. The alternative-theorem. Let us assume that the number of equations is equal to the number of unknowns,

$$\sum_\nu \alpha_\nu^\mu \xi^\nu = \eta^\mu \qquad (\mu = 1 \dots n). \tag{3.14}$$

Besides (3.14) consider the so-called "corresponding" homogeneous system

$$\sum_\nu \alpha_\nu^\mu \xi^\nu = 0 \qquad (\mu = 1 \dots n). \tag{3.15}$$

The mapping φ introduced in sec. 3.5 is now a linear mapping of the n-dimensional space E into a space of the same dimension. Hence we may apply the result of sec. 2.34 and obtain the following *alternative-theorem:*

If the homogeneous system possesses only the trivial solution (0...0), *the inhomogeneous system has a solution* $(\xi^1 \dots \xi^n)$ *for every choice of the right-hand side. If the homogeneous system has non-trivial solutions, then the inhomogeneous one is not solvable for every choice of the* $\eta^\nu (\nu = 1 \dots n)$.

From the last statement of section 3.5 it follows immediately that in the first case the solution of (3.14) is uniquely determined while in the second case the system (3.14) has – if it is solvable at all – infinitely many solutions.

3.8. The main-theorem. We now proceed to the general case of an arbitrary system

$$\sum_\nu \alpha_\nu^\mu \xi^\nu = \eta^\mu \qquad (\mu = 1 \dots m) \tag{3.16}$$

of m linear equations in n unknowns. As stated before, this system has a solution if and only if the vector

$$y = \sum_\mu \eta^\mu y_\mu$$

is contained in the image-space Im φ. In sec. 2.35 it has been shown that the space Im φ is the orthogonal complement of the kernel of the dual mapping $\varphi^*: F^* \to E^*$. In other words, the system (3.16) is solvable if and only if the right-hand side η^μ $(\mu = 1 \dots m)$ satisfies the conditions

$$\sum_\mu \eta^\mu \eta_\mu^* = 0 \tag{3.17}$$

for all solutions η_μ^* $(\mu = 1 \dots m)$ of the system

$$\sum_\mu \alpha_\nu^\mu \eta_\mu^* = 0 \quad (\nu = 1 \dots n). \tag{3.18}$$

We formulate this result in the following

Main-theorem: An inhomogeneous system of n equations in m unknowns has a solution if and only if every solution η^*_μ $(\mu=1\dots m)$ *of the transposed homogeneous system* (3.18) *satisfies the orthogonality-relation* (3.17).

Problems

1. Find the matrices corresponding to the following mappings $\varphi: E\to E$:

a) $\varphi x=0$.

b) $\varphi x=x$.

c) $\varphi x=\lambda x$.

d) $\varphi x=\sum_{\nu=1}^{m}\xi^\nu e_\nu$ where $e_\nu(\nu=1,\dots,n)$ is a given basis and $m\leqq n$ is a given number and $x=\sum_{\nu=1}^{n}\xi^\nu e_\nu$.

2. Consider a system of two equations in n unknowns

$$\sum_{\nu=1}^{n}\alpha_\nu\xi^\nu=\alpha \quad \sum_{\nu=1}^{n}\beta_\nu\xi^\nu=\beta\,.$$

Find the solutions of the corresponding transposed homogeneous system.

3. Prove the following statement:

The general solution of the inhomogeneous system is equal to the sum of any particular solution of this system and the general solution of the corresponding homogeneous system.

4. Let x_ν and $\bar{x}_\nu$ be two bases of E and A be the matrix of the basis-transformation $x_\nu\to\bar{x}_\nu$. Define the automorphism α of E by $\alpha x_\nu=\bar{x}_\nu$. Prove that A is the matrix of α as well with respect to the basis x_ν as with respect to the basis $\bar{x}_\nu$.

5. Show that a necessary and sufficient condition for the $n\times n$-matrix $A=(\alpha_\nu^\mu)$ to have rank $\leqq 1$ is that there exist elements $\alpha_1,\alpha_2,\dots,\alpha_n$ and $\beta^1,\beta^2,\dots,\beta^n$ such that

$$\alpha_\nu^\mu=\alpha_\nu\beta^\mu \qquad (\nu=1,2,\dots,n;\ \mu=1,2,\dots,n)\,.$$

If $A\neq 0$, show that the elements α_ν and β^μ are uniquely determined up to constant factors λ and μ respectively, where $\lambda\mu=1$.

6. Given a basis a_ν of a linear space E, define the mapping $\varphi: E\to E$ as

$$\varphi a_\nu=\sum_\mu a_\mu\,.$$

Find the matrix of the dual mapping relative to the dual basis.

7. Verify that the system of three equations:

$$\begin{aligned}\xi+\eta+\zeta&=3\,,\\ \xi-\eta-\zeta&=4\,,\\ \xi+3\eta+3\zeta&=1\end{aligned}$$

has no solution. Find a solution of the transposed homogeneous system which is not orthogonal to the vector (3, 4, 1). Replace the number 1 on the right-hand side of the third equation in such a way that the resulting system is solvable.

8. Let an inhomogeneous system of linear equations be given,

$$\sum_{\nu} \alpha_{\nu}^{\mu} \xi^{\nu} = \eta^{\mu} \qquad (\mu = 1, \ldots, m).$$

The *augmented matrix* of the system is defined as the $m \times (n+1)$-matrix obtained from the matrix α_{ν}^{μ} by adding the column $(\eta^1, \ldots, \eta^m)$. Prove that the above system has a solution if and only if the augmented matrix has the same rank as the matrix (α_{ν}^{μ}).

§ 2. Multiplication of matrices

3.9. The linear space of the $n \times m$ matrices. Consider the space $L(E; F)$ of all linear mappings $\varphi: E \to F$ and the set $M^{n \times m}$ of all $n \times m$-matrices. Once bases have been chosen in E and in F there is a 1-1 correspondence between the mappings $\varphi: E \to F$ and the $n \times m$-matrices defined by

$$\varphi \to M(\varphi, x_{\nu}, y_{\mu}). \tag{3.19}$$

This correspondence suggests defining a linear structure in the set $M^{n \times m}$ such that the mapping (3.19) becomes an isomorphism.

We define the *sum* of two $n \times m$-matrices

$$A = (\alpha_{\nu}^{\mu}) \quad \text{and} \quad B = (\beta_{\nu}^{\mu})$$

as the $n \times m$-matrix

$$A + B = (\alpha_{\nu}^{\mu} + \beta_{\nu}^{\mu})$$

and the product of a scalar λ and a matrix A as the matrix

$$\lambda A = (\lambda a_{\nu}^{\mu}).$$

It is immediately apparent that with these operations the set $M^{n \times m}$ is a linear space. The zero-vector in this linear space is the matrix which has only zero-entries.

Furthermore, it follows from the above definitions that

$$M(\lambda \varphi + \mu \psi) = \lambda M(\varphi) + \mu M(\psi) \qquad \varphi, \psi \in L(E; F)$$

i.e., that the mapping (3.19) defines an isomorphism between $L(E; F)$ and the space $M^{n \times m}$.

3.10. Product of matrices. Assume that

$$\varphi: E \to F \quad \text{and} \quad \psi: F \to G$$

are linear mappings between three linear spaces E, F, G of dimensions n, m and l, respectively. Then $\psi \circ \varphi$ is a linear mapping of E into G. Select a basis $x_\nu (\nu = 1 \dots n)$, $y_\mu (\mu = 1 \dots m)$ and $z_\lambda (\lambda = 1 \dots l)$ in each of the three spaces. Then the mappings φ and ψ determine two matrices (α_ν^μ) and (β_μ^λ) by the relations

$$\varphi x_\nu = \sum_\mu \alpha_\nu^\mu y_\mu$$

and

$$\psi y_\mu = \sum_\lambda \beta_\mu^\lambda z_\lambda .$$

These two equations yield

$$(\psi \circ \varphi) x_\nu = \sum_{\mu, \lambda} \alpha_\nu^\mu \beta_\mu^\lambda z_\lambda .$$

Consequently, the matrix of the mapping $\psi \circ \varphi$ relative to the bases x_ν and z_λ is given by

$$\gamma_\nu^\lambda = \sum_\mu \alpha_\nu^\mu \beta_\mu^\lambda . \tag{3.20}$$

The $n \times l$-matrix (3.20) is called the *product* of the $n \times m$-matrix $A = (\alpha_\nu^\mu)$ and the $m \times l$-matrix $B = (\beta_\nu^\mu)$ and is denoted by AB. It follows immediately from this definition that

$$M(\psi \circ \varphi) = M(\varphi) M(\psi) . \tag{3.21}$$

Note that the matrix $M(\psi \circ \varphi)$ of the product-mapping $\psi \circ \varphi$ is the product of the matrices $M(\varphi)$ and $M(\psi)$ in reversed order of the factors.

It follows immediately from (3.21) and the formulas of sec. 2.10 that the matrix-multiplication has the following properties (cf. sec. 2.25):

$$\begin{aligned} A(\lambda B_1 + \mu B_2) &= \lambda A B_1 + \mu A B_2 \\ (\lambda A_1 + \mu A_2) B &= \lambda A_1 B + \mu A_2 B \\ (AB)C &= A(BC) \\ (AB)^* &= B^* A^* . \end{aligned}$$

3.11. Automorphisms and regular matrices. An $n \times n$-matrix A is called *regular* if it has the maximal rank n. Let φ be an automorphism of the n-dimensional linear space E and $A = M(\varphi)$ the corresponding $n \times n$-matrix relative to a basis $x_\nu (\nu = 1 \dots n)$. By the result of section 3.4 the rank of φ is equal to the rank of the matrix A. Consequently, the matrix A is regular. Conversely, every linear transformation $\varphi: E \to E$ having a regular matrix is an automorphism.

To every regular matrix A there exists an *inverse matrix*, i.e., a matrix A^{-1} such that

$$AA^{-1} = A^{-1}A = J,$$

where J denotes the *unit matrix* whose entries are δ^{μ}_{ν}. In fact, let φ be the automorphism of E such that $M(\varphi) = A$ and let φ^{-1} be the inverse automorphism. Then

$$\varphi^{-1} \circ \varphi = \varphi \circ \varphi^{-1} = \iota,$$

whence

$$M(\varphi)M(\varphi^{-1}) = M(\varphi^{-1} \circ \varphi) = M(\iota) = J$$

and

$$M(\varphi^{-1})M(\varphi) = M(\varphi \circ \varphi^{-1}) = M(\iota) = J.$$

These equations show that the matrix

$$A^{-1} = M(\varphi^{-1})$$

is the inverse of the matrix A.

Problems

1. Verify the following properties:

a) $(A + B)^* = A^* + B^*$.
b) $(\lambda A)^* = \lambda A^*$.
c) $(A^{-1})^* = (A^*)^{-1}$.

2. A square-matrix is called *upper (lower) triangular* if all the elements below (above) the main diagonal are zero. Prove that sum and product of triangular matrices are again triangular.

3. Let φ be linear transformation such that $\varphi^2 = \varphi$. Show that there exists a basis in which φ is represented by a matrix of the form:

$$\underbrace{\begin{pmatrix} 1 & & & 0 \dots 0 \\ & 1 & & \cdot \quad \cdot \\ & & \cdot & \cdot \quad \cdot \\ & & 1 & 0 \dots 0 \\ 0 & \dots & \dots & \dots 0 \\ \cdot & & & \cdot \\ \cdot & & & \cdot \\ \cdot & & & \cdot \\ 0 & \dots & \dots & \dots 0 \end{pmatrix}}_{n} \begin{matrix} \left.\vphantom{\begin{matrix}1\\1\\ \cdot\\1\end{matrix}}\right\} m \\ \left.\vphantom{\begin{matrix}0\\ \cdot\\ \cdot\\ \cdot\\0\end{matrix}}\right\} n-m \end{matrix}$$

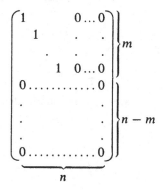

4. Denote by A_{ij} the matrix having the entry 1 at the place (i,j) and zero elsewhere. Verify the formula

$$A_{ij} \cdot A_{jk} = A_{ik}.$$

Prove that the matrices $\{A_{ij} | i, j=1, \ldots, n\}$ form a basis of the space $M^{n \times n}$.

§ 3. Basis-transformation

3.12. Definition. Consider two bases x_ν and $\bar{x}_\nu (\nu=1 \ldots n)$ of the space E. Then every vector $\bar{x}_\nu (\nu=1 \ldots n)$ can be written as

$$\bar{x}_\nu = \sum_\mu \alpha_\nu^\mu x_\mu. \tag{3.22}$$

Similarly,

$$x_\nu = \sum_\mu \check{\alpha}_\nu^\mu \bar{x}_\mu. \tag{3.23}$$

The two $n \times n$-matrices defined by (3.22) and (3.23) are inverse to each other. In fact, combining (3.22) and (3.23) we obtain

$$\bar{x}_\nu = \sum_{\mu, \lambda} \alpha_\nu^\mu \check{\alpha}_\mu^\lambda \bar{x}_\lambda.$$

This is equivalent to

$$\sum_\lambda \Big(\sum_\mu \alpha_\nu^\mu \check{\alpha}_\mu^\lambda - \delta_\nu^\lambda\Big) \bar{x}_\lambda = 0$$

and hence it implies that

$$\sum_\mu \alpha_\nu^\mu \check{\alpha}_\mu^\lambda = \delta_\nu^\lambda.$$

In a similar way the relations

$$\sum_\mu \check{\alpha}_\nu^\mu \alpha_\mu^\lambda = \delta_\nu^\lambda$$

are proved. Thus, any two bases of E are connected by a pair of inverse matrices.

Conversely, given a basis $x_\nu (\nu=1 \ldots n)$ and a regular $n \times n$-matrix (α_ν^μ), another basis can be obtained by

$$\bar{x}_\nu = \sum_\mu \alpha_\nu^\mu x_\mu.$$

To show that the vectors $\bar{x}_\nu$ are linearly independent, assume that

$$\sum_\nu \lambda^\nu \bar{x}_\nu = 0.$$

Then

$$\sum_{\nu, \mu} \lambda^\nu \alpha_\nu^\mu x_\mu = 0$$

and hence, in view of the linear independence of the vectors x_μ,

$$\sum_\nu \lambda^\nu \alpha_\nu^\mu = 0 \qquad (\mu = 1 \dots n).$$

Multiplication with the inverse matrix $(\check{\alpha}_\mu^\nu)$ yields

$$\sum_{\nu,\mu} \lambda^\nu \alpha_\nu^\mu \check{\alpha}_\mu^\kappa = \sum_\nu \lambda^\nu \delta_\nu^\kappa = \lambda^\kappa = 0 \qquad (\kappa = 1 \dots n).$$

3.13. Transformation of the dual basis. Let E^* be a dual space of E, $x^{*\nu}$ the dual basis of x_ν and $\bar{x}^{*\nu}$ the dual of the basis $\bar{x}_\nu (\nu = 1 \dots n)$. Then

$$\bar{x}^{*\varrho} = \sum_\sigma \beta_\sigma^\varrho x^{*\sigma}, \tag{3.24}$$

where β_σ^ϱ is a regular $n \times n$-matrix. Relations (3.23) and (3.24) yield

$$\sum_\sigma \beta_\sigma^\varrho \langle x^{*\sigma}, x_\nu \rangle = \sum_\mu \check{\alpha}_\nu^\mu \langle \bar{x}^{*\varrho}, \bar{x}_\mu \rangle. \tag{3.25}$$

Now

$$\langle x^{*\sigma}, x_\nu \rangle = \delta_\nu^\sigma \quad \text{and} \quad \langle \bar{x}^{*\varrho}, \bar{x}_\mu \rangle = \delta_\mu^\varrho.$$

Substituting this in (3.25) we obtain

$$\beta_\nu^\varrho = \check{\alpha}_\nu^\varrho.$$

This shows that the matrix of the basis-transformation $x^{*\nu} \to \bar{x}^{*\nu}$ is the inverse of the matrix of the transformation $x_\nu \to \bar{x}_\nu$. The two basis-transformations

$$\bar{x}_\nu = \sum_\mu \alpha_\nu^\mu x_\mu \quad \text{and} \quad \bar{x}^{*\nu} = \sum_\mu \check{\alpha}_\mu^\nu x^{*\mu} \tag{3.26}$$

are called *contragradient* to each other.

The relations (3.26) permit the derivation of the transformation-law for the components of a vector $x \in E$ under the basis-transformation $x_\nu \to \bar{x}_\nu$. Decomposing x relative to the bases x_ν and $\bar{x}_\nu$ we obtain

$$x = \sum_\nu \xi^\nu x_\nu \quad \text{and} \quad x = \sum_\nu \bar{\xi}^\nu \bar{x}_\nu.$$

From the two above equations we obtain in view of (3.26).

$$\bar{\xi}^\nu = \sum_\mu \check{\alpha}_\mu^\nu \langle x^{*\mu}, x \rangle = \sum_\mu \check{\alpha}_\mu^\nu \xi^\mu. \tag{3.27}$$

Comparing (3.27) with the second equation (3.26) we see that the components of a vector are transformed in exactly the same way as the vectors of the dual basis.

3.14. The transformation of the matrix of a linear mapping. In this section it will be investigated how the matrix of a linear mapping $\varphi: E \to F$ is changed under a basis-transformation in E as well as in F. Let $M(\varphi; x_\nu, y_\mu) = (\gamma_\nu^\mu)$ and $M(\varphi; \bar{x}_\nu, \bar{y}_\mu) = (\bar{\gamma}_\nu^\mu)$ be the $n \times m$-matrices of φ relative to the bases x_ν, y_μ and $\bar{x}_\nu, \bar{y}_\mu$ $(\nu = 1 \dots n,\ \mu = 1 \dots m)$, respectively. Then

$$\varphi x_\nu = \sum_\mu \gamma_\nu^\mu y_\mu \quad \text{and} \quad \varphi \bar{x}_\nu = \sum_\mu \bar{\gamma}_\nu^\mu \bar{y}_\mu \qquad (\nu = 1 \dots n). \tag{3.28}$$

Introducing the matrices

$$A = (\alpha_\nu^\lambda) \quad \text{and} \quad B = (\beta_\mu^\kappa)$$

of the basis-transformations $x_\nu \to \bar{x}_\nu$ and $y_\mu \to \bar{y}_\mu$ and their inverse matrices, we then have the relations

$$\begin{aligned} \bar{x}_\nu &= \sum_\lambda \alpha_\nu^\lambda x_\lambda & x_\nu &= \sum_\lambda \check{\alpha}_\nu^\lambda \bar{x}_\lambda \\ \bar{y}_\mu &= \sum_\kappa \beta_\mu^\kappa y_\kappa & y_\mu &= \sum_\kappa \check{\beta}_\mu^\kappa \bar{y}_\kappa . \end{aligned} \tag{3.29}$$

Equations (3.28) and (3.29) yield

$$\varphi \bar{x}_\nu = \sum_\lambda \alpha_\nu^\lambda \varphi x_\lambda = \sum_{\lambda,\mu} \alpha_\nu^\lambda \gamma_\lambda^\mu y_\mu = \sum_{\lambda,\mu,\kappa} \alpha_\nu^\lambda \gamma_\lambda^\mu \check{\beta}_\mu^\kappa \bar{y}_\kappa$$

and we obtain the following relation between the matrices (γ_ν^μ) and $(\bar{\gamma}_\nu^\mu)$:

$$\bar{\gamma}_\nu^\kappa = \sum_{\lambda,\mu} \alpha_\nu^\lambda \gamma_\lambda^\mu \check{\beta}_\mu^\kappa . \tag{3.30}$$

Using capital letters for the matrices we can write the transformation formula (3.30) in the form

$$M(\varphi; \bar{x}_\nu, \bar{y}_\mu) = A\, M(\varphi; x_\nu, y_\mu)\, B^{-1}.$$

It shows that all possible matrices of the mapping φ are obtained from a particular matrix by left-multiplication with a regular $n \times n$-matrix and right-multiplication with a regular $m \times m$-matrix.

Problems

1. Let f be a function defined in the set of all $n \times n$-matrices such that

$$f(T A T^{-1}) = f(A)$$

for every regular matrix T. Define the function F in the space $L(E; E)$ by

$$F(\varphi) = f(M(\varphi; x_\nu, x_\mu))$$

where E is an n-dimensional linear space and $x_\nu\,(\nu=1\ldots n)$ is a basis of E. Prove that the function F does not depend on the choice of the basis x_ν.

2. Assume that φ is a linear transformation $E\to E$ having the same matrix relative to every basis $x_\nu\,(\nu=1\ldots n)$. Prove that $\varphi=\lambda\iota$ where λ is a scalar.

3. Given the basis transformation

$$\begin{aligned} \bar{x}_1 &= 2x_1 - x_2 - x_3 \\ \bar{x}_2 &= -\,x_2 \\ \bar{x}_3 &= 2x_2 + x_3 \end{aligned}$$

find all the vectors which have the same components with respect to the bases x_μ and $\bar{x}_\mu$ $(\mu=1,2,3)$.

§ 4. Elementary transformations

3.15. Definition. Consider a linear mapping $\varphi: E\to F$. Then there exists a basis $a_\nu\,(\nu=1,\ldots,n)$ of E and a basis $b_\mu\,(\mu=1,\ldots,m)$ of F such that the corresponding matrix of φ has the following *normal-form:*

$$\begin{pmatrix} \underbrace{\begin{matrix} 1 & & \\ & \ddots & \\ & & 1 \end{matrix}}_{r} & & 0 \\ & 0 & \\ & & \ddots \\ 0 & & & 0 \end{pmatrix} \tag{3.31}$$

where r is the rank of φ. In fact, let $a_\nu\,(\nu=1,\ldots,n)$ be a basis of E such that the vectors $a_{r+1}\ldots a_n$ form a basis of the kernel. Then the vectors $b_\varrho=\varphi a_\varrho\,(\varrho=1,\ldots,r)$ are linearly independent and hence this system can be extended to a basis $(b_1,\ldots,b_m)$ of F. It follows from the construction of the bases a_ν and b_μ that the matrix of φ has the form (3.31).

Now let $x_\nu\,(\nu=1,\ldots,n)$ and $y_\mu\,(\mu=1,\ldots,m)$ be two arbitrary bases of E and F. It will be shown that the corresponding matrix $M(\varphi;x_\nu,y_\mu)$ can be converted into the normal-form (3.31) by a number of elementary basis-transformations. These transformations are:

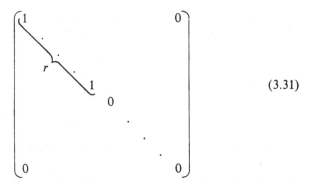

(I.1.) Interchange of two vectors x_i and $x_j (i \neq j)$.
(I.2.) Interchange of two vectors y_k and $y_l (k \neq l)$.
(II.1.) Adding to a vector x_i an arbitrary multiple of a vector $x_j (j \neq i)$.
(II.2.) Adding to a vector y_k an arbitrary multiple of a vector $y_l (l \neq k)$.

It is easy to see that the four above transformations have the following effect on the matrix $M(\varphi)$:

(I.1.) Interchange of the rows i and j.
(I.2.) Interchange of the columns k and l.
(II.1.) Replacement of the row-vector a_i by $a_i + \lambda a_j (j \neq i)$.
(II.2.) Replacement of the column-vector b_k by $b_k + \lambda b_l (l \neq k)$.

It remains to be shown that every $n \times m$-matrix can be converted into the normal form (3.31) by a sequence of these elementary matrix-transformations and the operations $a_\nu \to \lambda a_\nu$, $b_\mu \to \lambda b_\mu$ $(\lambda \neq 0)$.

3.16. Reduction to the normal-form. Let (γ_ν^μ) be the given $n \times m$-matrix. It is no restriction to assume that at least one $\gamma_\nu^\mu \neq 0$, otherwise the matrix is already in the normal-form. By the operations (I.1.) and (I.2.) this element can be moved to the place (1, 1). Then $\gamma_1^1 \neq 0$ and it is no restriction to assume that $\gamma_1^1 = 1$. Now, by adding proper multiples of the first row to the other rows we can obtain a matrix whose first column consists of zeros except for γ_1^1. Next, by adding certain multiples of the first column to the other columns this matrix can be converted into the form

$$\begin{pmatrix} 1 & 0 \ldots 0 \\ 0 & * \quad * \\ \cdot & \\ \cdot & \\ 0 & * \quad * \end{pmatrix} \tag{3.32}$$

If all the elements γ_ν^μ $(\nu = 2 \ldots n,\ \mu = 2 \ldots m)$ are zero, (3.32) is the normal-form. Otherwise there is an element $\gamma_\nu^\mu \neq 0 (2 \leqq \nu \leqq m,\ 2 \leqq \mu \leqq m)$. This can be moved to the place (2,2) by the operations (I.1. and (I.2.). Hereby the first row and the first column are not changed. Dividing the second row by γ_2^2 and applying the operations (II.1.) and (II.2.) we can obtain a matrix of the form

$$\begin{pmatrix} 1 & 0 & \ldots 0 \\ 0 & 1 & 0 \ldots 0 \\ \cdot & 0 & * \quad * \\ \cdot & & \\ 0 & 0 & * \quad * \end{pmatrix}$$

In this way the original matrix is ultimately converted into the form (3.31.).

3.17. The Gaussian elimination. The technique described in sec. 3.16 can be used to solve a system of linear equations by successive elimination. Let

$$\begin{aligned} \alpha_1^1 \xi^1 + \cdots \alpha_n^1 \xi^n &= \eta^1 \\ &\vdots \\ \alpha_1^m \xi^1 + \cdots \alpha_n^m \xi^n &= \eta^m \end{aligned} \tag{3.33}$$

be a system of m linear equations in n unknowns. Before starting the elimination we perform the following reductions:

If all coefficients in a certain row, say in the i-th row, are zero, consider the corresponding number η^i on the right hand-side. If $\eta^i \neq 0$, the i-th equation contains a contradiction and the system (3.33) has no solution. If $\eta^i = 0$, the i-th equation is an identity and can be omitted.

Hence, we can assume that at least one coefficient in every equation is different from zero. Rearranging the unknowns we can achieve that $\alpha_1^1 \neq 0$. Multiplying the first equation by $-(\alpha_1^1)^{-1}\alpha_1^\mu$ and adding it to the μ-th equation we obtain a system of the form

$$\begin{aligned} \alpha_1^1 \xi^1 + \alpha_2^1 \xi^2 + \cdots \alpha_n^1 \xi^n &= \zeta^1 \\ \beta_2^2 \xi^2 + \cdots \beta_n^2 \xi^n &= \zeta^2 \\ &\vdots \\ \beta_2^m \xi^2 + \cdots \beta_n^m \xi^n &= \zeta^m \end{aligned} \tag{3.34}$$

which is equivalent to the system (3.33).

Now apply the above reduction to the last $(m-1)$ equations of the system (3.34). If one of these equations contains a contradiction, the system (3.34) has no solutions. Then the equivalent system (3.33) does not have a solution either. Otherwise eliminate the next unknown, say ξ^2, from the reduced system.

Continue this process until either a contradiction arises at a certain step or until no equations are left after the reduction. In the first case, (3.33) does not have a solution. In the second case we finally obtain a triangular system

$$\begin{aligned} \alpha_1^1 \xi^1 + \alpha_2^1 \xi^2 + \cdots \alpha_n^1 \xi^n &= \omega^1 \quad \alpha_1^1 \neq 0 \\ \beta_2^2 \xi^2 + \cdots \beta_n^2 \xi^n &= \omega^2 \quad \beta_2^2 \neq 0 \\ &\cdot \\ &\cdot \\ \kappa_r^r \xi^r + \cdots \kappa_n^r \xi^n &= \omega^r \quad \kappa_r^r \neq 0 \end{aligned} \tag{3.35}$$

which is equivalent to the original system *).

---

*) If no equations are left after the reduction, then every n-tuple $(\xi^1 \dots \xi^n)$ is a solution of (3.33).

The system (3.35) can be solved in a step by step manner beginning with ξ^r,

$$\xi^r = -(\kappa^r)^{-1}\left(\omega^r - \sum_{\nu=r+1}^{n} \kappa_\nu^r \xi^\nu\right). \tag{3.36}$$

Inserting (3.36) into the first $(r-1)$ equations we can reduce the system to a triangular one of $r-1$ equations. Continuing this way we finally obtain the solution of (3.33) in the form

$$\xi^\nu = \sum_{\mu=r+1}^{n} \lambda_\mu^\nu \xi^\mu + \varrho^\nu \qquad (\nu = 1 \dots r)$$

where the $\xi^\nu\,(\nu = r+1 \dots n)$ are arbitrary parameters.

Problems

1. Two $n \times m$-matrices C and C' are called *equivalent* if there exists a regular $n \times n$-matrix A and a regular $m \times m$-matrix B such that $C' = A\,C\,B$. Prove that two matrices are equivalent if and only if they have the same rank.

2. Apply the Gauss elimination to the following systems:

a)
$$\begin{aligned} \xi^1 - \xi^2 + 2\xi^3 &= 1, \\ 2\xi^1 \qquad + 2\xi^3 &= 1, \\ \xi^1 - 3\xi^2 + 4\xi^3 &= 2. \end{aligned}$$

b)
$$\begin{aligned} \eta^1 + 2\eta^2 + 3\eta^3 + 4\eta^4 &= 5, \\ 2\eta^1 + \eta^2 + 4\eta^3 + \eta^4 &= 2, \\ 3\eta^1 + 4\eta^2 + \eta^3 + 5\eta^4 &= 6, \\ 2\eta^1 + 3\eta^2 + 5\eta^3 + 2\eta^4 &= 3. \end{aligned}$$

c)
$$\begin{aligned} \varepsilon^1 + \varepsilon^2 + \varepsilon^3 &= 1, \\ 3\varepsilon^1 + \varepsilon^2 - \varepsilon^3 &= 0. \\ 2\varepsilon^1 + \varepsilon^2 \qquad &= 1, \end{aligned}$$

Chapter IV

Determinants

In this chapter, except for the last paragraph, all vector spaces will be defined over a fixed but arbitrarily chosen field Γ of characteristic 0.

§ 1. Determinant functions

4.1. Even and odd permutations. Let X be an arbitrary set and denote for each $p \geqq 1$ by X^p the set of ordered p-tuples $(x_1, \ldots, x_p)$, $x_i \in X$. Let

$$\Phi: X^p \to Y$$

be a map from X^p to a second set, Y. Then every permutation $\sigma \in S_p$ determines a new map

$$\sigma \Phi: X^p \to Y$$

defined by

$$(\sigma \Phi)(x_1, \ldots, x_p) = \Phi(x_{\sigma(1)}, \ldots, x_{\sigma(p)}).$$

It follows immediately from the definitions that

$$\tau(\sigma \Phi) = (\tau \sigma) \Phi \qquad \sigma, \tau \in S_p \tag{4.1}$$

and

$$\iota \Phi = \Phi \tag{4.2}$$

where ι is the identity permutation.

Now let $X = \mathbb{Z}$, $Y = \mathbb{Z}$ and define Φ by

$$\Phi(x_1, \ldots, x_p) = \prod_{i<j} (x_i - x_j) \qquad x_i \in \mathbb{Z}.$$

It is easily checked that for every $\sigma \in S_p$

$$\sigma \Phi = \varepsilon_\sigma \cdot \Phi$$

where $\varepsilon_\sigma = \pm 1$. Formulae (4.1) and (4.2) imply that

$$\varepsilon_{\tau\sigma} = \varepsilon_\tau \cdot \varepsilon_\sigma$$

and

$$\varepsilon_\iota = 1.$$

Thus ε is a homomorphism from S_p to the multiplicative group $\{-1, +1\}$. A permutation σ is called *even* (respectively *odd*) if $\varepsilon_\sigma = 1$ (respectively $\varepsilon_\sigma = -1$).

Since for a transposition τ, $\varepsilon_\tau = -1$, the transpositions are odd permutations.

4.2. p-linear maps. Let E and F be vector spaces. A *p-linear map from E to F* is a map $\Phi: E^p \to F$ which is linear with respect to each argument; i.e.,

$$\begin{aligned}\Phi(x_1, \dots, \lambda x_i + \mu y_i, \dots, x_p) \\ = \lambda \Phi(x_1, \dots, x_i, \dots, x_p) + \mu \Phi(x_1, \dots, y_i, \dots, x_p)\end{aligned} \qquad \lambda, \mu \in \Gamma.$$

A p-linear map from E to Γ is called a *p-linear function in E*. As an example, let $f_1, \dots, f_p$ be linear functions in E and define Φ by

$$\Phi(x_1, \dots, x_p) = f_1(x_1) \dots f_p(x_p) \qquad x_i \in E.$$

A p-linear map $\Phi: E^p \to F$ is called *skew symmetric*, if for every permutation σ

$$\sigma \Phi = \varepsilon_\sigma \cdot \Phi;$$

that is,

$$\Phi(x_{\sigma(1)}, \dots, x_{\sigma(p)}) = \varepsilon_\sigma \Phi(x_1, \dots, x_p)$$

Every p-linear map $\Phi: E^p \to F$ determines a skew symmetric p-linear map, Ψ, given by

$$\Psi = \sum_\sigma \varepsilon_\sigma \cdot \sigma \Phi.$$

In fact, let τ be an arbitrary permutation. Then formula (4.1) yields

$$\begin{aligned}\tau \Psi &= \sum_\sigma \varepsilon_\sigma \, \tau(\sigma \Phi) = \sum_\sigma \varepsilon_\sigma (\tau \sigma) \Phi \\ &= \varepsilon_\tau \sum_\sigma \varepsilon_\tau \varepsilon_\sigma (\tau \sigma) \Phi = \varepsilon_\tau \sum_\sigma \varepsilon_{\tau\sigma} (\tau \sigma \Phi) \\ &= \varepsilon_\tau \sum_\varrho \varepsilon_\varrho (\rho \Phi) = \varepsilon_\tau \cdot \Psi\end{aligned}$$

and so Ψ is skew symmetric.

Proposition I: Let Φ be a p-linear map from E to F. Then the following conditions are equivalent:

(i) Φ is skew symmetric.

(ii) $\Phi(x_1, \dots, x_p) = 0$ whenever $x_i = x_j$ for some pair $i \neq j$.

(iii) $\Phi(x_1, \dots, x_p) = 0$ whenever the vectors $x_1, \dots, x_p$ are linearly dependent.

Proof: (i)$\Leftrightarrow$(ii). Assume that Φ is skew symmetric and that $x_i = x_j$ $(i \neq j)$.

Denote by τ the transposition interchanging i and j. Then, since $x_i = x_j$,

$$(\tau\Phi)(x_1, \ldots, x_p) = \Phi(x_1, \ldots, x_p).$$

On the other hand, since Φ is skew symmetric,

$$(\tau\Phi)(x_1, \ldots, x_p) = -\Phi(x_1, \ldots, x_p).$$

These relations imply that

$$2\Phi(x_1, \ldots, x_p) = 0$$

and so, since Γ has characteristic zero,

$$\Phi(x_1, \ldots, x_p) = 0.$$

Conversely, assume that Φ satisfies (ii). To show that Φ is skew symmetric, fix a pair i, j $(i<j)$ and set

$$\Psi(x, y) = \Phi(x_1, \ldots, x, \ldots, y, \ldots, x_p)$$

where the vectors $x_\nu (\nu \neq i, j)$ are fixed. Then

$$\Psi(x, x) = 0 \qquad x \in E.$$

It follows that

$$\Psi(x, y) + \Psi(y, x) = \Psi(x+y, x+y) - \Psi(x, x) - \Psi(y, y) = 0$$

whence

$$\Psi(y, x) = -\Psi(x, y) \qquad x, y \in E.$$

This relation shows that

$$\tau\Phi = -\Phi$$

for any transposition. Since every permutation is a product of transpositions, it follows that

$$\sigma\Phi = \varepsilon_\sigma \cdot \Phi \qquad \sigma \in S_p;$$

i.e., Φ is skew symmetric.

(ii)$\Leftrightarrow$(iii). Assume that Φ satisfies (ii) and let $x_1, \ldots, x_p$ be linearly dependent vectors. We may assume that

$$x_p = \sum_{\nu=1}^{p-1} \lambda^\nu x_\nu.$$

Then, by (ii),

$$\Phi(x_1, \ldots, x_p) = \sum_{\nu=1}^{p-1} \lambda^\nu \Phi(x_1, \ldots, x_{p-1}, x_\nu) = 0$$

and so Φ satisfies (iii). Obviously, (iii) implies (ii) and so the proof is complete.

Corollary: If $\dim E = n$, then every skew symmetric p-linear map Φ from E to F is zero if $p > n$.

Proposition II: Assume that E is an n-dimensional vector space and let Φ be a skew symmetric n-linear map from E to a vector space F. Then Φ is completely determined by its value on a basis of E. In particular, if Φ vanishes on a basis, then $\Phi = 0$.

Proof: In fact, choose a basis $a_1, \ldots, a_n$ of E and write

$$x_\lambda = \sum_{\nu=1}^{n} \xi_\lambda^\nu a_\nu \qquad (\lambda = 1 \ldots n).$$

Then

$$\begin{aligned}\Phi(x_1, \ldots, x_n) &= \sum_{\nu_1, \ldots, \nu_n} \xi_1^{\nu_1} \ldots \xi_n^{\nu_n} \Phi(a_{\nu_1}, \ldots, a_{\nu_n}) \\ &= \sum_\sigma \xi_1^{\sigma(1)} \ldots \xi_n^{\sigma(n)} \Phi(a_{\sigma(1)}, \ldots, a_{\sigma(n)}) \\ &= \Big(\sum_\sigma \varepsilon_\sigma \xi_1^{\sigma(1)} \ldots \xi_n^{\sigma(n)}\Big) \cdot \Phi(a_1, \ldots, a_n).\end{aligned}$$

Since the first factor does not depend on Φ, the proposition follows.

4.3. Determinant functions. A *determinant function* in an n-dimensional vector space E is a skew symmetric n-linear function from E to Γ.

In every n-dimensional vector space E $(n \geqq 1)$ there are determinant functions which are not identically zero. In fact, choose a basis, $f_1, \ldots, f_n$ of the space $L(E)$ and define the n-linear function Φ by

$$\Phi(x_1, \ldots, x_n) = f_1(x_1) \ldots f_n(x_n) \qquad x_\nu \in E.$$

Then, if $a_1, \ldots, a_n$ is the dual basis in E,

$$\Phi(a_{\sigma(1)}, \ldots, a_{\sigma(n)}) = \begin{cases} 1 & \sigma = \iota \\ 0 & \sigma \neq \iota. \end{cases} \tag{4.3}$$

Now set

$$\Delta = \sum_\sigma \varepsilon_\sigma (\sigma \Phi).$$

Then Δ is skew symmetric. Moreover, relation (4.3) yields

$$\Delta(a_1, \ldots, a_n) = 1$$

and so $\Delta \neq 0$.

Proposition III: Let E be an n-dimensional vector space and fix a non-zero determinant function Δ in E. Then every skew symmetric n-linear map φ from E to a vector space F determines a unique vector $b \in F$ such that

$$\varphi(x_1, \dots, x_n) = \Delta(x_1, \dots, x_n) \cdot b. \tag{4.4}$$

Proof: Choose a basis $a_1, \dots, a_n$ of E so that $\Delta(a_1, \dots, a_n) = 1$ and set $b = \varphi(a_1, \dots, a_n)$. Then the n-linear map $\psi: E^n \to F$ given by

$$\psi(x_1, \dots, x_n) = \Delta(x_1, \dots, x_n) \cdot b \tag{4.5}$$

agrees with φ on this basis. Thus, by Proposition II, $\psi = \varphi$, i.e., $\varphi = \Delta \cdot b$.

Clearly, the vector b is uniquely determined by relation (4.4).

Corollary: Let Δ be a non-zero determinant function in E. Then every determinant function is a scalar multiple of Δ.

Proposition IV: Let Δ be a determinant function in E ($\dim E = n$). Then, the following identity holds:

$$\sum_{j=1}^{n} (-1)^{j-1} \Delta(x, x_1, \dots, \hat{x}_j, \dots, x_n) \cdot x_j = \Delta(x_1, \dots, x_n) \cdot x \quad x_\nu \in E,\ x \in E, \tag{4.6}$$

where $\hat{x}_j$ means that the j-th argument is to be deleted.

Proof: If the vectors $x_1, \dots, x_n$ are linearly dependent a simple calculation shows that the left hand side of the above equation is zero. On the other hand, by sec. 4.2, $\Delta(x_1, \dots, x_n) = 0$. Thus we may assume that the vectors $x_1, \dots, x_n$ form a basis of E. Then, writing

$$x = \sum_{\nu=1}^{n} \xi^\nu x_\nu,$$

we have

$$\begin{aligned} &\sum_{\nu, j} (-1)^{j-1} \xi^\nu \Delta(x_\nu, x_1, \dots, \hat{x}_j, \dots, x_n)\, x_j \\ &= \sum_{\nu} (-1)^{\nu-1} \xi^\nu \Delta(x_\nu, x_1, \dots, \hat{x}_\nu, \dots, x_n)\, x_\nu \\ &= \Delta(x_1, \dots, x_n) \cdot \sum_{\nu} \xi^\nu x_\nu = \Delta(x_1, \dots, x_n) \cdot x. \end{aligned} \tag{4.7}$$

Problem

Let E^*, E be a pair of dual spaces and $\Delta \neq 0$ be a determinant function in E. Define the function Δ^* of n vectors in E^* as follows:

If the vectors $x^{*\nu}$ $(\nu = 1 \dots n)$ are linearly dependent, then $\Delta^*(x^{*1} \dots x^{*n}) = 0$.

If the vectors $x^{*\nu}$ $(\nu=1\dots n)$ are linearly independent, then $\Delta^*(x^{*1}\dots x^{*n})=\Delta(x_1\dots x_n)^{-1}$ where x_ν $(\nu=1\dots n)$ is the dual basis. Prove that Δ^* is a determinant function in E^*.

§ 2. The determinant of a linear transformation

4.4. Definition. Let φ be a linear transformation of the n-dimensional linear space E. To define the determinant of φ choose a non-trivial determinant function Δ. Then the function Δ_φ, defined by

$$\Delta_\varphi(x_1\dots x_n)=\Delta(\varphi x_1\dots\varphi x_n)$$

obviously is again a determinant function. Hence, by the uniqueness-theorem of section 4.3,

$$\Delta_\varphi=\alpha\Delta,$$

where α is a scalar. This scalar does not depend on the choice of Δ. In fact, if Δ' is another non-trivial determinant function, then $\Delta'=\lambda\Delta$ and consequently

$$\Delta'_\varphi=\lambda\Delta_\varphi=\lambda\alpha\Delta=\alpha\Delta'.$$

Thus, the scalar α is uniquely determined by the transformation φ. It is called the *determinant of* φ and it will be denoted by $\det\varphi$. So we have the following equation of definition:

$$\Delta_\varphi=\det\varphi\cdot\Delta,$$

where Δ is an arbitrary non-trivial determinant function. In a less condensed form this equation reads

$$\Delta(\varphi x_1\dots\varphi x_n)=\det\varphi\,\Delta(x_1\dots x_n). \tag{4.8}$$

In particular, if $\varphi=\lambda\iota$, then

$$\Delta_\varphi=\lambda^n\Delta$$

and hence

$$\det(\lambda\iota)=\lambda^n.$$

It follows from the above equation that the determinant of the identity-map is 1 and the determinant of the zero-map is zero.

4.5. Properties of the determinant. A linear transformation φ is regular if and only if its determinant is different from zero. To prove this, select a basis e_ν $(\nu=1\dots n)$ of E. Then

$$\Delta(\varphi e_1\dots\varphi e_n)=\det\varphi\,\Delta(e_1\dots e_n). \tag{4.9}$$

If φ is regular, the vectors $\varphi e_\nu (\nu=1\dots n)$ are linearly independent; hence

$$\Delta(\varphi e_1 \dots \varphi e_n) \neq 0. \tag{4.10}$$

Relations (4.9) and (4.10) imply that

$$\det \varphi \neq 0.$$

Conversely, assume that $\det \varphi \neq 0$. Then it follows from (4.9) that

$$\Delta(\varphi e_1 \dots \varphi e_n) \neq 0.$$

Hence the vectors $\varphi e_\nu (\nu=1\dots n)$ are linearly independent and φ is regular.

Consider two linear transformations φ and ψ of E. Then

$$\det(\psi \circ \varphi) = \det \psi \det \varphi. \tag{4.11}$$

In fact,

$$\begin{aligned} \Delta(\psi \varphi x_1 \dots \psi \varphi x_n) &= \det \psi\, \Delta(\varphi x_1 \dots \varphi x_n) \\ &= \det \psi \det \varphi\, \Delta(x_1 \dots x_n), \end{aligned}$$

whence (4.11). In particular, if φ is a linear automorphism and φ^{-1} is the inverse automorphism, we obtain

$$\det \varphi^{-1} \det \varphi = \det \iota = 1.$$

4.6. The classical adjoint. Let E be an n-dimensional vector space and let $\Delta \neq 0$ be a determinant function in E. Let $\varphi \in L(E; E)$. Then an n-linear map

$$\Phi: E^n \to L(E; E)$$

is given by

$$\Phi(x_1, \dots, x_n)x = \sum_{j=1}^{n} (-1)^{j-1} \Delta(x, \varphi x_1, \dots, \widehat{\varphi x_j}, \dots, \varphi x_n) \cdot x_j \qquad x_\nu \in E.$$

It is easy to check that Φ is skew symmetric. Thus, by Proposition III, there is a unique linear transformation, $\operatorname{ad}(\varphi)$, of E such that

$$\Phi(x_1, \dots, x_n) = \Delta(x_1, \dots, x_n) \operatorname{ad}(\varphi) \qquad x_\nu \in E;$$

i.e.,

$$\sum_{j=1}^{n} (-1)^{j-1} \Delta(x, \varphi x_1, \dots, \widehat{\varphi x_j}, \dots, \varphi x_n)\, x_j = \Delta(x_1, \dots, x_n) \operatorname{ad}(\varphi) x \qquad x, x_\nu \in E. \tag{4.12}$$

This equation shows that the element $\operatorname{ad}(\varphi) \in L(E; E)$ is independent of the choice of Δ and hence it is uniquely determined by φ. It is called the *classical adjoint* of φ.

Proposition V: The classical adjoint satisfies the relations

$$\operatorname{ad}(\varphi)\circ\varphi = \iota\cdot\det\varphi$$

and

$$\varphi\circ\operatorname{ad}(\varphi) = \iota\cdot\det\varphi .$$

Proof: Replacing x by φx in (4.12) we obtain

$$\sum_{j=1}^{n}(-1)^{j-1}\Delta(\varphi x, \varphi x_1, \ldots, \widehat{\varphi x_j}, \ldots, \varphi x_n)\, x_j = \Delta(x_1, \ldots, x_n)\operatorname{ad}\varphi(\varphi x).$$

Now observe that, in view of the definition of the determinant and Proposition IV,

$$\begin{aligned}\sum_{j=1}^{n}&(-1)^{j-1}\Delta(\varphi x, \varphi x_1, \ldots, \widehat{\varphi x_j}, \ldots, \varphi x_n)\, x_j\\ &= \det\varphi\cdot\sum_{j=1}^{n}(-1)^{j-1}\Delta(x, x_1, \ldots, \hat{x}_j, \ldots, x_n)\, x_j\\ &= \det\varphi\cdot\Delta(x_1, \ldots, x_n)\cdot x \qquad x_\nu\in E,\ x\in E.\end{aligned}$$

Composing these two relations yields

$$\Delta(x_1, \ldots, x_n)\operatorname{ad}\varphi(\varphi x) = \det\varphi\cdot\Delta(x_1, \ldots, x_n)\cdot x$$

and so the first relation of the proposition follows. The second relation is established by applying φ to (4.12) and using the identity in Proposition IV, with x_i replaced by φx_i $(i=1 \ldots n)$.

Corollary: If φ is a linear transformation with $\det\varphi \neq 0$ then φ has an inverse and the inverse is given by

$$\varphi^{-1} = \frac{1}{\det\varphi}\operatorname{ad}(\varphi).$$

Thus a linear transformation of E is a linear isomorphism if and only if its determinant is non-zero.

4.7. Stable subspaces. Let $\varphi: E\to E$ be a linear transformation and assume that E is the direct sum of two stable subspaces,

$$E = E_1 \oplus E_2 .$$

Then linear transformations

$$\varphi_1: E_1 \to E_1 \quad\text{and}\quad \varphi_2: E_2 \to E_2$$

are induced by φ. It will be shown that

$$\det \varphi = \det \varphi_1 \det \varphi_2 .$$

Define the transformations $\psi_1 : E \to E$ and $\psi_2 : E \to E$ by

$$\psi_1 = \begin{cases} \varphi_1 \text{ in } E_1 \\ \iota \text{ in } E_2 \end{cases} \qquad \psi_2 = \begin{cases} \iota \text{ in } E_1 \\ \varphi_2 \text{ in } E_2 \end{cases}$$

Then

$$\varphi = \psi_2 \circ \psi_1$$

and so

$$\det \varphi = \det \psi_2 \cdot \det \psi_1 .$$

Hence it is sufficient to prove that

$$\det \psi_1 = \det \varphi_1 \quad \text{and} \quad \det \psi_2 = \det \varphi_2 . \tag{4.13}$$

Let $\Delta \neq 0$ be a determinant function in E and $b_1 \ldots b_q$ be a basis of E_2. Then the function Δ_1, defined by

$$\Delta_1(x_1 \ldots x_p) = \Delta(x_1 \ldots x_p, b_1 \ldots b_q), \qquad x_i \in E_1 \quad p = \dim E_1, \tag{4.14}$$

is a non-trivial determinant function in E_1. Hence

$$\Delta_1(\varphi_1 x_1 \ldots \varphi_1 x_p) = \det \varphi_1 \, \Delta_1(x_1 \ldots x_p) .$$

On the other hand we obtain from (4.14)

$$\begin{aligned} \Delta_1(\varphi_1 x_1 \ldots \varphi_1 x_p) &= \Delta(\psi_1 x_1 \ldots \psi_1 x_p, \psi_1 b_1 \ldots \psi_1 b_q) \\ &= \det \psi_1 \, \Delta(x_1 \ldots x_p, b_1 \ldots b_q) \\ &= \det \psi_1 \, \Delta_1(x_1 \ldots x_p) . \end{aligned}$$

These relations yield

$$\det \varphi_1 = \det \psi_1 .$$

The second formula (4.13) is proved in the same way.

Problems

1. Consider the linear transformation $\varphi : E \to E$ defined by

$$\varphi e_\nu = \lambda_\nu e_\nu \qquad (\nu = 1 \ldots n)$$

where $e_\nu (\nu = 1 \ldots n)$ is a basis of E. Show that

$$\det \varphi = \lambda_1 \ldots \lambda_n .$$

2. Let $\varphi: E \to E$ be a linear transformation and assume that E_1 is a stable subspace. Consider the induced transformations $\varphi_1: E_1 \to E_1$ and $\bar{\varphi}: E/E_1 \to E/E_1$. Prove that

$$\det \varphi = \det \varphi_1 \cdot \det \bar{\varphi}.$$

3. Let $\alpha: E \to F$ be a linear isomorphism and φ be a linear transformation of E. Prove that

$$\det(\alpha \circ \varphi \circ \alpha^{-1}) = \det \varphi.$$

4. Let E be a vector space of dimension n and consider the space $L(E; E)$ of linear transformations.

a) Assume that F is a function in $L(E; E)$ satisfying

$$F(\psi \circ \varphi) = F(\psi) F(\varphi)$$

and

$$F(\iota) = 1.$$

Prove that F can be written in the form

$$F(\varphi) = f(\det \varphi)$$

where $f: \Gamma \to \Gamma$ is a mapping such that

$$f(\lambda \mu) = f(\lambda) f(\mu).$$

b) Suppose that F satisfies the additional condition that

$$F(\lambda \iota) = \lambda^n.$$

Then, if E is a real vector space of even dimension,

$$F(\varphi) = \det \varphi \quad \text{or} \quad F(\varphi) = |\det \varphi|$$

and if E is a real vector space of odd dimension or a complex vector space

$$F(\varphi) = \det \varphi.$$

Hint for part a): Let e_i $(i = 1 \ldots n)$ be a basis for E and define the transformations ψ_{ij} and φ_i by

$$\psi_{ij} e_\nu = \begin{cases} e_\nu & \nu \neq i \\ e_i + \lambda e_j & \nu = i \end{cases} \qquad i, j = 1 \ldots n$$

and

$$\varphi_i e_\nu = \begin{cases} e_\nu & \nu \neq i \\ \lambda e_i & \nu = i \end{cases} \qquad i = 1 \ldots n.$$

Show first that

$$F(\psi_{ij}) = 1$$

and that $F(\varphi_i)$ is independent of i.

4. Let E be a vector space with a countable basis and assume that a function F is given in $L(E; E)$ which satisfies the conditions of problem 4a). Prove that

$$F(\varphi) = 1 \qquad \varphi \in L(E; E).$$

Hint: Construct an injective mapping φ and a surjective mapping ψ such that

$$\psi \circ \varphi = 0.$$

5. Let ω be a linear transformation of E such that $\omega^2 = \iota$. Show that $\det \omega = (-1)^r$ where r is the rank of the map $\iota - \omega$.

6. Let j be a linear transformation of E such that $j^2 = -\iota$. Show that then the dimension of E must be even. Prove that $\det j = 1$.

7. Prove the following properties of the classical adjoint:

(i) $\operatorname{ad}(\psi \circ \varphi) = \operatorname{ad}(\varphi) \circ \operatorname{ad}(\psi)$.

(ii) $\det \operatorname{ad}(\varphi) = (\det \varphi)^{n-1}$.

(iii) If φ has rank $n-1$, then $\operatorname{Im} \operatorname{ad}(\varphi) = \ker \varphi$.

(iv) If φ has rank $\leqq n-2$, then $\operatorname{ad}(\varphi) = 0$.

(v) $\operatorname{ad}(\operatorname{ad} \varphi) = (\det \varphi)^{n-2} \cdot \varphi$.

(vi) $\det \operatorname{ad}(\operatorname{ad} \varphi) = (\det \varphi)^{(n-1)^2}$.

8. If $n = 2$ show that

$$\operatorname{ad}(\varphi) = \iota \cdot \operatorname{tr} \varphi - \varphi.$$

§ 3. The determinant of a matrix

4.8. Definition. Let φ be a linear transformation of E and (α_ν^μ) the corresponding matrix relative to a basis e_ν $(\nu = 1 \ldots n)$. Then

$$\varphi e_\nu = \sum_\mu \alpha_\nu^\mu e_\mu .$$

Substituting $x_\nu = e_\nu$ in (4.8) we obtain

$$\Delta(\varphi e_1, \ldots \varphi e_n) = \det \varphi \, \Delta(e_1 \ldots e_n).$$

The left-hand side of this equation can be written as

$$\begin{aligned} \Delta(\varphi e_1, \ldots \varphi e_n) &= \Delta\Big(\sum_\mu \alpha_1^\mu e_\mu, \ldots \sum_\mu \alpha_n^\mu e_\mu\Big) \\ &= \sum_\sigma \varepsilon_\sigma \alpha_1^{\sigma(1)} \ldots \alpha_n^{\sigma(n)} \cdot \Delta(e_1 \ldots e_n). \end{aligned}$$

We thus obtain

$$\det \varphi = \sum_\sigma \varepsilon_\sigma \alpha_1^{\sigma(1)} \ldots \alpha_n^{\sigma(n)} . \tag{4.15}$$

This formula shows how the determinant of φ is expressed in terms of the corresponding matrix.

We now define the *determinant of an* $n\times n$*-matrix* $A=(\alpha_\nu^\mu)$ by

$$\det A = \sum_\sigma \varepsilon_\sigma \alpha_1^{\sigma(1)} \dots \alpha_n^{\sigma(n)}. \tag{4.16}$$

Then equation (4.15) can be written as

$$\det \varphi = \det M(\varphi). \tag{4.17}$$

Now let A and B be two $n\times n$-matrices. Then

$$\det(AB) = \det A \, \det B. \tag{4.18}$$

In fact, let E be an n-dimensional vector space and define the linear transformations φ and ψ of E such that (with respect to a given basis)

$$M(\varphi) = A \quad \text{and} \quad M(\psi) = B.$$

Then

$$\begin{aligned}\det(AB) &= \det M(\varphi) M(\psi) = \det M(\psi\circ\varphi) = \det(\psi\circ\varphi)\\ &= \det\varphi\cdot\det\psi = \det M(\varphi)\det M(\psi) = \det A\cdot\det B.\end{aligned}$$

Formula (4.18) yields for two inverse matrices

$$\det A\cdot\det(A^{-1}) = \det J = 1 \quad (J \text{ unit-matrix})$$

showing that

$$\det(A^{-1}) = (\det A)^{-1}.$$

Finally note that if an $(n\times n)$-matrix A is of the form

$$A = \begin{pmatrix} A_1 & 0 \\ 0 & A_2 \end{pmatrix}$$

then

$$\det A = \det A_1 \cdot \det A_2 \tag{4.19}$$

as follows from sec. 4.7.

4.9. The determinant considered as a function of the rows. If the rows $a_\nu = (\alpha_\nu^1 \dots \alpha_\nu^n)$ of the matrix A are considered as vectors of the space Γ^n the determinant $\det A$ appears as a function of the n vectors a_ν $(\nu = 1\dots n)$. To investigate this function define a linear transformation φ of Γ^n by

$$\varphi e_\nu = a_\nu \qquad (\nu = 1\dots n)$$

where the vectors e_ν are the n-tuples

$$e_\nu = (\underbrace{0 \dots 1}_{\nu} \dots 0) \qquad (\nu = 1 \dots n).$$

Then A is the matrix of φ relative to the basis e_ν. Now let Δ be the determinant function in Γ^n which assumes the value one at the basis e_ν ($\nu = 1 \dots n$),

$$\Delta(e_1 \dots e_n) = 1.$$

Then

$$\Delta(a_1 \dots a_n) = \Delta(\varphi e_1 \dots \varphi e_n) = \det\varphi\, \Delta(e_1 \dots e_n) = \det\varphi$$

and hence

$$\det A = \Delta(a_1 \dots a_n). \tag{4.20}$$

This formula shows that the determinant of A considered as a function of the row-vectors has the following properties:

1. The determinant is linear with respect to every row-vector.
2. If two row-vectors are interchanged the determinant changes the sign.
3. The determinant does not change if to a row-vector a multiple of another row-vector is added.
4. The determinant is different from zero if and only if the row-vectors are linearly independent.

An argument similar to the one above shows that

$$\det A = \Delta(b^1, \dots b^n)$$

where the b^ν are the column-vectors of A. It follows that the properties 1–4 remain true if the determinant of A is considered as a function of the column-vectors.

Problems

1. Let $A = (\alpha_\nu^\mu)$ be a matrix such that $\alpha_\nu^\mu = 0$ if $\nu < \mu$. Prove that

$$\det A = \alpha_1^1 \dots \alpha_n^n.$$

2. Prove that the determinant of the $n \times n$-matrix

$$\alpha_\nu^\mu = 1 - \delta_\nu^\mu$$

is equal to $(n-1)(-1)^{n-1}$.

Hint: Consider the mapping $\varphi: E \to E$ defined by

$$\varphi e_\nu = \sum_\mu e_\mu - e_\nu \qquad (\nu = 1 \dots n).$$

3. Given an $n\times n$-matrix $A=(\alpha_\nu^\mu)$ define the matrix $B=(\beta_\nu^\mu)$ by

$$\beta_\nu^\mu = (-1)^{\nu+\mu}\alpha_\nu^\mu .$$

Prove that

$$\det B = \det A .$$

4. Given n complex numbers α_ν prove that

$$\det\begin{pmatrix}\alpha_1 & \alpha_2\dots\alpha_{n-1} & \alpha_n\\ \alpha_2 & \alpha_3\dots\alpha_n & \alpha_1\\ \vdots & & \\ \alpha_n & \alpha_1\dots\alpha_{n-2} & \alpha_{n-1}\end{pmatrix} = (-1)^{\frac{n(n-1)}{2}}\beta_1\dots\beta_n$$

where the numbers β_k are defined by

$$\beta_k = \sum_\nu \varepsilon_k^\nu \alpha_\nu \quad \varepsilon_k = \cos\frac{2\pi k}{n} + i\sin\frac{2\pi k}{n} \quad (k=1\dots n).$$

Hint: Multiply the above matrix by the matrix

$$\begin{pmatrix}\varepsilon_1 \dots \varepsilon_n\\ \varepsilon_1^2\dots\varepsilon_n^2\\ \dots\dots\dots\\ \dots\dots\dots\\ \varepsilon_1^n\dots\varepsilon_n^n\end{pmatrix}.$$

§ 4. Dual determinant functions

4.10. Let E^*, E be a pair of dual vector spaces and $\Delta^*\neq 0$, $\Delta\neq 0$ be determinant functions in E^* and E. It will be shown that

$$\Delta^*(x^{*1}\dots x^{*n})\Delta(x_1\dots x_n) = \alpha\det(\langle x^{*i}, x_j\rangle), \quad x^{*i}\in E^*, x_j\in E, \qquad (4.21)$$

where $\alpha\in\Gamma$ is a constant scalar. Consider the function Ω of $2n$ vectors defined by

$$\Omega(x^{*1}\dots x^{*n}; x_1\dots x_n) = \det(\langle x^{*i}, x_j\rangle).$$

Then it follows from the properties of the determinant of a matrix that Ω is linear with respect to each argument. Moreover, Ω is skew symmetric with respect to the vectors $x_i (i=1\dots n)$. Hence the uniqueness theorem (sec. 4.3) implies that Ω can be written as

$$\Omega(x^{*1},\dots x^{*n}; x_1\dots x_n) = \Phi(x^{*1}\dots x^{*n})\Delta(x_1\dots x_n) \qquad (4.22)$$

where Φ depends only on the vectors x^{*i}. Replacing the x_i in (4.21) by a basis e_i of E we obtain

$$\Omega(x^{*1}, \dots x^{*n}; e_1 \dots e_n) = \Phi(x^{*1} \dots x^{*n}) \Delta(e_1 \dots e_n).$$

This relation shows that Φ is linear with respect to every argument and skew symmetric. Applying the uniqueness theorem again we find that

$$\Phi(x^{*1} \dots x^{*n}) = \beta \Delta^*(x^{*1} \dots x^{*n}), \qquad \beta \in \Gamma. \tag{4.23}$$

Combining (4.22) and (4.23) we obtain

$$\Omega(x^{*1} \dots x^{*n}; x_1 \dots x_n) = \beta \Delta^*(x^{*1} \dots x^{*n}) \Delta(x_1 \dots x_n). \tag{4.24}$$

Now let e^{*i}, $e_i (i = 1 \dots n)$ be a pair of dual bases. Then (4.24) yields

$$1 = \beta \Delta^*(e^{*1} \dots e^{*n}) \Delta(e_1 \dots e_n)$$

and so $\beta \neq 0$. Multiplying (4.24) by $\alpha = \beta^{-1}$ we obtain the relation (4.21).

The determinant functions Δ^* and Δ are called *dual* if the factor α in (4.21) is equal to 1; i.e.,

$$\Delta^*(x^{*1} \dots x^{*n}) \Delta(x_1 \dots x_n) = \det(\langle x^{*i}, x_j \rangle). \tag{4.25}$$

To every determinant function $\Delta \neq 0$ in E there exists precisely one dual determinant function Δ^* in E^*. In fact, let $\Delta \neq 0$ be an arbitrary determinant function in E^* and set $\Delta_0^* = \alpha^{-1} \Delta$ where α is the scalar in (4.21). Then Δ_0^* and Δ are dual. To prove the uniqueness, assume that Δ_1^* and Δ_2^* are dual determinant functions to Δ. Then we have that

$$[\Delta_1^*(x^{*1} \dots x^{*n}) - \Delta_2^*(x^{*1} \dots x^{*n})] \Delta(x_1 \dots x_n) = 0 \qquad x^{*i} \in E^*, \ x_i \in E$$

whence $\Delta_1^* = \Delta_2^*$.

4.11. The determinant of dual transformations. Let $\varphi : E \to E$ and $\varphi^* : E^* \leftarrow E^*$ be two dual linear transformations. Then

$$\det \varphi^* = \det \varphi .$$

To prove this, let Δ^*, Δ be a pair of dual determinant functions in E^* and E. Then we have in view of (4.25)

$$\Delta^*(x^{*1}, \dots x^{*n}) \Delta(x_1 \dots x_n) = \det(\langle x^{*i}, x_j \rangle).$$

This relation yields

$$\Delta^*(\varphi^* x^{*1} \dots \varphi^* x^{*n}) \Delta(x_1 \dots x_n) = \det(\langle \varphi^* x^{*i}, x_j \rangle)$$

and

$$\Delta^*(x^{*1}\dots x^{*n})\,\Delta(\varphi x_1\dots\varphi x_n) = \det(\langle x^{*i},\varphi x_j\rangle).$$

Since

$$\langle\varphi^* x^{*i}, x_j\rangle = \langle x^{*i},\varphi x_j\rangle \qquad (i,j = 1\dots n)$$

it follows that

$$\Delta^*(\varphi^* x^{*1},\dots\varphi^* x^{*n})\,\Delta(x_1\dots x_n) = \Delta^*(x^{*1}\dots x^{*n})\,\Delta(\varphi x_1\dots\varphi x_n).$$

But

$$\Delta^*(\varphi^* x^{*1},\dots\varphi^* x^{*n}) = \det\varphi^*\cdot\Delta^*(x^{*1}\dots x^{*n})$$

and

$$\Delta(\varphi x_1\dots\varphi x_n) = \det\varphi\,\Delta(x_1\dots x_n)$$

and so we obtain

$$(\det\varphi^* - \det\varphi)\,\Delta^*(x^{*1}\dots x^{*n})\,\Delta(x_1\dots x_n) = 0$$

whence the result.

The above result implies that transposed $n\times n$-matrices have the same determinant. In fact, let A be an $n\times n$-matrix and let φ be a linear transformation of an n-dimensional vector space such that (with respect to a given basis) $M(\varphi)=A$. Then it follows that

$$\det A^* = \det M(\varphi)^* = \det M(\varphi^*) = \\ = \det\varphi^* = \det\varphi = \det M(\varphi) = \det A.$$

Problems

1. Show that the determinant functions, Δ, Δ^* of the problem in § 1 are dual.
2. Using the expansion formula (4.16) prove that,

$$\det A^* = \det A.$$

§ 5. The adjoint matrix

4.12. Definition. Let $\varphi\colon E\to E$ $(\dim E = n)$ be a linear transformation and let $\operatorname{ad}(\varphi)$ be the adjoint transformation (cf. sec. 4.6). We shall express the matrix of $\operatorname{ad}(\varphi)$ in terms of the matrix of φ. It is called the *adjoint* of the matrix of φ.

Let $a_1,\dots,a_n$ be a basis of E such that $\Delta(a_1,\dots,a_n)=1$ and write

$$\operatorname{ad}(\varphi)\,a_j = \sum_i \beta^i_j\,a_i.$$

Then, setting $x_\nu = a_\nu$ $(\nu = 1 \dots n)$ and $x = a_j$ in the identity (4.12) we obtain

$$\sum_i (-1)^{i-1} \Delta(a_j, \varphi a_1, \dots, \widehat{\varphi a_i}, \dots, \varphi a_n)\, a_i = \sum_i \beta_j^i a_i$$

whence

$$\begin{aligned}\beta_j^i &= (-1)^{i-1} \Delta(a_j, \varphi a_1, \dots, \varphi a_i, \dots, \varphi a_n) \\ &= \Delta(\varphi a_1, \dots, \varphi a_{i-1}, a_j, \varphi a_{i+1}, \dots, \varphi a_n) \qquad (i, j = 1 \dots n).\end{aligned}$$

This equation shows that, for $i, j = 1, \dots, n$,

$$\beta_j^i = \det \varphi_j^i$$

where $\varphi_j^i: E \to E$ is the linear transformation given by

$$\begin{aligned}\varphi_j^i(a_\nu) &= \varphi a_\nu \qquad (\nu \neq i) \\ \varphi_j^i a_i &= a_j.\end{aligned}$$

Since the matrix of φ_j^i with respect to the basis $a_1, \dots, a_n$ is given by

$$C_i^j = \begin{pmatrix} \alpha_1^1 & \dots \alpha_1^{j-1} & \alpha_1^j & \alpha_1^{j+1} & \dots \alpha_1^n \\ \vdots & \vdots & & \vdots & \vdots \\ \alpha_{i-1}^1 & \dots \alpha_{i-1}^{j-1} & \alpha_{i-1}^j & \alpha_{i-1}^{j+1} & \dots \alpha_{i-1}^n \\ 0 & \dots\ 0 & 1 & 0 & \dots\ 0 \\ \alpha_{i+1}^1 & \dots \alpha_{i+1}^{j-1} & \alpha_{i+1}^j & \alpha_{i+1}^{j+1} & \dots \alpha_{i+1}^n \\ \vdots & \vdots & & \vdots & \vdots \\ \alpha_n^1 & \dots \alpha_n^{j-1} & \alpha_n^j & \alpha_n^{j+1} & \dots \alpha_n^n \end{pmatrix} \qquad (i, j = 1 \dots n)$$

we have

$$\beta_j^i = \det C_i^j.$$

Definition: The determinant of C_i^j is called the *cofactor* of α_i^j and is denoted by $\operatorname{cof} \alpha_i^j$. Thus we can write

$$\beta_j^i = \operatorname{cof} \alpha_i^j \qquad (i, j = 1 \dots n);$$

i.e., the adjoint matrix is the transpose of the matrix of cofactors.

4.13. Cramer's formula. In view of the result above we obtain from Proposition V, sec. 4.6, the relations

$$\sum_j \alpha_i^j \beta_j^k = \delta_i^k \cdot \det A \tag{4.26}$$

and

$$\sum_j \beta_i^j \alpha_j^k = \delta_i^k \cdot \det A. \tag{4.27}$$

Setting $k=i$ in (4.26) we obtain the expansion formula by cofactors,

$$\det A = \sum_j \alpha_i^j \beta_j^i \qquad (i = 1 \dots n). \tag{4.28}$$

Now assume that $\det A \neq 0$ and define a matrix $\check{\alpha}_i^j$ by setting

$$\check{\alpha}_i^j = \frac{1}{\det A} \beta_i^j .$$

Then the equations (4.26) and (4.27) yield

$$\sum_j \alpha_i^j \check{\alpha}_j^k = \sum_j \check{\alpha}_i^j \alpha_j^k = \delta_i^k$$

showing that the matrix $(\check{\alpha}_i^j)$ is the inverse of the matrix (α_i^j).

From (4.26) we obtain *Cramers solution formula* of a system of n linear equations in n unknowns whose determinant is non-zero. In fact, let

$$\sum_{k=1}^{n} \alpha_k^j \xi^k = \eta^j \qquad j = 1 \dots n \tag{4.29}$$

be such a system. Then we have, in view of (4.26),

$$\xi^i \cdot \det A = \sum_j \beta_j^i \eta^j$$

whence

$$\xi^i = \frac{1}{\det A} \sum_j \beta_j^i \eta^j = \frac{1}{\det A} \sum_j \operatorname{cof}(\alpha_i^j)\, \eta^j .$$

This formula expresses the (unique) solution of (4.29) in terms of the cofactors of the matrix A and the determinant of A.

4.14. The submatrices S_i^j. Given an $(n \times n)$-matrix denote for each pair (i,j) $(i=1 \dots n,\ j=1 \dots n)$ by S_i^j the $(n-1) \times (n-1)$-matrix obtained from A by deleting the i-th row and the j-th column. We shall show that

$$\operatorname{cof} \alpha_i^j = (-1)^{i+j} \det S_i^j \qquad (i,j = 1 \dots n). \tag{4.30}$$

In fact, by $(i-1)$ interchanges of rows and $(j-1)$ interchanges of columns we can transform C_i^j into the matrix (cf. sec. 4.12)

$$B_i^j = \begin{pmatrix} 1 & 0 \; . \; . \; . \; . \; 0 \\ \alpha_1^1 & \\ \cdot & \\ \hat{\alpha}_i^1 & S_i^j \\ \cdot & \\ \cdot & \\ \alpha_n^1 & \end{pmatrix} .$$

Thus

$$\operatorname{cof}\alpha_i^j = \det C_i^j = (-1)^{i+j}\det B_i^j.$$

On the other hand we have, in view of the expansion formula (4.28), with $i = i$

$$\det B_i^j = \det S_i^j$$

and so (4.30) follows.

4.15. Expansion by cofactors. From the relations (4.28) and (4.30) we obtain the *expansion-formula* of the determinant with respect to the ith row,

$$\det A = \sum_j (-1)^{i+j}\alpha_i^j \det S_i^j \qquad (i = 1 \dots n). \tag{4.31}$$

By this formula the evaluation of the determinant of n rows is reduced to the evaluation of n determinants of $n-1$ rows.

In the same way the expansion-formula with respect to the jth column is proved:

$$\det A = \sum_i (-1)^{i+j}\alpha_i^j \det S_i^j \qquad (j = 1 \dots n). \tag{4.32}$$

4.16. Minors. Let $A = (\alpha_\nu^\mu)$ be a given $n \times m$-matrix. For every system of indices

$$1 \leqq i_1 < i_2 < \cdots < i_k \leqq n \quad \text{and} \quad 1 \leqq j_1 < j_2 < \cdots < j_k \leqq m$$

denote by $A_{i_1 \dots i_k}^{j_1 \dots j_k}$ the submatrix of A, consisting of the rows $i_1 \dots i_k$ and the columns $j_1 \dots j_k$. The determinant of $A_{i_1 \dots i_k}^{j_1 \dots j_k}$ is called a *minor of order k* of the matrix A. It will be shown that in a matrix of rank r there is always a minor of order r which is different from zero, whereas all minors of order $k > r$ are zero. Let $A_{i_1 \dots i_k}^{j_1 \dots j_k}$ be a minor of order $k > r$. Then the row-vectors $a_{i_1} \dots a_{i_k}$ of A are linearly dependent. This implies that the rows of the matrix $A_{i_1 \dots i_k}^{j_1 \dots j_k}$ are also linearly dependent and thus the determinant must be zero.

It remains to be shown that there is a minor of order r which is different from zero. Since A has rank r, there are r linearly independent row-vectors $a_{i_1} \dots a_{i_r}$. The submatrix consisting of these row-vectors has again the rank r. Therefore it must contain r linearly independent column-vectors $b^{j_1} \dots b^{j_r}$ (cf. sec. 3.4). Consider the matrix $A_{i_1 \dots i_r}^{j_1 \dots j_r}$. Its column-vectors are linearly independent, whence

$$\det A_{i_1 \dots i_r}^{j_1 \dots j_r} \neq 0.$$

If A is a square-matrix, the minors

$$\det A_{i_1 \dots i_k}^{i_1 \dots i_k}$$

are called the *principal minors* of order k.

Problems

1. Prove the *Laplace expansion* formula for the determinant of an $(n \times n)$-matrix: Let p $(1 \leqq p \leqq n-1)$ be a fixed integer. Then

$$\det A = \sum_{\nu_1 < \cdots < \nu_n} \varepsilon(\nu_1, \ldots, \nu_p) \det A^{\nu_1 \ldots \nu_p} \det B^{\nu_{p+1} \ldots \nu_n}$$

where

$$A^{\nu_1 \ldots \nu_p} = A^{\nu_1 \ldots \nu_p}_{1 \ldots p}$$

$$B^{\nu_{p+1} \ldots \nu_n} = A^{\nu_{p+1} \ldots \nu_n}_{p+1 \ldots n}$$

and

$$\varepsilon(\nu_1 \ldots \nu_p) = (-1)^{\sum_{i=1}^{p} (\nu_i - i)}.$$

2. Let $a_1, \ldots, a_n$ and $b_1, \ldots, b_n$ be two bases of a vector space E and let $p (1 \leqq p \leqq n-1)$ be given. Show that the vectors b_ν can be reordered such that

(i) $a_1, \ldots, a_p, b_{\nu_{p+1}}, \ldots, b_{\nu_n}$ is a basis of E,
(ii) $b_{\nu_1}, \ldots, b_{\nu_p}, a_{p+1}, \ldots, a_n$ is also a basis of E.

3. Compute the inverse of the following matrices.

$$A = \frac{1}{2}\begin{pmatrix} 1 & 1 & 1 & 1 \\ 1 & 1 & -1 & -1 \\ 1 & -1 & 1 & -1 \\ 1 & -1 & -1 & 1 \end{pmatrix}$$

$$B = \begin{pmatrix} \lambda & 1 & & & \\ & \lambda & 1 & & \\ & & \ddots & \ddots & \\ & & & \ddots & 1 \\ & & & & \lambda \end{pmatrix} \qquad (\lambda \neq 0)$$

4. Show that

a) $$\det \begin{pmatrix} x & a & a & \ldots & a \\ a & x & a & \ldots & a \\ \vdots & & & & \vdots \\ a & \ldots & & a & x \end{pmatrix} = [x + (n-1)a](x-a)^{n-1}$$

and that

b) $$\det\begin{pmatrix} 1 & 1 & \dots 1 \\ \lambda_1 & \lambda_2 & \dots \lambda_n \\ \lambda_1^2 & \lambda_2^2 & \dots \lambda_n^2 \\ \dots & \dots & \dots \\ \lambda_1^{n-1} & \lambda_2^{n-1} & \dots \lambda_n^{n-1} \end{pmatrix} = \prod_{i>j}(\lambda_i - \lambda_j).$$

(*Vandermonde* determinant.)

5. Define

$$\Delta_n = \det\begin{pmatrix} x_1 & 1 & & & & \\ -1 & x_2 & 1 & & & \\ & -1 & x_3 & 1 & & \\ & & \ddots & \ddots & \ddots & \\ & & & & x_{n-1} & 1 \\ & & & & -1 & x_n \end{pmatrix}$$

Show that $\Delta_n = x_n\Delta_{n-1} + \Delta_{n-2} \quad (n>2)$

$$\Delta_1 = x_1; \quad \Delta_2 = x_1 x_2 + 1.$$

6. Verify the following formula for a *quasi-triangular* determinant:

$$\det\begin{pmatrix} x_{11} \dots x_{1p} & 0 \dots 0 \\ \vdots \quad\quad \vdots & \vdots \quad \vdots \\ x_{p1} \dots x_{pp} & 0 \dots 0 \\ x_{p+1\,1} \dots\dots & x_{p+1\,n} \\ \vdots & \vdots \\ x_{n1} \dots\dots & x_{nn} \end{pmatrix} = \det\begin{pmatrix} x_{11} \dots x_{1p} \\ \vdots \quad\quad \vdots \\ x_{p1} \dots x_{pp} \end{pmatrix} \cdot \det\begin{pmatrix} x_{p+1\,p+1} \dots x_{p+1\,n} \\ \vdots \quad\quad \vdots \\ x_{n\,p+1} \dots x_{nn} \end{pmatrix}.$$

7. Prove that the operation $A \to \operatorname{ad} A$ (cf. sec. 4.12) has the following properties:

a) $\operatorname{ad}(AB) = \operatorname{ad} B \cdot \operatorname{ad} A$.
b) $\det \operatorname{ad} A = (\det A)^{n-1}$.
c) $\operatorname{ad}\operatorname{ad} A = (\det A)^{n-2} \cdot A$.
d) $\det \operatorname{ad}\operatorname{ad} A = (\det A)^{(n-1)^2}$.

§ 6. The characteristic polynomial

4.17. Eigenvectors. Consider a linear transformation φ of an n-dimensional linear space E. A vector $a \neq 0$ of E is called an *eigenvector* of φ if

$$\varphi a = \lambda a .$$

The scalar λ is called the corresponding *eigenvalue*. A linear transformation φ need not have eigenvectors. As an example let E be a real linear space of two dimensions and define φ by

$$\varphi x_1 = x_2 \quad \varphi x_2 = -x_1$$

where the vectors x_1 and x_2 form a basis of E. This mapping does not have eigenvectors. In fact, assume that

$$a = \xi^1 x_1 + \xi^2 x_2$$

is an eigenvector. Then $\varphi a = \lambda a$ and hence

$$\xi^1 = \lambda \xi^2, \quad \xi^2 = -\lambda \xi^1 .$$

These equations yield

$$(\xi^1)^2 + (\xi^2)^2 = 0$$

whence $\xi^1 = 0$ and $\xi^2 = 0$.

4.18. The characteristic equation. Assume that a is an eigenvector of φ and that λ is the corresponding eigenvalue. Then

$$\varphi a = \lambda a , \qquad a \neq 0 .$$

This equation can be written as

$$(\varphi - \lambda \iota) a = 0 \tag{4.33}$$

showing that $\varphi - \lambda \iota$ is not regular. This implies that

$$\det(\varphi - \lambda \iota) = 0 . \tag{4.34}$$

Hence, every eigenvalue of φ satisfies the equation (4.34). Conversely, assume that λ is a solution of the equation (4.34). Then $\varphi - \lambda \iota$ is not regular. Consequently there is a vector $a \neq 0$ such that

$$(\varphi - \lambda \iota) a = 0 ,$$

whence $\varphi a = \lambda a$.

Thus, the eigenvalues of φ are the solutions of the equation (4.34). This equation is called the *characteristic equation* of the linear transformation φ.

4.19. The characteristic polynomial. To obtain a more explicit expression for the characteristic equation choose a determinant function $\Delta \neq 0$ in E. Then

$$\Delta(\varphi x_1 - \lambda x_1 \dots \varphi x_n - \lambda x_n) = \det(\varphi - \lambda \iota)\Delta(x_1 \dots x_n) \quad x_\nu \in E(\nu = 1 \dots n). \tag{4.35}$$

Expanding the left hand-side we obtain a sum of 2^n terms of the form

$$\Delta(z_1 \dots z_n),$$

where every argument is either φx_ν or $-\lambda x_\nu$. Denote by $S_p (0 \leqq p \leqq n)$ the sum of all terms in which p arguments are equal to φx_ν and $n-p$ arguments are equal to $-\lambda x_\nu$. Collect in each term of S_p the indices $\nu_1 \dots \nu_p$ $(\nu_1 < \dots < \nu_p)$ such that

$$z_{\nu_1} = \varphi x_{\nu_1} \dots z_{\nu_p} = \varphi x_{\nu_p}$$

and the indices $\nu_{p+1} \dots \nu_n (\nu_{p+1} < \dots < \nu_n)$ such that

$$z_{\nu_{p+1}} = -\lambda x_{\nu_{p+1}} \dots z_{\nu_n} = -\lambda x_{\nu_n}.$$

Introducing the permutation σ by

$$\sigma(i) = \nu_i \qquad (i = 1 \dots n)$$

we can write

$$\begin{aligned} \Delta(z_1 \dots z_n) &= \varepsilon_\sigma \Delta(z_{\sigma(1)} \dots z_{\sigma(n)}) \\ &= \varepsilon_\sigma \Delta(\varphi x_{\sigma(1)} \dots \varphi x_{\sigma(p)}, -\lambda x_{\sigma(p+1)} \dots -\lambda x_{\sigma(n)}) \\ &= (-\lambda)^{n-p} \varepsilon_\sigma \Delta(\varphi x_{\sigma(1)} \dots \varphi x_{\sigma(p)}, x_{\sigma(p+1)} \dots x_{\sigma(n)}). \end{aligned}$$

Thus,

$$S_p = (-\lambda)^{n-p} \sum_\sigma \varepsilon_\sigma \Delta(\varphi x_{\sigma(1)} \dots \varphi x_{\sigma(p)}, x_{\sigma(p+1)} \dots x_{\sigma(n)}) \tag{4.36}$$

where the sum is extended over all permutations σ subject to the conditions

$$\sigma(1) < \dots < \sigma(p) \quad \text{and} \quad \sigma(p+1) < \dots < \sigma(n).$$

Observing the skew symmetry of Δ we obtain from (4.36)

$$S_p = \frac{(-\lambda)^{n-p}}{p!(n-p)!} \sum_\sigma \varepsilon_\sigma \Delta(\varphi x_{\sigma(1)} \dots \varphi x_{\sigma(p)}, x_{\sigma(p+1)} \dots x_{\sigma(n)}) \tag{4.37}$$

where the sum on the right hand-side is taken over all permutations. Let Φ_p be the function defined by

$$\Phi_p(x_1 \dots x_n) = \sum_\sigma \varepsilon_\sigma \Delta(\varphi x_{\sigma(1)} \dots \varphi x_{\sigma(p)}, x_{\sigma(p+1)} \dots x_{\sigma(n)}) \quad (0 \leqq p \leqq n)$$

and τ be an arbitrary permutation of $(1 \ldots n)$. Then

$$\begin{aligned}\Phi_p(x_{\tau(1)} \ldots x_{\tau(n)}) &= \sum_\sigma \varepsilon_\sigma \Delta(\varphi x_{\tau\sigma(1)} \ldots \varphi x_{\tau\sigma(p)}, x_{\tau\sigma(p+1)} \ldots x_{\tau\sigma(n)}) \\ &= \varepsilon_\tau \sum_\sigma \varepsilon_{\tau\sigma} \Delta(\varphi x_{\tau\sigma(1)} \ldots \varphi x_{\tau\sigma(p)}, x_{\tau\sigma(p+1)} \ldots x_{\tau\sigma(n)}) \\ &= \varepsilon_\tau \sum_\sigma \varepsilon_\varrho \Delta(\varphi x_{\varrho(1)} \ldots \varphi x_{\varrho(p)}, x_{\varrho(p+1)} \ldots x_{\varrho(n)}) \\ &= \varepsilon_\tau \Phi_p(x_1 \ldots x_n).\end{aligned}$$

This equation shows that Φ_p is skew symmetric with respect to all arguments. This implies that

$$\Phi_p = (-1)^{n-p} p!(n-p)! \alpha_p \cdot \Delta \tag{4.38}$$

where α_p is a scalar. Inserting (4.38) into (4.37) we obtain

$$S_p = \alpha_p \lambda^{n-p} \cdot \Delta .$$

Hence, the left hand-side of (4.35) can be written as

$$\Delta(\varphi x_1 - \lambda x_1, \ldots \varphi x_n - \lambda x_n) = \Delta(x_1 \ldots x_n) \sum_{p=0}^{n} \alpha_p \lambda^{n-p} . \tag{4.39}$$

Now equations (4.35) and (4.39) yield

$$\det(\varphi - \lambda \iota) = \sum_{p=0}^{n} \alpha_p \lambda^{n-p}$$

showing that the determinant of $\varphi - \lambda \iota$ is a polynomial of degree n in λ. This polynomial is called the *characteristic polynomial* of the linear transformation φ. The coefficients of the characteristic polynomial are determined by equation (4.38), and are called the *characteristic coefficients*.

These relations yield for $p=0$ and $p=n$

$$\alpha_0 = (-1)^n \quad \text{and} \quad \alpha_n = \det \varphi$$

respectively.

4.20. Existence of eigenvalues. Combining the results of sec. 4.18 and 4.19, we see that the eigenvalues of φ are the roots of the characteristic polynomial

$$f(\lambda) = \sum_{\nu=0}^{n} \alpha_\nu \lambda^{n-\nu} .$$

This shows that *a linear transformation of an n-dimensional linear space has at most n different eigenvalues.*

Assume that E is a complex linear space. Then, according to the fundamental theorem of algebra, the polynomial f has at least one zero. Consequently, *every linear transformation of a complex linear space has at least one eigenvalue.*

If E is a real linear space, this does not generally hold, as has been shown in the beginning of this paragraph.

Now assume that the dimension of E is odd. Then

$$\lim_{\lambda\to\infty} f(\lambda) = -\infty \quad \text{and} \quad \lim_{\lambda\to-\infty} f(\lambda) = +\infty$$

and thus the polynomial $f(\lambda)$ must have at least one zero. This proves that a *linear transformation of an odd-dimensional real linear space has at least one eigenvalue.* Observing that

$$f(0) = \alpha_n = \det\varphi$$

we see that a linear transformation of positive determinant has at least one positive eigenvalue and a linear transformation of negative determinant has at least one negative eigenvalue, provided that E has odd dimension.

If the dimension of E is even we have the relations

$$\lim_{\lambda\to\infty} f(\lambda) = \infty \quad \text{and} \quad \lim_{\lambda\to-\infty} f(\lambda) = \infty$$

and hence nothing can be said if $\det\varphi > 0$. However, if $\det\varphi < 0$, there exists at least one positive and one negative eigenvalue.

4.21. The characteristic polynomial of the inverse mapping. It follows (cf. sec. 4.11) that the characteristic polynomial of the dual transformation φ^* coincides with the characteristic polynomial of φ.

Suppose now that $E = E_1 \oplus E_2$ where E_1 and E_2 are stable subspaces. Then the result of sec. 4.7 implies that the characteristic polynomial of φ is the product of the characteristic polynomials of the induced transformations $\varphi_1 : E_1 \to E_1$ and $\varphi_2 : E_2 \to E_2$.

Finally, let $\varphi : E \to E$ be a regular linear transformation and consider the inverse transformation φ^{-1}. The characteristic polynomial of φ^{-1} is defined by

$$F(\lambda) = \det(\varphi^{-1} - \lambda\iota).$$

Now,

$$\varphi^{-1} - \lambda\iota = \varphi^{-1}\circ(\iota - \lambda\varphi) = -\lambda\varphi^{-1}\circ(\varphi - \lambda^{-1}\iota),$$

whence

$$\det(\varphi^{-1} - \lambda\iota) = (-\lambda)^n \det\varphi^{-1} \cdot \det(\varphi - \lambda^{-1}\iota).$$

This equation shows that the characteristic polynomials of φ and of φ^{-1} are related by

$$F(\lambda) = (-\lambda)^n \det\varphi^{-1} f(\lambda^{-1}).$$

Expanding $F(\lambda)$ as

$$F(\lambda) = \sum_{\nu=0}^{n} \beta_\nu \lambda^{n-\nu}$$

we obtain the following relations between the coefficients of f and of F:

$$\beta_\nu = (-1)^n \det\varphi^{-1} \alpha_{n-\nu} \qquad (\nu = 0 \dots n).$$

4.22. The characteristic polynomial of a matrix. Let $e_\nu (\nu = 1 \dots n)$ be a basis of E and $A = M(\varphi)$ be the matrix of the linear transformation φ relative to this basis. Then

$$M(\varphi - \lambda\iota) = M(\varphi) - \lambda M(\iota) = A - \lambda J$$

whence

$$\det(\varphi - \lambda\iota) = \det M(\varphi - \lambda\iota) = \det(A - \lambda J).$$

Thus, the characteristic polynomial of φ can be written as

$$f(\lambda) = \det(A - \lambda J). \tag{4.40}$$

The polynomial (4.40) is called the *characteristic polynomial of the matrix A*. The roots of the polynomial f are called the *eigenvalues of the matrix A*.

Problems

1. Compute the eigenvalues of the matrix

$$\begin{pmatrix} 1 & 0 & 3 \\ 3 & -2 & -1 \\ 1 & -1 & 1 \end{pmatrix}.$$

2. Show that the eigenvalues of the real matrix

$$\begin{pmatrix} \alpha & \beta \\ \beta & \delta \end{pmatrix} \quad \text{are real,} \qquad \alpha, \beta, p, \delta \in \mathbb{R}.$$

3. Prove that the characteristic polynomial of a projection $\pi: E \to E$ (see Chapter II, sec. 2.19) is given by

$$f(\lambda) = (-1)^{n-p} \lambda^{n-p} (1 - \lambda)^p$$

where $n = \dim E$ and $p = \dim \operatorname{Im} \pi$.

4. Show that the coefficients of the characteristic polynomial of an involution satisfy the relations

$$\alpha_p = \varepsilon \alpha_{n-p} \quad \varepsilon = \pm 1 \qquad (p = 0 \dots n).$$

5. Consider a direct decomposition $E = E_1 \oplus E_2$. Given linear transformations $\varphi_i: E_i \to E_i$ $(i = 1, 2)$ consider the linear transformation $\varphi = \varphi_1 \oplus \varphi_2: E \to E$. Prove that the characteristic polynomial of φ is the product of the characteristic polynomials of φ_1 and of φ_2.

6. Let $\varphi: E \to E$ be a linear transformation and assume that E_1 is a stable subspace. Consider the induced transformations $\varphi_1: E_1 \to E_1$ and $\bar{\varphi}: E/E_1 \to E/E_1$. Prove that

$$\chi = \chi_1 \bar{\chi}$$

where χ, χ_1 and $\bar{\chi}$ denote the characteristic polynomials of φ, φ_1 and $\bar{\varphi}$ respectively. In particular show that

$$\chi(\lambda) = (-\lambda)^s \bar{\chi}(\lambda)$$

where $\bar{\varphi}$ is the induced transformation of $E/\ker \varphi$ and s denotes the dimension of $\ker \varphi$.

7. A linear transformation, φ, of E is called *nilpotent* if $\varphi^k = 0$ for some k. Prove that φ is nilpotent if and only if the characteristic polynomial has the form

$$\chi(\lambda) = (-\lambda)^n.$$

Hint: Use problem 6.

8. Given two linear transformations φ and ψ of E show that $\det(\varphi - \lambda\psi)$ is a polynomial in λ.

9. Let φ and ψ be two linear transformations. Prove that $\varphi \circ \psi$ and $\psi \circ \varphi$ have the same characteristic polynomial.

Hint: Consider first the case that ψ is regular.

§ 7. The trace

4.23. The trace of a linear transformation. In a similar way as the determinant, another scalar can be associated with a given linear transformation φ. Let $\Delta \neq 0$ be a determinant function in E. Consider the sum

$$\sum_{i=1}^{n} \Delta(x_1 \ldots \varphi x_i \ldots x_n).$$

This sum obviously is again a determinant function and thus it can be written as

$$\sum_{i=1}^{n} \Delta(x_1 \ldots \varphi x_i \ldots x_n) = \alpha \cdot \Delta(x_1 \ldots x_n) \tag{4.41}$$

where α is a scalar. This scalar which is uniquely determined by φ is called the *trace* of φ and will be denoted by $\operatorname{tr}\varphi$. It follows immediately that the trace depends linearly on φ,

$$\operatorname{tr}(\lambda\varphi + \mu\psi) = \lambda \operatorname{tr}\varphi + \mu \operatorname{tr}\psi .$$

Next we show that

$$\operatorname{tr}(\psi \circ \varphi) = \operatorname{tr}(\varphi \circ \psi) \tag{4.42}$$

for any two linear transformations φ and ψ. The trace of $\psi \circ \varphi$ is defined by the equation

$$\sum_i \Delta(x_1 \ldots (\psi \circ \varphi) x_i \ldots x_n) = \operatorname{tr}(\psi \circ \varphi) \Delta(x_1 \ldots x_n) \quad x_\nu \in E .$$

Replacing the vectors x_ν by ψx_ν $(\nu = 1 \ldots n)$ we obtain

$$\begin{aligned} &\sum_i \Delta(\psi x_1 \ldots (\psi \circ \varphi \circ \psi) x_i \ldots \psi x_n) \\ &= \operatorname{tr}(\psi \circ \varphi) \Delta(\psi x_1 \ldots \psi x_n) = \operatorname{tr}(\psi \circ \varphi) \det\psi\, \Delta(x_1 \ldots x_n). \end{aligned} \tag{4.43}$$

The left hand-side of this equation can be written as

$$\begin{aligned} \sum_i \Delta(\psi x_1 \ldots (\psi \circ \varphi \circ \psi) x_i \ldots \psi x_n) &= \det\psi \sum_i \Delta(x_1 \ldots (\varphi \circ \psi) x_i \ldots x_n) \\ &= \det\psi \cdot \operatorname{tr}(\varphi \circ \psi) \Delta(x_1 \ldots x_n) \end{aligned}$$

and thus (4.43) implies that

$$\det\psi \operatorname{tr}(\varphi \circ \psi) = \operatorname{tr}(\psi \circ \varphi) \det\psi . \tag{4.44}$$

If ψ is regular, this equation may be divided by $\det\psi$ yielding (4.42). If ψ is non-regular, consider the mapping $\psi - \lambda\iota$ where λ is different from all eigenvalues of ψ. Then $\psi-\lambda\iota$ is regular, whence

$$\operatorname{tr}[(\psi-\lambda\iota)\circ\varphi]=\operatorname{tr}[\varphi\circ(\psi-\lambda\iota)].$$

In view of the linearity of the trace-operator this equation yields

$$\operatorname{tr}(\psi\circ\varphi)-\lambda\operatorname{tr}\varphi=\operatorname{tr}(\varphi\circ\psi)-\lambda\operatorname{tr}\varphi$$

whence (4.42).

Finally it will be shown that the coefficient of λ^{n-1} in the characteristic polynomial of φ can be written as

$$\alpha_1=(-1)^{n-1}\operatorname{tr}\varphi. \tag{4.45}$$

Formula (4.38) yields for $p=1$

$$\sum_\sigma \varepsilon_\sigma \Delta(\varphi x_{\sigma(1)}, x_{\sigma(2)}\dots x_{\sigma(n)})=(-1)^{n-1}\alpha_1\Delta(x_1\dots x_n) \tag{4.46}$$

the sum being taken over all permutations σ subject to the restrictions

$$\sigma(2)<\dots<\sigma(n).$$

This sum can be written as

$$\sum_{i=1}^n(-1)^{i-1}\Delta(\varphi x_i, x_1\dots\hat{x}_i\dots x_n)=\sum_{i=1}^n\Delta(x_1\dots x_{i-1},\varphi x_i, x_{i+1}\dots x_n).$$

We thus obtain from (4.46)

$$\sum_i\Delta(x_1\dots\varphi x_i\dots x_n)=(-1)^{n-1}\alpha_1\Delta(x_1\dots x_n). \tag{4.47}$$

Comparing the relations (4.47) and (4.41) we find (4.45).

4.24. The trace of a matrix. Let $e_\nu(\nu=1\dots n)$ be a basis of E. Then φ determines an $n\times n$-matrix α_ν^μ by the equations

$$\varphi e_\nu=\sum_\mu\alpha_\nu^\mu e_\mu. \tag{4.48}$$

Inserting $x_\nu=e_\nu$ $(\nu=1\dots n)$ in (4.41) we find

$$\sum_i\Delta(e_1\dots\varphi e_i\dots e_n)=\operatorname{tr}\varphi\,\Delta(e_1\dots e_n). \tag{4.49}$$

Equations (4.48) and (4.49) imply that

$$\Delta(e_1\dots e_n)\sum_{i=1}^n\alpha_i^i=\Delta(e_1\dots e_n)\operatorname{tr}\varphi$$

whence

$$\operatorname{tr}\varphi=\sum_i\alpha_i^i. \tag{4.50}$$

Observing that

$$\alpha_\nu^\mu=\langle e^{*\mu},\varphi e_\nu\rangle,$$

where $e^{*\nu}$ $(\nu=1\dots n)$ is the dual basis of e_ν, we can rewrite equation (4.50) as

$$\operatorname{tr}\varphi = \sum_i \langle e^{*i}, \varphi e_i\rangle . \tag{4.51}$$

Formula (4.50) shows that the trace of a linear transformation is equal to the sum of all entries in the main-diagonal of the corresponding matrix. For any $n\times n$-matrix $A=(\alpha_\nu^\mu)$ this sum is called the *trace* of A and will be denoted by $\operatorname{tr} A$,

$$\operatorname{tr} A = \sum_i \alpha_i^i . \tag{4.52}$$

Now equation (4.50) can be written in the form

$$\operatorname{tr}\varphi = \operatorname{tr} M(\varphi).$$

4.25. The duality of $L(E;F)$ and $L(F;E)$. Now consider two linear spaces E and F and the spaces $L(E;F)$ and $L(F;E)$ of all linear mappings $\varphi: E\to F$ and $\psi: F\to E$. With the help of the trace a scalar product can be introduced in these spaces in the following way:

$$\langle\varphi,\psi\rangle = \operatorname{tr}(\varphi\circ\psi) \quad \varphi\in L(E;F), \quad \psi\in L(F;E). \tag{4.53}$$

The function defined by (4.53) is obviously bilinear. Now assume that

$$\langle\varphi,\psi\rangle = 0 \tag{4.54}$$

for a fixed mapping $\varphi\in L(E;F)$ and all linear mappings $\psi\in L(F;E)$. It has to be shown that this implies that $\varphi=0$. Assume that $\varphi\neq 0$. Then there exists a vector $a\in E$ such that $\varphi a\neq 0$. Extend the vector $b_1=\varphi a$ to a basis $(b_1\dots b_m)$ of F and define the linear mapping $\psi: F\to E$ by

$$\psi b_1 = a, \quad \psi b_\mu = 0 \qquad (\mu = 2\dots m).$$

Then

$$(\varphi\circ\psi) b_1 = b_1, \quad (\varphi\circ\psi) b_\mu = 0 \qquad (\mu = 2\dots m),$$

whence

$$\langle\varphi,\psi\rangle = \operatorname{tr}(\varphi\circ\psi) = 1 .$$

This is in contradiction with (4.54). Interchanging E and F we see that the relation

$$\langle\varphi,\psi\rangle = 0$$

for a fixed mapping $\psi\in L(F;E)$ and all mappings $\varphi\in L(E;F)$ implies that $\psi=0$. Hence, a scalar-product is defined in $L(E;F)$ and $L(F;E)$ by (4.53) (cf. problem 14 below).

Problems

1. Show that the characteristic polynomial of a linear transformation φ of a 2-dimensional linear space can be written as

$$f(\lambda) = \lambda^2 - \lambda \operatorname{tr} \varphi + \det \varphi .$$

Verify that every such φ satisfies its characteristic equation,

$$\varphi^2 - \varphi \cdot \operatorname{tr} \varphi + \iota \cdot \det \varphi = 0 .$$

2. Given three linear transformations φ, ψ, χ of E show that

$$\operatorname{tr}(\chi \circ \psi \circ \varphi) \neq \operatorname{tr}(\chi \circ \varphi \circ \psi)$$

in general.

3. Show that the trace of a projection operator $\pi: E \to E$ (see Chapter II sec. 2.19) is equal to the dimension of $\operatorname{Im} \pi$.

4. Consider two pairs of dual spaces E^*, E and F^*, F. Prove that the spaces $L(E; F)$ and $L(E^*; F^*)$ are dual with respect to the scalar-product defined by

$$\langle \varphi, \psi \rangle = \operatorname{tr}(\varphi^* \circ \psi) \quad \varphi \in L(E; F) \quad \psi \in L(E^*; F^*) .$$

5. Let f be a linear function in the space $L(E; E)$. Show that f can be written as

$$f(\varphi) = \operatorname{tr}(\varphi \circ \alpha)$$

where α is a fixed linear transformation in E. Prove that α is uniquely determined by f.

6. Assume that f is a linear function in the space $L(E; E)$ such that

$$f(\psi \circ \varphi) = f(\varphi \circ \psi) .$$

Prove that

$$f(\varphi) = \lambda \cdot \operatorname{tr} \varphi$$

where λ is a scalar.

7. Let φ and ψ be two linear transformations of E. Consider the sum

$$\sum_{i \neq j} \Delta(x_1 \ldots \varphi x_i \ldots \psi x_j \ldots x_n)$$

where $\Delta \neq 0$ is a determinant function in E. This sum is again a determinant function and hence it can be written as

$$\sum_{i \neq j} \Delta(x_1 \ldots \varphi x_i \ldots \psi x_j \ldots x_n) = B(\varphi, \psi) \Delta(x_1 \ldots x_n) .$$

By the above relation a bilinear function B is defined in the space $L(E; E)$. Prove:

a) $B(\varphi, \psi) = \operatorname{tr}\varphi \operatorname{tr}\psi - \operatorname{tr}(\psi\circ\varphi)$.

b) $\frac{1}{2}B(\varphi, \varphi) = (-1)^n \alpha_2$ where α_2 is the coefficient of λ^{n-2} in the characteristic polynomial of φ.

c) $\alpha_2 = \dfrac{(-1)^n}{2}[(\operatorname{tr}\varphi)^2 - \operatorname{tr}(\varphi^2)]$.

8. Consider two $n\times n$-matrices A and B. Prove the relation

$$\operatorname{tr}(AB) = \operatorname{tr}(BA)$$

a) by direct computation.

b) using the relation $\operatorname{tr}\varphi = \operatorname{tr} M(\varphi)$.

9. If φ and ψ are two linear transformations of a 2-dimensional linear space prove the relation

$$\psi\circ\varphi + \varphi\circ\psi = \varphi \operatorname{tr}\psi + \psi\operatorname{tr}\varphi + \iota(\operatorname{tr}(\psi\circ\varphi) - \operatorname{tr}\varphi\operatorname{tr}\psi).$$

10. Let $A: L(E; E)\to L(E; E)$ be a linear transformation such that

$$A(\varphi\circ\psi) = A(\varphi)\circ A(\psi) \qquad \varphi,\psi\in L(E;E)$$

and

$$A(\iota) = \iota.$$

Prove that $\operatorname{tr} A(\varphi) = \operatorname{tr}\varphi$.

11. Let E be a 2-dimensional vector space and φ be a linear transformation of E. Prove that φ satisfies the equation $\varphi^2 = -\lambda\iota$, $\lambda > 0$ if and only if

$$\det\varphi > 0 \quad \text{and} \quad \operatorname{tr}\varphi = 0.$$

12. Let $\varphi: E_1\to E_1$ and $\varphi_2: E_2\to E_2$ be linear transformations. Consider

$$\varphi = \varphi_1\oplus\varphi_2: E_1\oplus E_2\to E_1\oplus E_2.$$

Prove that $\operatorname{tr}\varphi = \operatorname{tr}\varphi_1 + \operatorname{tr}\varphi_2$.

13. Let $\varphi: E\to E$ be a linear transformation and assume that there is a decomposition $E = E_1\oplus\cdots\oplus E_r$ into subspaces such that $E_i\cap\varphi E_i = 0$ $(i = 1\ldots r)$. Prove that $\operatorname{tr}\varphi = 0$.

14. Let φ: $E\to F$ and ψ: $E\leftarrow F$ be linear maps between finite dimensional vector spaces. Show that $\operatorname{tr}(\varphi\circ\psi) = \operatorname{tr}(\psi\circ\varphi)$.

15. Show that the trace of $\operatorname{ad}(\varphi)$ (cf. sec. 4.6) is the negative $(n-1)$-th characteristic coefficient of φ, i.e., $\operatorname{tr}\operatorname{ad}(\varphi) = -\alpha_{n-1}$.

§ 8. Oriented vector spaces

In this paragraph E will be a real vector space of dimension $n \geqq 1$.

4.26. Orientation by a determinant function. Let $\Delta_1 \neq 0$ and $\Delta_2 \neq 0$ be two determinant functions in E. Then $\Delta_2 = \lambda \Delta_1$ where $\lambda \neq 0$ is a real number. Hence we can introduce an equivalence relation in the set of all determinant functions $\Delta \neq 0$ as follows:

$$\Delta_1 \sim \Delta_2 \quad \text{if} \quad \lambda > 0 .$$

It is easy to verify that this is indeed an equivalence. Hence a decomposition of all determinant functions $\Delta \neq 0$ into two equivalence classes is induced. Each of these classes is called an *orientation* of E. If (Δ) is an orientation and $\Delta \in (\Delta)$ we shall say that Δ *represents* the given orientation. Since there are two equivalence classes of determinant functions the vector space E can be oriented in two different ways.

A basis $e_\nu (\nu = 1 \ldots n)$ of an oriented vector space is called *positive* if

$$\Delta(e_1 \ldots e_n) > 0$$

where Δ is a representing determinant function. If $(e_1 \ldots e_n)$ is a positive basis and σ is a permutation of the numbers $(1 \ldots n)$ then the basis $(e_{\sigma(1)} \ldots e_{\sigma(n)})$ is positive if and only if the permutation σ is even.

Suppose now that E^* is a dual space of E and that an orientation is defined in E. Then the dual determinant function (cf. sec. 4.10) determines an orientation in E^*. It is clear that this orientation depends only on the orientation of E. Hence, an orientation in E^* is induced by the orientation of E.

4.27. Orientation preserving linear mappings. Let E and F be two oriented vector spaces of the same dimension n and $\varphi: E \to F$ be a linear isomorphism. Given two representing determinant functions Δ_E and Δ_F in E and F consider the function Δ_φ defined by

$$\Delta_\varphi(x_1 \ldots x_n) = \Delta_F(\varphi x_1, \ldots, \varphi x_n) .$$

Clearly Δ_φ is again a determinant function in E and hence we have that

$$\Delta_\varphi = \lambda \Delta_E$$

where $\lambda \neq 0$ is a real number. The sign of λ depends only on φ and on the given orientations (and not on the choice of the representing determinant functions). The linear isomorphism φ is called *orientation preserving* if

$\lambda>0$. The above argument shows that, given a linear isomorphism $\varphi:E\to F$ and an orientation in E, then there exists precisely one orientation in F such that φ preserves the orientation. This orientation will be called the orientation *induced by* φ.

Now let φ be a linear *automorphism*; i.e., $F=E$. Then we have $\Delta_F=\Delta_E$ and hence it follows that

$$\lambda = \det\varphi .$$

This relation shows that a linear automorphism $\varphi:E\to E$ preserves the orientation if and only if $\det\varphi>0$.

As an example consider the mapping $\varphi=-\iota$. Since

$$\det(-\iota)=(-1)^n$$

it follows that φ preserves the orientation if and only if the dimension of n is even.

4.28. Factor spaces. Let E be an orientated vector space and F be an oriented subspace. Then an orientation is induced in the factor space E/F in the following way: Let Δ be a representing determinant function in E and $a_1 \dots a_p$ be a positive basis of F. Then the function

$$\Delta(a_1 \dots a_p, x_{p+1} \dots x_n), \qquad x_i\in E$$

depends only on the classes $\bar{x}_i$. In fact, assume for instance that y_{p+1} and x_{p+1} are equivalent mod F.
Then

$$y_{p+1} = x_{p+1} + \sum_{\nu=1}^{p} \lambda^\nu a_\nu$$

and we obtain

$$\Delta(a_1 \dots a_p, y_{p+1} \dots x_n) = \Delta(a_1 \dots a_p, x_{p+1} \dots x_n) + \\ + \sum_{\nu=1}^{p} \lambda^\nu \Delta(a_1 \dots a_p, a_\nu \dots x_n) = \Delta(a_1 \dots a_p, x_{p+1} \dots x_n).$$

Hence a function $\bar{\Delta}$ of $(n-p)$ vectors in E/F is well defined by

$$\bar{\Delta}(\bar{x}_{p+1} \dots \bar{x}_n) = \Delta(a_1 \dots a_p, x_{p+1} \dots x_n). \tag{4.55}$$

It is clear that $\bar{\Delta}$ is linear with respect to every argument and skew symmetric. Hence $\bar{\Delta}$ is a determinant function in E/F. It will now be shown that the orientation defined in E/F by $\bar{\Delta}$ depends only on the orientations of E and F. Clearly, if Δ' is another representing determinant function in

E we have that $\Delta' = \lambda\Delta$, $\lambda > 0$ and hence $\bar{\Delta}' = \lambda\bar{\Delta}$. Now let $(a'_1 \dots a'_p)$ be another positive basis of F. Then we have that

$$a'_\nu = \sum_\mu \alpha^\mu_\nu a_\mu, \quad \det(\alpha^\mu_\nu) > 0$$

whence

$$\Delta(a'_1 \dots a'_p, x_{p+1} \dots x_n) = \det(\alpha^\mu_\nu)\,\Delta(a_1 \dots a_p, x_{p+1} \dots x_n).$$

It follows that the function $\bar{\Delta}'$ obtained from the basis $a'_1 \dots a'_p$ is a positive multiple of the function $\bar{\Delta}$ obtained from the basis $a_1 \dots a_p$.

4.29. Direct decompositions. Consider a direct decomposition

$$E = E_1 \oplus E_2 \tag{4.56}$$

and assume that orientations are defined in E_1 and E_2. Then an orientation is induced in E as follows: Let $a_i (i=1 \dots p)$ and $b_j (j=1 \dots q)$ be positive bases of E_1 and E_2 respectively. Then choose the orientation of E such that the basis $a_1 \dots a_p, b_1 \dots b_q$ is positive. To prove that this orientation depends only on the orientations of E_1 and E_2 let $\bar{a}_i (i=1 \dots p)$ and $\bar{b}_j (j=1 \dots q)$ be two other positive bases of E_1 and E_2. Consider the linear transformations $\varphi: E_1 \to E_1$ and $\psi: E_2 \to E_2$ defined by

$$\varphi a_i = \bar{a}_i \quad (i = 1 \dots p) \quad \text{and} \quad \psi b_j = \bar{b}_j \quad (j = 1 \dots q).$$

Then the transformation $\varphi \oplus \psi$ carries the basis $(a_1 \dots a_p, b_1 \dots b_q)$ into the basis $(\bar{a}_1 \dots \bar{a}_p, \bar{b}_1 \dots \bar{b}_q)$. Since $\det \varphi > 0$ and $\det \psi > 0$ it follows from sec. 4.7 that

$$\det(\varphi \oplus \psi) = \det \varphi \det \psi > 0$$

and hence $(\bar{a}_1 \dots \bar{a}_p, \bar{b}_1 \dots \bar{b}_q)$ is again a positive basis of E.

Suppose now that in the direct decomposition (4.56) orientations are given in E and E_1. Then an orientation is induced in E_2. In fact, consider the projection $\pi: E \to E_2$ defined by the decomposition (4.56). It induces an isomorphism

$$\varphi: E/E_1 \xrightarrow{\cong} E_2.$$

In view of sec. 4.28 an orientation in E/E_1 is determined by the orientations of E and E_1. Hence an orientation is induced in E_2 by φ. To describe this orientation explicitly let Δ be a representing determinant function in E and $e_1, \dots, e_p$ be a positive basis of E_1. Then formula (4.55) implies that the induced orientation in E_2 is represented by the determinant function

$$\Delta_2(y_{p+1} \dots y_n) = \Delta(e_1 \dots e_p, y_{p+1} \dots y_n), \quad y_i \in E_2. \tag{4.57}$$

Now let $e_{p+1}, \ldots, e_n$ be a positive basis of E_2 with respect to the induced orientation. Then we have

$$\Delta_2(e_{p+1} \ldots e_n) > 0$$

and hence formula (4.57) implies that

$$\Delta(e_1 \ldots e_p, e_{p+1} \ldots e_n) > 0.$$

It follows that the basis $e_1 \ldots e_p, e_{p+1} \ldots e_n$ of E is positive. In other words, the orientation induced in E by E_1 and E_2 coincides with the original orientation.

The space E_2 in turn induces an orientation in E_1. It will be shown that this orientation coincides with the original orientation of E_1 if and only if $p(n-p)$ is even. The induced orientation of E_1 is represented by the determinant function

$$\Delta_1(x_1 \ldots x_p) = \Delta(e_{p+1} \ldots e_n, x_1 \ldots x_p) \tag{4.58}$$

where $e_\lambda (\lambda = p+1, \ldots n)$ is a positive basis of E_2. Substituting $x_\nu = e_\nu$ $(\nu = 1 \ldots n)$ in equation (4.58) we find that

$$\Delta_1(e_1 \ldots e_p) = \Delta(e_{p+1} \ldots e_n, e_1 \ldots e_p) = (-1)^{p(n-p)} \Delta_2(e_{p+1} \ldots e_n). \tag{4.59}$$

But $e_\lambda (\lambda = p+1, \ldots, n)$ is a positive basis of E_2 whence

$$\Delta_2(e_{p+1} \ldots e_n) > 0. \tag{4.60}$$

It follows from (4.59) and (4.60) that

$$\Delta_1(e_1 \ldots e_p) \begin{cases} > 0 & \text{if } p(n-p) \text{ is even} \\ < 0 & \text{if } p(n-p) \text{ is odd.} \end{cases} \tag{4.61}$$

Since the basis $(e_1 \ldots e_p)$ of E_1 is positive with respect to the original orientation, relation (4.61) shows that the induced orientation coincides with the original orientation if and only if $p(n-p)$ is even.

4.30. Example. Consider a 2-dimensional linear space E. Given a basis (e_1, e_2) we choose the orientation of E in which the basis e_1, e_2 is positive. Then the determinant function Δ, defined by

$$\Delta(e_1, e_2) = 1$$

represents this orientation. Now consider the subspace $E_j (j = 1,2)$ generated by $e_j (j = 1,2)$ with the orientation defined by e_j. Then E_1 induces in E_2 the given orientation, but E_2 induces in E_1 the inverse orientation.

In fact, defining the determinant-functions Δ_1 and Δ_2 in E_1 and in E_2 by

$$\Delta_1(x) = \Delta(e_2, x) \quad x \in E_1, \quad \text{and} \quad \Delta_2(x) = \Delta(e_1, x) \quad x \in E_2$$

we find that

$$\Delta_2(e_2) = \Delta(e_1, e_2) = 1 \quad \text{and} \quad \Delta_1(e_1) = \Delta(e_2, e_1) = -1.$$

4.31. Intersections. Let E_1 and E_2 be two subspaces of E such that

$$E = E_1 + E_2 \tag{4.62}$$

and assume that orientations are given in E_1, E_2 and E. It will be shown that then an orientation is induced in the intersection $E_{12} = E_1 \cap E_2$. Setting

$$\dim E_1 = p, \quad \dim E_2 = q, \quad \dim E_{12} = r$$

we obtain from (4.62) and (1.32) that

$$r = p + q - n.$$

Now consider the isomorphisms

$$\varphi: E/E_1 \xrightarrow{\cong} E_2/E_{12}$$

and

$$\psi: E/E_2 \xrightarrow{\cong} E_1/E_{12}.$$

Since orientations are induced in E/E_1 and E/E_2 these isomorphisms determine orientations in E_2/E_{12} and in E_1/E_{12} respectively. Now choose two positive bases $\bar{a}_{r+1} \dots \bar{a}_p$ and $\bar{b}_{r+1} \dots \bar{b}_q$ in E_1/E_{12} and E_2/E_{12} respectively and let $a_i \in E_1$ and $b_j \in E_2$ be vectors such that

$$\pi_1 a_i = \bar{a}_i \quad \text{and} \quad \pi_2 b_j = \bar{b}_j$$

where π_1 and π_2 denote the canonical projections

$$\pi_1: E_1 \to E_1/E_{12} \quad \text{and} \quad \pi_2: E_2 \to E_2/E_{12}.$$

Now define the function Δ_{12} by

$$\Delta_{12}(z_1 \dots z_r) = \Delta(z_1 \dots z_r, a_{r+1} \dots a_p, b_{r+1} \dots b_q). \tag{4.63}$$

In a similar way as in sec. 4.28 it is shown that the orientation defined in E_{12} by Δ_{12} depends only on the orientations of E_1, E_2 and E (and not on the choice of the vectors a_i and b_j). Hence an orientation is induced in E_{12}.

Interchanging E_1 and E_2 in (4.63) we obtain

$$\Delta_{21}(z_1 \dots z_r) = \Delta(z_1 \dots z_r, b_{r+1} \dots b_q, a_{r+1} \dots a_p). \tag{4.64}$$

Hence it follows that

$$\Delta_{21} = (-1)^{(p-r)(q-r)} \Delta_{12} = (-1)^{(n-p)(n-q)} \Delta_{12}. \tag{4.65}$$

Now consider the special case of a direct decomposition. Then $p+q=n$ and $E_{12}=(0)$. The function Δ_{12} reduces to the scalar

$$\alpha_{12} = \Delta(a_1 \dots a_p, b_1 \dots b_q). \tag{4.66}$$

It follows from (4.66) that $\alpha_{12} \neq 0$. Moreover the number $\frac{\alpha_{12}}{|\alpha_{12}|}$ depends only on the orientations of E_1, E_2 and E. It is called the *intersection number* of the oriented subspaces E_1 and E_2. From (4.65) we obtain the relation

$$\alpha_{21} = (-1)^{p(n-p)} \alpha_{12}.$$

4.32. Basis deformation. Let a_ν and $b_\nu (\nu=1 \dots n)$ be two bases of E. Then the basis a_ν is called *deformable* into the basis b_ν if there exist n continuous mappings

$$x_\nu : t \to x_\nu(t) \quad t_0 \leqq t \leqq t_1$$

satisfying the conditions

1. $x_\nu(t_0) = a_\nu$ and $x_\nu(t_1) = b_\nu$
2. The vectors $x_\nu(t)(\nu=1 \dots n)$ are linearly independent for every fixed t.

The deformability of two bases is obviously an equivalence relation. Hence, the set of all bases of E is decomposed into classes of deformable bases. We shall now prove that there are precisely two such classes. This is a consequence of the following

Theorem: Two bases a_ν and $b_\nu (\nu=1 \dots n)$ are deformable into each other if and only if the linear transformation $\varphi: E \to E$ defined by $\varphi a_\nu = b_\nu$ has positive determinant.

Proof: Let $\Delta \neq 0$ be an arbitrary determinant function. Then formula 4.16 together with the observation that the components $\xi_\nu^i (i=1 \dots n)$ are continuous functions of x_ν shows that the mapping $\underbrace{E \times \cdots \times E}_{n} \to \mathbb{R}$ defined by Δ is continuous.

Now assume that $t \to x_\nu(t)$ is a deformation of the basis a_ν into the basis b_ν. Consider the real valued function

$$\Phi(t) = \Delta(x_1(t) \dots x_n(t)).$$

The continuity of the function Δ and the mappings $t \to x_\nu(t)$ implies that the function Φ is continuous. Furthermore,

$$\Phi(t) \neq 0 \qquad (t_0 \leqq t \leqq t_1)$$

because the vectors $x_\nu(t)(\nu = 1 \dots n)$ are linearly independent. Thus the function Φ assumes the same sign at $t = t_0$ and at $t = t_1$. But

$$\Phi(t_1) = \Delta(b_1 \dots b_n) = \Delta(\varphi a_1 \dots \varphi a_n) = \det\varphi\, \Delta(a_1 \dots a_n) = \det\varphi \cdot \Phi(t_0)$$

whence

$$\det\varphi > 0$$

and so the first part of the theorem is proved.

4.33. Conversely, assume that the linear transformation $a_\nu \to b_\nu$ has positive determinant. To construct a deformation $(a_1 \dots a_n) \to (b_1 \dots b_n)$ assume first that the vector n-tuple

$$(a_1 \dots a_i, b_{i+1} \dots b_n) \tag{4.67}$$

is linearly independent for every $i (1 \leqq i \leqq n-1)$. Then consider the decomposition

$$b_n = \sum_\nu \beta^\nu a_\nu .$$

By the above assumption the vectors $(a_1 \dots a_{n-1}, b_n)$ are linearly independent, whence $\beta^n \neq 0$. Define the number ε_n by

$$\varepsilon_n = \begin{cases} +1 & \text{if} \quad \beta^n > 0 \\ -1 & \text{if} \quad \beta^n < 0. \end{cases}$$

It will be shown that the n mappings

$$\begin{cases} x_\nu(t) = a_\nu (\nu = 1 \dots n-1) \\ x_n(t) = (1-t) a_n + t \varepsilon_n b_n \end{cases} \qquad (0 \leqq t \leqq 1)$$

define a deformation

$$(a_1 \dots a_n) \to (a_1 \dots a_{n-1}, \varepsilon_n b_n).$$

Let $\Delta \neq 0$ be a determinant function in E. Then

$$\Delta(x_1(t) \dots x_n(t)) = ((1-t) + \varepsilon_n \beta^n t) \Delta(a_1 \dots a_n).$$

Since $\varepsilon_n \beta^n > 0$, it follows that

$$1 - t + \varepsilon_n \beta^n t > 0 \qquad (0 \leqq t \leqq 1)$$

whence

$$\Delta(x_1(t)\dots x_n(t)) \neq 0 \qquad (0 \leqq t \leqq 1).$$

This implies the linear independence of the vectors $x_\nu(t)(\nu=1\dots n)$ for every t.

In the same way a deformation

$$(a_1 \dots a_{n-1}, \varepsilon_n b_n) \to (a_1 \dots a_{n-2}, \varepsilon_{n-1} b_{n-1}, \varepsilon_n b_n)$$

can be constructed where $\varepsilon_{n-1} = \pm 1$. Continuing this way we finally obtain a deformation

$$(a_1 \dots a_n) \to (\varepsilon_1 b_1 \dots \varepsilon_n b_n) \quad \varepsilon_\nu = \pm 1 \qquad (\nu = 1 \dots n).$$

To construct a deformation

$$(\varepsilon_1 b_1 \dots \varepsilon_n b_n) \to (b_1 \dots b_n)$$

consider the linear transformations

$$\varphi : a_\nu \to \varepsilon_\nu b_\nu \qquad (\nu = 1 \dots n)$$

and

$$\psi : \varepsilon_\nu b_\nu \to b_\nu \qquad (\nu = 1 \dots n).$$

The product of these linear transformations is given by

$$\psi \circ \varphi : a_\nu \to b_\nu \qquad (\nu = 1 \dots n).$$

By hypothesis,

$$\det(\psi \circ \varphi) > 0 \tag{4.68}$$

and by the result of sec. 4.32

$$\det \varphi > 0. \tag{4.69}$$

Relations (4.68), and (4.69) imply that

$$\det \psi > 0.$$

But

$$\det \psi = \varepsilon_1 \dots \varepsilon_n$$

whence

$$\varepsilon_1 \dots \varepsilon_n = +1.$$

Thus, the number of ε_ν equal to -1 is even. Rearranging the vectors b_ν $(\nu=1\dots n)$ we can achieve that

$$\varepsilon_\nu = \begin{cases} -1 & (\nu = 1 \dots 2p) \\ +1 & (\nu = 2p + 1 \dots n). \end{cases}$$

Then a deformation

$$(\varepsilon_1 b_1 \dots \varepsilon_n b_n) \to (b_1 \dots b_n)$$

is defined by the mappings

$$\left.\begin{aligned} x_{2\nu-1}(t) &= -b_{2\nu-1}\cos t + b_{2\nu}\sin t \\ x_{2\nu}(t) &= -b_{2\nu-1}\sin t - b_{2\nu}\cos t \end{aligned}\right\} (\nu = 1 \dots p) \qquad 0 \leqq t \leqq \pi .$$
$$x_\nu(t) = b_\nu \qquad (\nu = 2p+1 \dots n)$$

4.34. The case remains to be considered that not all the vector n-tuples (4.67) are linearly independent. Let $\Delta \neq 0$ be a determinant function. The linear independence of the vectors $a_\nu\,(\nu = 1 \dots n)$ implies that

$$\Delta(a_1 \dots a_n) \neq 0 .$$

Since Δ is a continuous function, there exist spherical neighbourhoods U_{a_ν} of $a_\nu\,(\nu = 1 \dots n)$ such that

$$\Delta(x_1 \dots x_n) \neq 0 \quad \text{if} \quad x_\nu \in U_{a_\nu} \qquad (\nu = 1 \dots n) .$$

Choose a vector $a_1' \in U_{a_1}$ which is not contained in the $(n-1)$-dimensional subspace generated by the vectors $(b_2 \dots b_n)$. Then the vectors $(a_1', b_2 \dots b_n)$ are linearly independent. Next, choose a vector $a_2' \in U_{a_2}$ which is not contained in the $(n-1)$-dimensional subspace generated by the vectors $(a_1', b_3 \dots b_n)$. Then the vectors $(a_1', a_2', b_3 \dots b_n)$ are linearly independent. Going on this way we finally obtain a system of n vectors $a_\nu'\,(\nu = 1 \dots n)$ such that every n-tuple

$$(a_1' \dots a_i', b_{i+1} \dots b_n) \qquad (i = 1 \dots n)$$

is linearly independent. Since $a_\nu' \in U_{a_\nu}$, it follows that

$$\Delta(a_1' \dots a_n') \neq 0 .$$

Hence the vectors $a_\nu'\,(\nu = 1 \dots n)$ form a basis of E. The n mappings

$$x_\nu(t) = (1-t)\,a_\nu + t\,a_\nu' \qquad (0 \leqq t \leqq 1)$$

define a deformation

$$(a_1 \dots a_n) \to (a_1' \dots a_n') . \tag{4.70}$$

In fact, $x_\nu(t)\,(0 \leqq t \leqq 1)$ is contained in U_{a_ν} whence

$$\Delta(x_1(t) \dots x_n(t)) \neq 0 \qquad (0 \leqq t \leqq 1) .$$

This implies the linear independence of the vectors $x_\nu(t)\,(\nu = 1 \dots n)$.

By the result of sec. 4.33 there exists a deformation

$$(a'_1 \dots a'_n) \to (b_1 \dots b_n). \tag{4.71}$$

The two deformations (4.70) and (4.71) yield a deformation

$$(a_1 \dots a_n) \to (b_1 \dots b_n).$$

This completes the proof of the theorem in sec. 4.32.

4.35. Basis deformation in an oriented linear space. If an orientation is given in the linear space E, the theorem of sec. 4.32 can be formulated as follows: Two bases a_ν and b_ν $(\nu = 1 \dots n)$ can be deformed into each other if and only if they are both positive or both negative with respect to the given orientation. In fact, the linear transformation

$$\varphi : a_\nu \to b_\nu \qquad (\nu = 1 \dots n)$$

has positive determinant if and only if the bases a_ν and $b_\nu (\nu = 1 \dots n)$ are both positive or both negative.

Thus the two classes of deformable bases consist of all positive bases and all negative bases.

4.36. Complex vector spaces. The existence of two orientations in a real linear space is based upon the fact that every real number $\lambda \neq 0$ is either positive or negative. Therefore it is not possible to distinguish two orientations of a complex linear space. In this context the question arises whether any two bases of a complex linear space can be deformed into each other. It will be shown that this is indeed always possible.

Consider two bases a_ν and $b_\nu (\nu = 1 \dots n)$ of the complex linear space E. As in sec. 4.33 we can assume that the vector n-tuples

$$(a_1 \dots a_i, b_{i+1} \dots b_n)$$

are linearly independent for every $i (1 \leqq i \leqq n-1)$. It follows from the above assumption that the coefficient β^n in the decomposition

$$b_n = \sum_\nu \beta^\nu a_\nu$$

is different from zero. The complex number β^n can be written as

$$\beta^n = r\, e^{i\vartheta} \qquad (r > 0, 0 \leqq \vartheta < 2\pi).$$

Now choose a positive continuous function $r(t) (0 \leqq t \leqq 1)$ such that

$$r(0) = 1, \quad r(1) = r \tag{4.72}$$

and a continuous function $\vartheta(t)(0 \leqq t \leqq 1)$ such that

$$\vartheta(0) = 0, \quad \vartheta(1) = \vartheta. \tag{4.73}$$

Define mappings $x_\nu(t), (0 \leqq t \leqq 1)$ by

$$\left.\begin{aligned} x_\nu(t) &= a_\nu \quad (\nu = 1 \dots n-1) \\ \text{and} \qquad x_n(t) &= t \sum_{\nu=1}^{n-1} \beta^\nu a_\nu + r(t) e^{i\vartheta(t)} a_n . \end{aligned}\right\} 0 \leqq t \leqq 1. \tag{4.74}$$

Then the vectors $x_\nu(t)(\nu = 1 \dots n)$ are linearly independent for every t. In fact, assume a relation

$$\sum_{\nu=1}^{n} \lambda^\nu x_\nu(t) = 0 .$$

Then

$$\sum_{\nu=1}^{n-1} \lambda^\nu a_\nu + \lambda^n t \sum_{\nu=1}^{n-1} \beta^\nu a_\nu + \lambda^n r(t) e^{i\vartheta(t)} a_n = 0$$

whence

$$\lambda^\nu + \lambda^n t \beta^\nu = 0 \quad (\nu = 1 \dots n-1)$$

and

$$\lambda^n r(t) e^{i\vartheta(t)} = 0 .$$

Since $r(t) \neq 0$ for $0 \leqq t \leqq 1$, the last equation implies that $\lambda^n = 0$. Hence the first $(n-1)$ equations reduce to $\lambda^\nu = 0 (\nu = 1 \dots n-1)$.

It follows from (4.74), (4.72) and (4.73) that

$$x_n(0) = a_n \quad \text{and} \quad x_n(1) = b_n .$$

Thus the mappings (4.74) define a deformation

$$(a_1 \dots a_{n-1}, a_n) \to (a_1 \dots a_{n-1}, b_n) .$$

Continuing this way we obtain after n steps a deformation of the basis a_ν into the basis $b_\nu (\nu = 1 \dots n)$.

Problems

1. Let E be an oriented n-dimensional linear space and $x_\nu (\nu = 1 \dots n)$ be a positive basis; denote by E_i, the subspace generated by the vectors $(x_1, \dots \hat{x}_i \dots x_n)$. Prove that the basis $(x_1 \dots \hat{x}_i \dots x_n)$ is positive with respect to the orientation induced in E_i by the vector $(-1)^{i-1} x_i$.

2. Let E be an oriented vector space of dimension 2 and let a_1, a_2 be two linearly independent vectors. Consider the 1-dimensional subspaces E_1 and E_2 generated by a_1 and a_2 and define orientations in E_i such that the bases a_i are positive $(i=1,2)$. Show that the intersection number of E_1 and E_2 is $+1$ if and only if the basis a_1, a_2 of E is positive.

3. Let E be a vector space of dimension 4 and assume that e_ν $(\nu=1\dots4)$ is a basis of E. Consider the following quadruples of vectors:

I. $e_1+e_2,\ e_1+e_2+e_3,\ e_1+e_2+e_3+e_4,\ e_1-e_2+e_4$
II. $e_1+2e_3,\ e_2+e_4,\ e_2-e_1+e_4,\ e_2$
III. $e_1+e_2-e_3,\ e_2+e_4,\ e_3+e_2,\ e_2-e_1$
IV. $e_1+e_2-e_3,\ e_2-e_4,\ e_3+e_2,\ e_2-e_1$
V. $e_1-3e_3,\ e_2+e_4,\ e_2-e_1-e_4,\ e_2$.

a) Verify that each quadruple is a basis of E and decide for each pair of bases if they determine the same orientation of E.

b) If for any pair of bases, the two bases determine the same orientation, construct an explicit deformation.

c) Consider E as a subspace of a 5-dimensional vector space $\tilde{E}$ and assume that $e_\nu\,(\nu=1,\dots 5)$ is a basis of $\tilde{E}$. Extend each of the bases above to a basis of $\tilde{E}$ which determines the same orientation as the basis e_ν $(\nu=1,\dots,5)$. Construct the corresponding deformations explicitly.

4. Let E be an oriented vector space and let E_1, E_2 be two oriented subspaces such that $E=E_1+E_2$. Consider the intersection $E_1\cap E_2$ together with the induced orientation. Given a positive basis $(c_1,\dots,c_r)$ of $E_1\cap E_2$ extend it to a positive basis $(c_1,\dots,c_r,a_{r+1},\dots,a_p)$ of E_1 and to a positive basis $(c_1,\dots,c_r,b_{r+1},\dots,b_q)$ of E_2. Prove that then $(c_1,\dots,c_r,$ $a_{r+1},\dots,a_p,b_{r+1},\dots,b_q)$ is a positive basis of E.

5. *Linear isotopies.* Let E be an n-dimensional real vector space. Two linear automorphisms $\varphi\colon E\xrightarrow{\cong}E$ and $\psi\colon E\xrightarrow{\cong}E$ will be called *linearly isotopic* if there is a continuous map $\Phi\colon I\times E\to E$ (I the closed unit interval) such that $\Phi(0,x)=\varphi(x)$, $\Phi(1,x)=\psi(x)$ and such that for each $t\in I$ the map $\Phi_t\colon E\to E$ given by $\Phi_t(x)=\Phi(t,x)$ is a linear automorphism.

(i) Show that two automorphisms of E are linearly isotopic if and only if their determinants have the same sign.

(ii) Let $j\colon E\to E$ be a linear map such that $j^2=-\iota$. Show that j is linearly isotopic to the identity map and conclude that $\det j=1$.

6. *Complex structures.* A *complex structure* in real vector space E is a linear transformation $j\colon E\to E$ satisfying $j^2=-\iota$.

(i) Show that a complex structure exists in an n-dimensional vector space if and only if n is even.

(ii) Let j be a complex structure in E where $\dim E = 2n$. Let Δ be a determinant function. Show that for $x_\nu \in E$ $(\nu = 1 \ldots n)$ either

$$\Delta(x_1, \ldots, x_n, jx_1, \ldots, jx_n) \geqq 0$$

or

$$\Delta(x_1, \ldots, x_n, jx_1, \ldots, jx_n) \leqq 0.$$

The natural orientation of (E, j) is defined to be the orientation represented by a non-zero determinant function satisfying the first of these. Conclude that the underlying real vector space of a complex space carries a natural orientation.

(iii) Let (E, j) be a vector space with complex structure and consider the complex structure $(E, -j)$. Show that the natural orientations of (E, j) and $(E, -j)$ coincide if and only if n is even $(\dim E = 2n)$.

Chapter V

Algebras

In paragraphs one and two all vector spaces are defined over a fixed, but arbitrarily chosen field Γ of characteristic 0.

§ 1. Basic properties

5.1. Definition: An *algebra, A,* is a vector space together with a mapping $A \times A \to A$ such that the conditions (M_1) and (M_2) below both hold. The image of two vectors $x \in A$, $y \in A$, under this mapping is called the *product* of x and y and will be denoted by xy.

The mapping $A \times A \to A$ is required to satisfy:

$$(M_1) \qquad (\lambda x_1 + \mu x_2) y = \lambda (x_1 y) + \mu (x_2 y)$$

$$(M_2) \qquad x(\lambda y_1 + \mu y_2) = \lambda (x y_1) + \mu (x y_2).$$

As an immediate consequence of the definition we have that

$$0 \cdot x = x \cdot 0 = 0.$$

Suppose B is a second algebra. Then a linear mapping $\varphi: A \to B$ is called a *homomorphism* (of algebras) if φ preserves products; i.e.,

$$\varphi(x y) = \varphi x \cdot \varphi y. \tag{5.1}$$

A homomorphism that is injective (resp. surjective, bijective) is called a monomorphism (resp. epimorphism, isomorphism). If $B = A$, φ is called an endomorphism.

Note: To distinguish between mappings of vector spaces and mappings of algebras, we reserve the word *linear mapping* for a mapping between vector spaces satisfying (1.8), (1.9) and *homomorphism* for a linear mapping between algebras which satisfies (5.1).

Let A be a given algebra and let U, V be two subsets of A. We denote by UV, the set

$$\{x \in A \mid x = \sum_i u_i v_i, \quad u_i \in U, v_i \in V\}.$$

Every vector $a\in A$ induces a linear mapping

$$\mu(a): A\to A$$

defined by

$$\mu(a)x = a\,x \tag{5.2}$$

$\mu(a)$ is called the *multiplication operator* determined by a.

An algebra A is called *associative* if

$$x(y\,z)=(x\,y)\,z \qquad x,y,z\in A$$

and *commutative* if

$$x\,y = y\,x \qquad x,y\in A\,.$$

From every algebra A we can obtain a second algebra A^{opp} by defining

$$(x\,y)^{\text{opp}} = y\,x$$

A^{opp} is called the algebra *opposite to A*. It is clear that if A is associative then so is A^{opp}. If A is commutative we have $A^{\text{opp}}=A$.

If A is an associative algebra, a subset $S\subset A$ is called a *system of generators* of A if each vector $x\in A$ is a linear combination of products of elements in S,

$$x=\sum_{(\nu)}\lambda^{\nu_1\ldots\nu_p}x_{\nu_1}\ldots x_{\nu_p}\,, \qquad x_{\nu_i}\in S,\ \lambda^{\nu_1\ldots\nu_p}\in\Gamma\,.$$

A *unit element* (or identity) in an algebra is an element e such that for every x

$$x\,e = e\,x = x\,. \tag{5.3}$$

If A has a unit element, then it is unique. In fact, if e and e' are unit elements, we obtain from (5.3)

$$e = e\,e' = e'\,.$$

Let A be an algebra with unit element e_A and φ be an epimorphism of A onto a second algebra B. Then $e_B=\varphi\,e_A$ is the unit element of B. In fact, if $y\in B$ is arbitrary, there exists an element $x\in A$ such that $y=\varphi x$. This gives

$$y\,e_B=\varphi\,x\cdot\varphi\,e_A=\varphi(x\,e_A)=\varphi(x)=y\,.$$

In the same way it is shown that $e_B y=y$.

An algebra with unit element is called a *division algebra*, if to every element $a\neq 0$ there is an element a^{-1} such that $aa^{-1}=a^{-1}a=e$.

5.2. Examples: 1. Consider the space $L(E;E)$ of all linear transformations of a vector space E. Define the product of two transformations by

$$\psi\,\varphi=\psi\circ\varphi\,.$$

The relations (2.17) imply that the mapping $(\varphi, \psi) \rightarrow \psi\varphi$ satisfies (M_1) and (M_2) and hence $L(E; E)$ is made into an algebra. $L(E; E)$ together with this multiplication is called the *algebra of linear transformations of E* and is denoted by $A(E; E)$. The identity transformation ι acts as unit element in $A(E; E)$. It follows from (2.14) that the algebra $A(E; E)$ is associative.

However, it is not commutative if $\dim E \geqq 2$. In fact, write

$$E = (x_1) \oplus (x_2) \oplus F$$

where (x_1) and (x_2) are the one-dimensional subspaces generated by two linearly independent vectors x_1 and x_2, and F is a complementary subspace. Define linear transformations φ and ψ by

$$\varphi x_1 = 0, \quad \varphi x_2 = x_1; \qquad \varphi y = 0, y \in F$$

and

$$\psi x_1 = x_2, \quad \psi x_2 = 0; \qquad \psi y = 0, y \in F.$$

Then

$$\varphi \psi x_2 = \varphi 0 = 0$$

while

$$\psi \varphi x_2 = \psi x_1 = x_2$$

whence $\varphi\psi \neq \psi\varphi$.

Suppose now that A is an associative algebra and consider the linear mapping

$$\mu: A \rightarrow A(A; A)$$

defined by

$$\mu(a)x = a\,x. \tag{5.4}$$

Then we have that

$$\mu(a\,b)x = a\,b\,x = \mu(a)\mu(b)x$$

whence

$$\mu(a\,b) = \mu(a)\mu(b).$$

This relation shows that μ is a homomorphism of A into $A(A; A)$.

Example 2: Let $M^{n \times n}$ be the vector space of $(n \times n)$-matrices for a given integer n and define the product of two $(n \times n)$-matrices by formula (3.20). Then it follows from the results of sec. 3.10 that the space $M^{n \times n}$ is made into an associative algebra under this multiplication with the unit matrix J as unit element. Now consider a vector space E of dimension n with a distinguished basis e_ν $(\nu = 1 \ldots n)$. Then every linear transformation $\varphi: E \rightarrow E$ determines a matrix $M(\varphi)$. The correspondence $\varphi \rightarrow M(\varphi)$

determines a linear isomorphism of $A(E; E)$ onto $M^{n\times n}$. In view of sec. 3.10 we have that

$$M(\psi\circ\varphi) = M(\varphi)M(\psi). \tag{3.21}$$

This relation shows that M is an isomorphism of the algebra $A(E; E)$ onto the opposite algebra $(M^{n\times n})^{\mathrm{opp}}$.

Example 3: Suppose $\Gamma_1 \subset \Gamma$ is a subfield. Then Γ is an algebra over Γ_1. We show first that Γ is a vector space over Γ_1. In fact, consider the mapping $\Gamma_1 \times \Gamma \to \Gamma$ defined by

$$(\lambda, x) \to \lambda x, \qquad \lambda \in \Gamma_1, x \in \Gamma.$$

It satisfies the relations

$$\begin{aligned}(\lambda+\mu)x &= \lambda x + \mu x\\ \lambda(x+y) &= \lambda x + \lambda y\\ (\lambda\mu)x &= \lambda(\mu x)\\ 1x &= x\end{aligned}$$

where $\lambda, \mu \in \Gamma_1$, $x, y \in \Gamma$. Thus Γ is a vector space over Γ_1.

Define the multiplication in Γ by

$$(x, y) \to xy \quad \text{(field multiplication)}.$$

Then M_1 and M_2 follow from the distribution laws for field multiplication. Hence Γ is an associative commutative algebra over Γ_1 with 1 as unit element.

Example 4: Let C^r be the vector space of functions of a real variable t which have derivatives up to order r. Defining the product by

$$(fg)(t) = f(t)g(t)$$

we obtain an associative and commutative algebra in which the function $f(t) = 1$ acts as unit element.

5.3. Subalgebras and ideals. A *subalgebra*, A_1, of an algebra A is a linear subspace which is closed under the multiplication in A; that is, if x and y are arbitrary elements of A_1, then $xy \in A_1$. Thus A_1 inherits the structure of an algebra from A. It is clear that a subalgebra of an associative (commutative) algebra is itself associative (commutative).

Let S be a subset of A, and suppose that A is associative. Then the subspace $B \subset A$ generated (linearly) by elements of the form

$$s_1 \ldots s_r, \qquad s_i \in S$$

is clearly a subalgebra of A, called the *subalgebra generated by S.* It is easily verified that

$$B = \bigcap_\alpha A_\alpha$$

where the A_α are all the subalgebras of A containing S.

A *right (left) ideal* in an algebra A is a subspace I such that for every $x\in I$, and every $y\in A$, $xy\in I(yx\in I)$. A subspace that is both a right and left ideal is called a *two-sided ideal*, or simply an *ideal in* A. Clearly, every right (left) ideal is a subalgebra. As an example of an ideal, consider the subspace A^2 (linearly generated by the products xy). A^2 is clearly an ideal and is called the *derived algebra.*

The *ideal I generated by a set S* is the intersection of all ideals containing S. If A is associative, I is the subspace of A generated (linearly) by elements of the form

$$s, a s, s a \qquad s\in S, a\in A.$$

In particular every single element a generates an ideal I_a. I_a is called the *principal ideal* generated by a.

Example 5: Suppose A is an algebra with unit element e, and let $\varphi:\Gamma\to A$ be the linear mapping defined by

$$\varphi\lambda = \lambda e.$$

Considering Γ as an algebra over itself we have that

$$\varphi(\lambda\mu) = (\lambda\mu)e = (\lambda e)(\mu e) = \varphi(\lambda)\varphi(\mu).$$

Hence φ is a homomorphism. Moreover, if $\varphi\lambda=0$, then $\lambda e=0$ whence $\lambda=0$. It follows that φ is a monomorphism. Consequently we may identify Γ with its image under φ. Then Γ becomes a subalgebra of A and scalar multiplication coincides with algebra multiplication. In fact, if λ is any scalar, then

$$\lambda a = \lambda(e\cdot a) = (\lambda e)\cdot a = \varphi(\lambda)a.$$

Example 6: Given an element a of an associative algebra consider the set, N_a, of all elements $x\in A$ such that $ax=0$. If $x\in N_a$ then we have for every $y\in A$

$$a(xy) \equiv (ax)y = 0$$

and so $xy\in N_a$. This shows that N_a is a right ideal in A. It is called the *right annihilator of* a. Similarly the left annihilator of a is defined.

5.4. Factor algebras. Let A be an algebra and B be an arbitrary subspace of A. Consider the canonical projection

$$\pi: A\to A/B.$$

It will be shown that A/B admits a multiplication such that π is a homomorphism if and only if B is an ideal in A.

Assume first that there exists such a multiplication in A/B. Then for every $x \in A, y \in B$, we have

$$\pi(x\,y) = \pi\,x \cdot \pi\,y = \pi\,x \cdot 0 = 0$$

whence $x\,y \in B$.
Similarly it follows that $y\,x \in B$ and so B must be an ideal.

Conversely, assume B is an ideal. Then define the multiplication in A/B by

$$\bar{x}\,\bar{y} = \pi(x\,y) \qquad \bar{x}, \bar{y} \in A/B \tag{5.5}$$

where x and y are any representatives of $\bar{x}$ and $\bar{y}$ respectively.

It has to be shown that the above product does not depend on the choice of x and y. Let x' and y' be two other elements such that $\pi\,x' = \bar{x}$ and $\pi\,y' = \bar{y}$. Then

$$x' - x \in B \quad \text{and} \quad y' - y \in B.$$

Hence we can write

$$x' = x + b, \quad b \in B \quad \text{and} \quad y' = y + c, \quad c \in B.$$

It follows that

$$x'\,y' - x\,y = b\,y + x\,c + b\,c \in B$$

and so

$$\pi(x'\,y') = \pi(x\,y).$$

The multiplication in A/B clearly satisfies (M_1) and (M_2) as follows from the linearity of π. Finally, rewriting (5.5) in the form

$$\pi(x\,y) = \pi\,x \cdot \pi\,y$$

we see that π is a homomorphism and that the multiplication in A/B is uniquely determined by the requirement that π be a homomorphism.

The vector space A/B together with the multiplication (5.5) is called the *factor algebra* of A with respect to the ideal B. It is clear that if A is associative (commutative) then so is A/B. If A has a unit element e then $\bar{e} = \pi\,e$ is the unit element of the algebra A/B.

5.5. Homomorphisms. Suppose A and B are algebras and $\varphi: A \to B$ is a homomorphism. Then the kernel of φ is an ideal in A. In fact, if $x \in \ker \varphi$ and $y \in A$ are arbitrary we have that

$$\varphi(x\,y) = \varphi\,x \cdot \varphi\,y = 0 \cdot \varphi\,y = 0$$

whence $x\,y \in \ker \varphi$. In the same way it follows that $y\,x \in \ker \varphi$. Next consider the subspace $\operatorname{Im} \varphi \subset B$. Since for every two elements $x, y \in A$

$$\varphi\,x \cdot \varphi\,y = \varphi(x\,y) \in \operatorname{Im} \varphi$$

it follows that $\operatorname{Im} \varphi$ is a subalgebra of B.

Now let

$$\bar{\varphi}: A/\ker\varphi \to B$$

be the induced injective linear mapping. Then we have the commutative diagram

$$\begin{array}{ccc} A & \xrightarrow{\varphi} & B \\ {\scriptstyle\pi}\downarrow & \nearrow_{\bar{\varphi}} & \\ A/\ker\varphi & & \end{array}$$

and since π is a homomorphism, it follows that

$$\begin{aligned} \bar{\varphi}(\pi x \cdot \pi y) &= \bar{\varphi}\pi(x y) \\ &= \varphi(x y) \\ &= \varphi(x)\cdot\varphi(y) \\ &= \bar{\varphi}(\pi x)\cdot\bar{\varphi}(\pi y). \end{aligned}$$

This relation shows that $\bar{\varphi}$ is a homomorphism and hence a monomorphism. In particular, the induced mapping

$$\bar{\varphi}: A/\ker\varphi \xrightarrow{\cong} \operatorname{Im}\varphi$$

is an isomorphism.

Finally, assume that C is a third algebra, and let $\psi: B \to C$ be a homomorphism. Then the composition $\psi \circ \varphi: A \to C$ is again a homomorphism. In fact, we have

$$\begin{aligned} (\psi\circ\varphi)(x y) &= \psi(\varphi x \cdot \varphi y) \\ &= \psi\varphi x \cdot \psi\varphi y \\ &= (\psi\circ\varphi) x \cdot (\psi\circ\varphi) y. \end{aligned}$$

Let $\varphi: A \to B$ be any homomorphism of associative algebras and S be a system of generators for A. Then φ determines a set map $\varphi_0: S \to B$ by

$$\varphi_0 x = \varphi x, \qquad x \in S.$$

The homomorphism φ is completely determined by φ_0. In fact, if

$$x = \sum_{(\nu)} \lambda^{\nu_1 \ldots \nu_p} x_{\nu_1} \ldots x_{\nu_p}, \qquad x_{\nu_i} \in S, \lambda^{\nu_1 \ldots \nu_p} \in \Gamma$$

is an arbitrary element we have that

$$\begin{aligned} \varphi x &= \sum_{(\nu)} \lambda^{\nu_1 \ldots \nu_p} \varphi x_{\nu_1} \ldots \varphi x_{\nu_p} \\ &= \sum_{(\nu)} \lambda^{\nu_1 \ldots \nu_p} \varphi_0 x_{\nu_1} \ldots \varphi_0 x_{\nu_p} \end{aligned}$$

Proposition I: Let $\varphi_0: S \to B$ be an arbitrary set map. Then φ_0 can be can be extended to a homomorphism $\varphi: A \to B$ if and only if

$$\sum_{(\nu)} \lambda^{\nu_1 \ldots \nu_p} \varphi_0 x_{\nu_1} \ldots \varphi_0 x_{\nu_p} = 0 \quad \text{whenever} \quad \sum_{(\nu)} \lambda^{\nu_1 \ldots \nu_p} x_{\nu_1} \ldots x_{\nu_p} = 0. \tag{5.6}$$

Proof: It is clear that the above condition is necessary. Conversely, assume that (5.6) is satisfied. Then define a mapping $\varphi: A \to B$ by

$$\varphi \sum_{(\nu)} \xi^{\nu_1 \ldots \nu_p} x_{\nu_1} \ldots x_{\nu_p} = \sum_{(\nu)} \xi^{\nu_1 \ldots \nu_p} \varphi_0 x_{\nu_1} \ldots \varphi_0 x_{\nu_p}, \qquad x_{\nu_i} \in S. \tag{5.7}$$

To show that φ is, in fact, well defined we notice that if

$$\sum_{(\nu)} \xi^{\nu_1 \ldots \nu_p} x_{\nu_1} \ldots x_{\nu_p} = \sum_{(\mu)} \eta^{\mu_1 \ldots \mu_q} y_{\mu_1} \ldots y_{\mu_q}$$

then

$$\sum_{(\nu)} \xi^{\nu_1 \ldots \nu_p} x_{\nu_1} \ldots x_{\nu_p} - \sum_{(\mu)} \eta^{\mu_1 \ldots \mu_q} y_{\mu_1} \ldots y_{\mu_q} = 0.$$

In view of (5.6)

$$\sum_{(\nu)} \xi^{\nu_1 \ldots \nu_p} \varphi_0 x_{\nu_1} \ldots \varphi_0 x_{\nu_p} - \sum_{(\mu)} \eta^{\mu_1 \ldots \mu_q} \varphi_0 y_{\mu_1} \ldots \varphi_0 y_{\mu_q} = 0$$

and so

$$\sum_{\nu} \xi^{\nu_1 \ldots \nu_p} \varphi_0 x_{\nu_1} \ldots \varphi_0 x_{\nu_p} = \sum_{\mu} \eta^{\mu_1 \ldots \mu_q} \varphi_0 y_{\mu_1} \ldots \varphi_0 y_{\mu_q}.$$

It follows from (5.7) that

$$\varphi x = \varphi_0 x \qquad x \in S$$

$$\varphi(\lambda x + \mu y) = \lambda \varphi x + \mu \varphi y$$

and

$$\varphi(x y) = \varphi x \cdot \varphi y$$

and hence φ is a homomorphism.

Now suppose $\{e_\alpha\}$ is a basis for A and let $\varphi: A \to B$ be a linear map such that

$$\varphi(e_\alpha e_\beta) = \varphi e_\alpha \varphi e_\beta$$

for each α, β. Then φ is a homomorphism, as follows from the relation

$$\begin{aligned} \varphi(x y) &= \varphi\{(\sum_\alpha \xi^\alpha e_\alpha)(\sum_\beta \eta^\beta e_\beta)\} \\ &= \varphi(\sum_{\alpha,\beta} \xi^\alpha \eta^\beta e_\alpha e_\beta) = \sum_{\alpha,\beta} \xi^\alpha \eta^\beta \varphi(e_\alpha) \varphi(e_\beta) \\ &= (\sum_\alpha \xi^\alpha \varphi(e_\alpha))(\sum_\beta \eta^\beta \varphi(e_\beta)) = \varphi(x) \varphi(y). \end{aligned}$$

5.6. Derivations. A linear mapping $\theta: A \to A$ of an algebra into itself is called a *derivation* if

$$\theta(x\,y) = \theta x \cdot y + x \cdot \theta y \qquad x, y \in A. \tag{5.8}$$

As an example let A be the algebra of C^∞-functions $f: \mathbb{R} \to \mathbb{R}$ and define the mapping θ by $\theta: f \to f'$ where f' denotes the derivative of f. Then the elementary rules of calculus imply that θ is a derivation.

If A has a unit element e it follows from (5.8) that

$$\theta e = \theta e + \theta e$$

whence $\theta e = 0$. A derivation is completely determined by its action on a system of generators of A, as follows from an argument similar to that used to prove the same result for homomorphisms. Moreover, if $\theta: A \to A$ is a linear map such that

$$\theta(e_\alpha e_\beta) = \theta(e_\alpha) e_\beta + e_\alpha \theta(e_\beta)$$

where $\{e_\alpha\}$ is a basis for A, then θ is a derivation in A.

For every derivation θ we have the *Leibniz formula*

$$\theta^n(x\,y) = \sum_{r=0}^{n} \binom{n}{r} \theta^r x \cdot \theta^{n-r} y. \tag{5.9}$$

In fact, for $n=1$, (5.9) coincides with (5.8). Suppose now by induction that (5.9) holds for some n. Then

$$\begin{aligned}
\theta^{n+1}(x\,y) &= \theta\,\theta^n(x\,y) \\
&= \sum_{r=0}^{n} \binom{n}{r} \theta^{r+1} x \cdot \theta^{n-r} y + \sum_{r=0}^{n} \binom{n}{r} \theta^r x \cdot \theta^{n-r+1} y \\
&= x \cdot \theta^{n+1} y + \sum_{r=1}^{n} \left[\binom{n}{r} + \binom{n}{r-1}\right] \theta^r x \cdot \theta^{n+1-r} y + \theta^{n+1} x \cdot y \\
&= x \cdot \theta^{n+1} y + \sum_{r=1}^{n} \binom{n+1}{r} \theta^r x \cdot \theta^{n-r+1} y + \theta^{n+1} x \cdot y \\
&= \sum_{r=0}^{n+1} \binom{n+1}{r} \theta^r x \cdot \theta^{n+1-r} y.
\end{aligned}$$

and so the induction is closed.

The image of a derivation θ in A is of course a subspace of A, but it is in general not a subalgebra. Similarly, the kernel is a subalgebra, but it is not, in general, an ideal. To see that $\ker\theta$ is a subalgebra, we notice that for any two elements $x, y \in \ker\theta$

$$\theta(x\,y) = \theta x \cdot y + x \cdot \theta y = 0$$

whence $x y \in \ker\theta$.

It follows immediately from (5.8) that a linear combination of derivations $\theta_i: A \to A$ is again a derivation in A. But the product of two derivations θ_1, θ_2 satisfies

$$\begin{aligned}(\theta_1\theta_2)(x\,y) &= \theta_1(\theta_2 x \cdot y + x \cdot \theta_2 y)\\ &= \theta_1\theta_2 x \cdot y + \theta_2 x \cdot \theta_1 y + \theta_1 x \cdot \theta_2 y + x \cdot \theta_1\theta_2 y \qquad (5.10)\end{aligned}$$

and so is, in general, not a derivation. However, the commutator

$$[\theta_1, \theta_2] = \theta_1\theta_2 - \theta_2\theta_1$$

is again a derivation, as follows at once from (5.10).

5.7. φ-derivations. Let A and B be algebras and $\varphi: A \to B$ be a fixed homomorphism. Then a linear mapping $\theta: A \to B$ is called a φ-derivation if

$$\theta(x\,y) = \theta x \cdot \varphi y + \varphi x \cdot \theta y \qquad x, y \in A.$$

In particular, all derivations in A are ι-derivations where $\iota: A \to A$ denotes the identity map.

As an example of a φ-derivation, let A be the algebra of C^∞-functions $f: \mathbb{R} \to \mathbb{R}$ and let $B = \mathbb{R}$. Define the homomorphism φ to be the evaluation homomorphism

$$\varphi: f \to f(0)$$

and the mapping θ by

$$\theta: f \to f'(0).$$

Then it follows that

$$\begin{aligned}\theta(f g) &= (f g)'(0)\\ &= f'(0) g(0) + f(0) g'(0)\\ &= \theta f \cdot \varphi g + \varphi f \cdot \theta g\end{aligned}$$

and so θ is a φ-derivation.

More generally, if θ_A is any derivation in A, then $\theta = \varphi \circ \theta_A$ is a φ-derivation. In fact,

$$\begin{aligned}\theta(x\,y) &= \varphi\theta_A(x\,y)\\ &= \varphi(\theta_A x \cdot y + x \cdot \theta_A y)\\ &= \varphi\theta_A x \cdot \varphi y + \varphi x \cdot \varphi\theta_A y\\ &= \theta x \cdot \varphi y + \varphi x \cdot \theta y.\end{aligned}$$

Similarly, if θ_B is a derivation in B, then $\theta_B \circ \varphi$ is a φ-derivation.

5.8. Antiderivations. Recall that an involution in a linear space is a linear transformation whose square is the identity. Similarly we define an involution ω in an algebra A to be an *endomorphism* of A whose square is the identity map. Clearly the identity map of A is an involution. If A has a unit element e it follows from sec. 5.1 that $\omega e = e$.

Now let ω be a fixed involution in A. A linear transformation $\Omega: A \to A$ will be called an *antiderivation with respect to* ω if it satisfies the relation

$$\Omega(x\,y) = \Omega x \cdot y + \omega x \cdot \Omega y. \tag{5.11}$$

In particular, a derivation is an antiderivation with respect to the involution ι. As in the case of a derivation it is easy to show that an antiderivation is determined by its action on a system of generators for A and that $\ker \Omega$ is a subalgebra of A. Moreover, if A has a unit element e, then $\Omega e = 0$. It also follows easily that any linear combination of antiderivations with respect to a fixed involution ω is again an antiderivation with respect to ω.

Suppose next that Ω_1 and Ω_2 are antiderivations in A with respect to the involutions ω_1 and ω_2 and assume that $\omega_1 \circ \omega_2 = \omega_2 \circ \omega_1$. Then $\omega_1 \circ \omega_2$ is again an involution. The relations

$$(\Omega_1 \Omega_2)(x\,y) = \Omega_1(\Omega_2 x \cdot y + \omega_2 x \cdot \Omega_2 y) = \Omega_1 \Omega_2 x \cdot y + \\ + \omega_1 \Omega_2 x \cdot \Omega_1 y + \Omega_1 \omega_2 x \cdot \Omega_2 y + \omega_1 \omega_2 x \cdot \Omega_1 \Omega_2 y$$

and

$$(\Omega_2 \Omega_1)(x\,y) = \Omega_2(\Omega_1 x \cdot y + \omega_1 x \cdot \Omega_1 y) = \Omega_2 \Omega_1 x \cdot y + \\ + \omega_2 \Omega_1 x \cdot \Omega_2 y + \Omega_2 \omega_1 x \cdot \Omega_1 y + \omega_2 \omega_1 x \cdot \Omega_2 \Omega_1 y$$

yield

$$(\Omega_1 \Omega_2 \pm \Omega_2 \Omega_1)(x\,y) = \\ = (\Omega_1 \Omega_2 \pm \Omega_2 \Omega_1) x \cdot y + (\omega_1 \Omega_2 \pm \Omega_2 \omega_1) x \cdot \Omega_1 y + \\ + (\Omega_1 \omega_2 \pm \omega_2 \Omega_1) x \cdot \Omega_2 y + \omega_1 \omega_2 x \cdot (\Omega_1 \Omega_2 \pm \Omega_2 \Omega_1) y. \tag{5.12}$$

Now consider the following special cases:

1. $\omega_1 \Omega_2 = \Omega_2 \omega_1$ and $\omega_2 \Omega_1 = \Omega_1 \omega_2$ (this is trivially true if $\omega_1 = \pm\iota$ and $\omega_2 = \pm\iota$). Then the relation shows that $\Omega_1 \Omega_2 - \Omega_2 \Omega_1$ is an antiderivation with respect to the involution $\omega_1 \omega_2$. In particular, if Ω is an antiderivation with respect to ω and θ is a derivation such that $\omega\theta = \theta\omega$, then $\theta\Omega - \Omega\theta$ is again an antiderivation with respect to ω.

2. $\omega_1 \Omega_2 = -\Omega_2 \omega_1$ and $\omega_2 \Omega_1 = -\Omega_1 \omega_2$. Then $\Omega_1 \Omega_2 + \Omega_2 \Omega_1$ is an antiderivation with respect to the involution $\omega_1 \omega_2$.

Now let Ω_1 and Ω_2 be two antiderivations with respect to the same involution ω such that

$$\omega\Omega_i = -\Omega_i\omega \qquad (i = 1, 2).$$

Then it follows that $\Omega_1\Omega_2 + \Omega_2\Omega_1$ is a derivation. In particular, if Ω is any antiderivation such that

$$\omega\Omega = -\Omega\omega$$

then Ω^2 is a derivation.

Finally, let B be a second algebra, and let $\varphi: A \to B$ be a homomorphism. Assume that ω_A is an involution in A. Then a φ-antiderivation with respect to ω_A is a linear mapping $\Omega: A \to B$ satisfying

$$\Omega(xy) = \Omega x \cdot \varphi y + \varphi\omega_A x \cdot \Omega y. \qquad (5.13)$$

If ω_B is an involution in B such that

$$\varphi\omega_A = \omega_B\varphi$$

then equation (5.13) can be rewritten in the form

$$\Omega(xy) = \Omega x \cdot \varphi y + \omega_B\varphi x \cdot \Omega y.$$

Problems

1. Let A be an arbitrary algebra and consider the set $C(A)$ of elements $a \in A$ that commute with every element in A. Show that $C(A)$ is a subspace of A. If A is associative, prove that $C(A)$ is a subalgebra of A. $C(A)$ is called the *centre* of A.

2. If A is any algebra and θ is a derivation in A, prove that $C(A)$ and the derived algebra are stable under θ.

3. Construct an explicit example to prove that the sum of two endomorphisms is in general not an endomorphism.

4. Suppose $\varphi: A \to B$ is a homomorphism of algebras and let $\lambda \neq 0, 1$ be an arbitrarily chosen scalar. Prove that $\lambda\varphi$ is a homomorphism if and only if the derived algebra is contained in $\ker\varphi$.

5. Let C^1 and C denote respectively the algebras of continuously differentiable and continuous functions $f: \mathbb{R} \to \mathbb{R}$ (cf. Example 4). Consider the linear mapping

$$d: C^1 \to C$$

given by $df = f'$ where f' is the derivative of f.

a) Prove that this is an i-derivation where $i: C^1 \to C$ denotes the canonical injection.

b) Show that d is surjective and construct a right inverse for d.

c) Prove that d cannot be extended to a derivation in the algebra C.

6. Suppose A is an associative commutative algebra and θ is a derivation in A. Prove that

$$\theta x^p = p x^{p-1} \theta(x).$$

7. Suppose that θ is a derivation in an associative commutative algebra A with identity e and assume that $x \in A$ is invertible; i.e.; there exists an element x^{-1} such that

$$x x^{-1} = x^{-1} x = e.$$

Prove that $x^p (p \geqq 1)$ is invertible and that

$$(x^p)^{-1} = (x^{-1})^p.$$

Denoting the inverse of x^p by x^{-p} show that for every derivation θ

$$\theta(x^{-p}) = -p x^{-p-1} \theta(x).$$

8. Let L be an algebra in which the product of two elements x, y is denoted by $[x, y]$. Assume that

$$[x, y] + [y, x] = 0 \quad \text{(skew symmetry)}$$
$$[[x, y], z] + [[y, z], x] + [[z, x], y] = 0 \quad \text{(Jacobi identity)}$$

Then L is called a *Lie algebra.*
Let $\mathrm{Ad}(a)$ be the multiplication operator in the Lie algebra L. Prove that $\mathrm{Ad}(a)$ is a derivation.

9. Let A be an associative algebra with product xy. Show that the multiplication $(x, y) \to [x, y]$ where

$$[x, y] = x y - y x$$

makes A into a Lie algebra.

10. Let A be any algebra and consider the space $D(A)$ of derivations in A. Define a multiplication in $D(A)$ by setting

$$[\theta_1, \theta_2] = \theta_1 \theta_2 - \theta_2 \theta_1.$$

a) Prove that $D(A)$ is a Lie algebra.

b) Assume that A is a Lie algebra itself and consider the mapping $\varphi: A \to D(A)$ given by $\varphi: x \to \mathrm{Ad}\, x$. Show that φ is a homomorphism of Lie algebras. Determine the kernel of φ.

11. If L is a Lie algebra and I is an ideal in A, prove that the algebra L/I is again a Lie algebra.

12. Let E be a finite dimensional vector space. Show that the mapping

$$\Phi: A(E; E) \to A(E^*; E^*)^{\mathrm{opp}}$$

given by $\varphi \to \varphi^*$ is an isomorphism of algebras.

13. Let A be an associative algebra with identity and consider the multiplication operator

$$\mu: A \to A(A; A).$$

Show that μ is a monomorphism.

14. Let E be an n-dimensional vector space. Show that each basis e_i $(i=1\dots n)$ of E determines a basis $\varrho_{ij}(i,j=1\dots n)$ of $L(E; E)$ such that

(i) $\varrho_{ij}\varrho_{lk} = \delta_{jl}\varrho_{ik}$
(ii) $\sum_i \varrho_{ii} = \iota$.

Conversely, given n^2 linear transformations ϱ_{ij} of E satisfying i) and ii), prove that they form a basis of $L(E; E)$ and are induced by a basis of E.

Show that two bases e_i and e'_i of E determine the same basis of $L(E; E)$ if and only if $e'_i = \lambda e_i$, $\lambda \in \Gamma$.

15. Define an equivalence relation in the set of all linearly independent n^2-tuples $(\varphi_1 \dots \varphi_{n^2})$, $\varphi_\nu \in L(E; E)$, $(n = \dim E)$, in the following way:

$$(\varphi_1 \dots \varphi_{n^2}) \sim (\psi_1 \dots \psi_{n^2})$$

if and only if there exists an element $\chi \in GL(E)$ such that

$$\psi_\nu = \chi \varphi_\nu \chi^{-1} \quad (\nu = 1 \dots n^2).$$

Prove that

$$(\varphi_1, \dots \varphi_{n^2}) \sim (\lambda \varphi_1 \dots \lambda \varphi_{n^2}) \quad \lambda \in \Gamma$$

only if $\lambda = 1$.

16. Prove that the bases of $L(E; E)$ defined in problem 14 form an equivalence class under the equivalence relation of problem 15. Use this to show that every non-zero endomorphism $\Phi: A(E; E) \to A(E; E)$ is an inner automorphism; i.e., there exists a fixed linear automorphism α of E such that

$$\Phi(\varphi) = \alpha \varphi \alpha^{-1} \quad \varphi \in A(E; E).$$

17. Let A be an associative algebra, and let L denote the corresponding

Lie algebra (cf. problem 9). Show that a linear mapping $\theta: A \to A$ is a derivation in A only if it is a derivation in L.

18. Let E be a finite dimensional vector space and consider the mapping $\theta_\alpha: A(E;E) \to A(E;E)$ defined by

$$\theta_\alpha(\varphi) = \alpha\varphi - \varphi\alpha$$

Prove that θ_α is a derivation. Conversely, prove that every derivation in $A(E;E)$ is of this form.

Hint: Use problem 14.

§ 2. Ideals

5.9. The lattice of ideals. Let A be an algebra, and consider the set $\mathscr{I}$ of ideals in A. We order this set by inclusion; i.e., if I_1 and I_2 are ideals in A, then we write $I_1 \leqq I_2$ if and only if $I_1 \subset I_2$. The relation $\leqq$ is clearly a partial order in $\mathscr{I}$ (cf. sec. 0.6). Now let I_1 and I_2 be ideals in A. Then it is easily cheeked that $I_1 + I_2$ and $I_1 \cap I_2$ are again ideals, and are in fact the least upper bound and the greatest lower bound of I_1 and I_2. Hence, the relation $\leqq$ induces in $\mathscr{I}$ the structure of a lattice.

5.10. Nilpotent ideals. Let A be an associative algebra. Then an element $a \in A$ will be called *nilpotent* if for some k,

$$a^k = 0. \tag{5.14}$$

The least k for which (5.14) holds is called the *degree of nilpotency* of a. An ideal I will be called nilpotent if for some k,

$$I^k = 0. \tag{5.15}$$

The least k for which (5.15) holds is called the *degree of nilpotency of I* and will be denoted by deg I.

5.11.* Radicals. Let A be an associative commutative algebra. Then the nilpotent elements of A form an ideal. In fact, if x and y are nilpotent of degree p and q respectively we have that

$$\begin{aligned}(\lambda x + \mu y)^{p+q} &= \sum_{i=0}^{p+q} \binom{p+q}{i} \lambda^i \mu^{p+q-i} y^{p+q-i} x^i \\ &= \sum_{i=0}^{p+q} \alpha_i x^i y^{p+q-i} \\ &= \sum_{i=0}^{p} \alpha_i x^i y^{p+q-i} + \sum_{i=p+1}^{p+q} \alpha_i x^i y^{p+q-i} = 0\end{aligned}$$

and

$$(x y)^p = x^p y^p = 0.$$

The ideal consisting of the nilpotent elements is called the *radical* of A and will be denoted by rad A. (The definition of radical can be generalized to the non-commutative case; the theory is then much more difficult and belongs to the theory of rings and algebras. The reader is referred to [14]). It is clear that

$$\operatorname{rad}(\operatorname{rad} A) = \operatorname{rad} A.$$

The factor algebra $A/\operatorname{rad} A$ contains no non-zero nilpotent elements. To prove this assume that $\bar{x} \in A/\operatorname{rad} A$ is an element such that $\bar{x}^k = 0$ for some k. Then $x^k \in \operatorname{rad} A$ and hence the definition of rad A yields l such that

$$x^{kl} = (x^k)^l = 0.$$

It follows that $x \in \operatorname{rad} A$ whence $\bar{x} = 0$. The above result can be expressed by the formula

$$\operatorname{rad}(A/\operatorname{rad} A) = 0.$$

Now assume that the algebra A has dimension n. Then rad A is a nilpotent ideal, and

$$\deg(\operatorname{rad} A) \leqq \dim(\operatorname{rad} A) + 1 \leqq n + 1. \tag{5.16}$$

For the proof, we choose a basis $e_1, \ldots, e_r$ of rad A. Then each e_i is nilpotent. Let $k = \max_i(\deg e_i)$, and consider the ideal $(\operatorname{rad} A)^{rk}$. An arbitrary element in this ideal is a sum of elements of the form

$$e_1^{k_1} \ldots e_r^{k_r}$$

where

$$k_1 + \cdots + k_r = k r.$$

In particular, for some i, $k_i \geqq k$ and so $e_1^{k_1} \ldots e_r^{k_r} = 0$. This shows that

$$(\operatorname{rad} A)^{kr} = 0$$

and so rad A is nilpotent.

Now let s be the degree of nilpotency of rad A, and suppose that for some $m < s$,

$$(\operatorname{rad} A)^m = (\operatorname{rad} A)^{m+1}. \tag{5.17}$$

Then we obtain by induction that

$$(\operatorname{rad} A)^m = (\operatorname{rad} A)^{m+1} = (\operatorname{rad} A)^{m+2} = \cdots = (\operatorname{rad} A)^s = 0$$

which is a contradiction. Hence (5.17) is false and so in particular

$$\dim(\operatorname{rad} A)^m > \dim(\operatorname{rad} A)^{m+1}, \qquad m < s.$$

It follows at once that $s-1$ cannot be greater than the dimension of rad A, which proves (5.16).

As a corollary, we notice that for any nilpotent element $x \in A$, its degree of nilpotency is less than or equal to $n+1$,

$$\deg x \leqq n+1.$$

5.12.* Simple algebras. An algebra A is called *simple* if it has no proper non-trivial ideals and if $A^2 \neq 0$. As an example consider a field Γ as an algebra over a subfield Γ_1. Let $I \neq 0$ be an ideal in Γ. If x is a non-zero element of I, then

$$1 = x^{-1} x \in I$$

and it follows that

$$\Gamma = \Gamma \cdot 1 \subset I$$

whence $\Gamma = I$. Since $\Gamma^2 \neq 0$, Γ is simple.

As a second example consider the algebra $A(E; E)$ where E is a vector space of dimension n. Suppose I is a non-trivial ideal in $A(E; E)$ and let $\varphi \neq 0$ be an arbitrary element of I. Then there exists a vector $a \in E$ such that $\varphi a \neq 0$. Now define the linear transformations φ_i by

$$\varphi_i e_k = \delta^i_k a \qquad i, k = 1 \ldots n$$

where e_i $(i = 1 \cdots n)$ is a basis of E. Choose linear transformations ψ_i such that

$$\psi_i \varphi a = e_i \qquad i = 1 \ldots n.$$

Let $\psi \in A(E; E)$ be arbitrary and α^j_i let be the matrix of ψ with respect to the basis e_i. Then

$$\psi e_k = \sum_j \alpha^j_k e_j = \sum_j \alpha^j_k \psi_j \varphi a = \Big(\sum_{i,j} \alpha^j_i \psi_j \varphi \varphi_i\Big) e_k$$

whence

$$\psi = \sum_{i,j} \alpha^j_i \psi_j \varphi \varphi_i.$$

It follows that $\psi \in I$ and so $I = A(E; E)$. Since, (clearly) $A(E; E)^2 \neq 0$, $A(E; E)$ is a simple algebra.

Theorem I: Let A be a simple commutative associative algebra. Then A is a division algebra.

Proof: We show first that A has a unit element. Since $A^2 \neq 0$, there is an element $a \in A$ such that $I_a \neq 0$. Since A is simple, $I_a = A$. Thus there is an element $e \in A$ such that

$$a \cdot e = a. \tag{5.18}$$

Next we show that $e^2=e$. In fact, equation (5.18) implies that

$$a(e^2-e)=(a\,e)\,e-a\,e=a\,e-a\,e=0$$

and so e^2-e is contained in the annihilator of a, $e^2-e\in N_a$. Since N_a is an ideal and $N_a\neq A$ it follows that $N_a=0$ whence $e^2=e$.

Now let $x\in A$ be any element. Since $a=ae\in I_e$ we have $I_e\neq 0$ and so $I_e=A$. Thus we can write

$$x=e\,y \qquad \text{for some } y\in A.$$

It follows that $ex=e^2y=ey=x$ and so e is the unit element of A.

Thus every element $x\in A$ satisfies $x=x\cdot e$; i.e., $x\in I_x$. In particular, $I_x\neq 0$ if $x\neq 0$. Hence, if $x\neq 0$, $I_x=A$ and so there is an element $x^{-1}\in A$ such that $xx^{-1}=x^{-1}x=e$; i.e., A is a division algebra.

5.13.* Totally reducible algebras. An algebra A is called *totally reducible* if to every ideal I there is a complementary ideal I',

$$A=I\oplus I'.$$

Every ideal I in a totally reducible algebra is itself a totally reducible algebra. In fact, let I' be a complementary ideal. Then

$$I\cdot I'\subset I\cap I'=0.$$

Consequently, if J is an ideal in I, we have

$$J\cdot I\subset J \quad\text{and}\quad J\cdot I'\subset I\cdot I'=0$$

whence

$$J\cdot A\subset J.$$

It follows that J is an ideal in A. Let J' be a complementary ideal in A,

$$A=J\oplus J'.$$

Intersecting with I and observing that $J\subset I$ we obtain (cf. sec. 1.13)

$$I=J\oplus(I\cap J').$$

It follows that I is again totally reducible.

An algebra, A, is called *irreducible* if it cannot be written as the direct sum of two non-trivial ideals.

5.14.* Semisimple algebras. In this section A will denote an associative commutative algebra. A will be called *semisimple* if it is totally reducible and if for every non-zero ideal I, $I^2\neq 0$.

Proposition I: If A is totally reducible, then A is the direct sum of its radical and a semisimple ideal. The square of the radical is zero.

Proof: Let B denote a complementary ideal for rad A,

$$A = \operatorname{rad} A \oplus B.$$

Since $B \cong A/\operatorname{rad} A$ it follows that B contains no non-zero nilpotent elements and so $B^2 \neq 0$. It follows from sec. (5.13) that B is totally reducible and hence B is semisimple.

To show that the square of rad A is zero, let k be the degree of nilpotency of rad A, $(\operatorname{rad} A)^k = 0$. Then $(\operatorname{rad} A)^{k-1}$ is an ideal in rad A, and so there exists a complementary ideal J,

$$(\operatorname{rad} A)^{k-1} \oplus J = \operatorname{rad} A. \tag{5.19}$$

Now we have the relations

$$(\operatorname{rad} A^{k-1})^2 = \operatorname{rad} A^k = 0$$

$$J \cdot (\operatorname{rad} A)^{k-1} = 0$$

$$J^{k-1} \subset (\operatorname{rad} A)^{k-1} \cap J = 0.$$

These relations yield

$$(\operatorname{rad} A)^{\max(2,k-1)} = 0.$$

But $(\operatorname{rad} A)^{k-1} \neq 0$ and so

$$(\operatorname{rad} A)^2 = 0.$$

Corollary: A is semisimple if and only if A is totally reducible and rad $A = 0$.

Problems

1. Suppose that I_1, I_2 are ideals in an algebra A. Prove that

$$(I_1 + I_2)/I_1 \cong I_2/(I_1 \cap I_2).$$

2. Show that the algebra C^1 defined in Example 4, § 1, has no nilpotent elements $\neq 0$.

3. Consider the set S of step functions $f: [0,1] \to \mathbb{R}$. Show that the operations

$$(f+g)(t) = f(t) + g(t)$$
$$(\lambda f)(t) = \lambda f(t)$$
$$(fg)(t) = f(t)g(t)$$

make S into a commutative associative algebra with identity. (A function $f:[0,1]\to\mathbb{R}$ is called a *step function* if there exists a decomposition of the unit interval,

$$0=t_0<t_1<\cdots<t_n=1$$

such that f is constant in every interval $t_{i-1}<t<t_i\,(i=1\dots n)$.

4. Show that the algebra constructed in problem 3 has zero divisors, but no non-zero nilpotent elements.

5. Show that the algebra S of problem 3 has ideals which are not principal. Let $(a,b)\subset[0,1]$ be any open interval, and let f be a step function such that $f(t)=0$ if and only if $a<t<b$. Prove that the ideal generated by f is precisely the subset of functions g such that $g(t)=0$ for $a<t<b$.

6. Let I be any principal ideal in S (cf. problem 3). Show that there exists a complementary principal ideal I_1. Conversely, if $S=I\oplus I_1$ is a decomposition of S into ideals, prove that I and I_1 are principal.

7. Let E be an algebra with identity. Show that if E is totally reducible, then every ideal is principal.

8. Let E be an infinite dimensional vector space. Show that the linear transformations of E whose kernels have finite codimension form an ideal. Conclude that $A(E;E)$ is not simple.

Hint: See problem 11, chap. II, § 6 and problem 8, chap. I, § 4.

§ 3. Change of coefficient field of a vector space

5.15. Vector space over a subfield. Let E be a vector space over a field Γ and let Δ be a subfield of Γ. The vector space structure of E involves a mapping

$$\Gamma\times E\to E$$

satisfying the conditions (II.1), (II.2) and (II.3) of sec. 1.1. The restriction of this mapping to $\Delta\times E$ again satisfies these conditions, and so it determines on E the structure of a vector space over Δ. A subspace (factor space) of E considered as a Δ-vector space, will be called a Δ-subspace (Δ-factor space). Similarly we refer to Γ-subspaces and Γ-factor spaces. Clearly every Γ-subspace (factor space) is a Δ-subspace (Δ-factor space).

Now let F be a second vector space over Γ and suppose that $\varphi:E\to F$ is a Γ-linear mapping; i.e.,

$$\varphi(\lambda x+\mu y)=\lambda\varphi x+\mu\varphi y\quad x,y\in E,\lambda,\mu\in\Gamma.$$

Then φ is a Δ-linear mapping if E and F are considered as Δ-vector spaces.

As an example, consider the field Γ as a 1-dimensional vector space over itself. Then (cf. sec. 5.2 Example 3) Γ is an algebra over Δ.

5.16. Dimensions. To distinguish between the dimension of E over Γ and Δ we shall write $\dim_\Gamma E$ and $\dim_\Delta E$. Suppose now that Γ is finite dimensional over Δ. Assume further that the dimension of E over Γ is finite. It will be shown that

$$\dim_\Delta E = \dim_\Gamma E \cdot \dim_\Delta \Gamma .$$

Let $e_i\,(i=1\dots n)$ be a basis of E over Γ and consider the Γ-subspace E_i of E generated by e_i. Then there is a Γ-isomorphism $\varphi:\Gamma \xrightarrow{\cong} E_i$. But φ is also a Δ-isomorphism and hence it follows that

$$\dim_\Delta E_i = \dim_\Delta \Gamma, \qquad i=1\dots n .$$

Since the Δ-vector space E is the direct sum of the Δ-vector spaces E_i we obtain from this that

$$\dim_\Delta E = n \cdot \dim_\Delta \Gamma = \dim_\Gamma E \cdot \dim_\Delta \Gamma .$$

As an example let E be a complex vector space of dimension n. Then, since $\dim_{\mathbb{R}} \mathbb{C} = 2$, E, considered as a real vector space, has dimension $2n$. If $z_\nu\,(\nu=1\dots n)$ is a basis of the complex vector space E then the vectors $z_\nu,\ iz_\nu\,(\nu=1\dots n)$ form a basis of E, considered as a real vector space.

5.17. Algebras over subfields. Again let Δ be a subfield of Γ and let A be an algebra over Γ. Then A may be considered as a vector space over Δ, and it is clear that A, together with its Δ-vector space structure, is an algebra over Δ. We (in a way similar to the case of vector spaces) distinguish between Δ-subalgebras, Δ-homomorphisms and Γ-subalgebras, Γ-homomorphisms. Clearly every Γ-subalgebra, (Γ-homomorphism) is a Δ-subalgebra, (Δ-homomorphism).

5.18. Extension fields as subalgebras of $A_\Delta(E;E)$. Let E be a non-trivial vector space over Γ and $\Delta \subset \Gamma$ be a subfield. Then E can be considered as a vector space over Δ. Denote by $A_\Delta(E{:}\,E)$ the algebra (over Δ) of Δ-linear transformations of E. Define a mapping

$$\Phi: \Gamma \to A_\Delta(E;E)$$

by

$$\Phi(\alpha)x = \alpha x, \qquad \alpha \in \Gamma, x \in E$$

where αx is the ordinary scalar multiplication defined between Γ and E. Then

$$\Phi(\alpha\beta)x = (\alpha\beta)x = \alpha(\beta x) = \Phi(\alpha)\Phi(\beta)x$$

and

$$\Phi(\alpha+\beta)x = (\alpha+\beta)x = \alpha x + \beta x$$
$$= \Phi(\alpha)x + \Phi(\beta)x$$
$$= (\Phi(\alpha) + \Phi(\beta))x$$

whence

$$\Phi(\alpha+\beta) = \Phi(\alpha) + \Phi(\beta)$$

and

$$\Phi(\alpha\beta) = \Phi(\alpha)\Phi(\beta) \qquad \alpha, \beta \in \Gamma.$$

Since $\Delta \subset \Gamma$, it follows that Φ is a Δ-homomorphism. Moreover, Φ is injective. In fact, $\Phi(\alpha)=0$ implies that $\alpha x=0$ for every $x \in E$ whence $\alpha=0$.

Since Φ is a monomorphism we may identify Γ with the Δ-subalgebra $\operatorname{Im}\Phi$ of $A_\Delta(E; E)$.

Conversely, let E be a vector space over a field Δ and assume that $\Gamma \subset A_\Delta(E; E)$ is a field containing the identity. We may identify Δ with the subalgebra of Γ consisting of elements of the form $\lambda \cdot \iota$, $\lambda \in \Delta$, the identification map being given by $\lambda \to \lambda\iota$. Then we have $\Delta \subset \Gamma$; i.e., Δ is a subfield of Γ.

Now define a mapping $\Gamma \times E \to E$ by

$$(\varphi, x) \to \varphi x \qquad \varphi \in \Gamma, x \in E. \tag{5.20}$$

Then we have the relations

$$\varphi\psi(x) = \varphi(\psi x)$$
$$\varphi(x+y) = \varphi x + \varphi y$$
$$(\varphi+\psi)x = \varphi x + \psi x$$
$$\iota x = x \qquad x, y \in E;\ \varphi, \psi \in \Gamma,$$

and hence E is made into a vector space over Γ.

The restriction of the mapping (5.20) to Δ gives the original structure of E as a vector space over Δ while the mapping Φ restricted to Δ reduces to the canonical injection of Δ into $A_\Delta(E; E)$.

5.19. Linear transformations over extension fields. Let $\Delta \subset \Gamma$ be a subfield and E be a vector space over Γ. Then we have shown that $A_\Gamma(E; E) \subset A_\Delta(E; E)$. Now we shall prove the more precise

Proposition: $A_\Gamma(E; E)$ is the subalgebra of $A_\Delta(E; E)$ consisting of those Δ-linear transformations which commute with every Δ-linear transformation of the form

$$\varepsilon_\alpha : x \to \alpha x, \qquad \alpha \in \Gamma.$$

Proof: Let $\varphi \in A_\Gamma(E; E)$. Then

$$\varphi(\alpha x) = \alpha \varphi x \qquad x \in E, \alpha \in \Gamma$$

and so

$$\varphi \circ \varepsilon_\alpha = \varepsilon_\alpha \circ \varphi .$$

Conversely, if this holds, then by inverting the above argument we obtain that $\varphi \in A_\Gamma(E; E)$.

Corollary: Suppose E is a vector space over Δ and $\Gamma \subset A_\Delta(E; E)$ is a field such that $\iota \in \Gamma$. Then a transformation $\varphi \in A_\Delta(E; E)$ is contained in $A_\Gamma(E; E)$ if and only if it commutes with every Δ-linear transformation in Γ.

Problems

1. Suppose $\Delta \subset \Gamma$ is a subfield of Γ such that Γ has finite dimension over Δ. Suppose further that $A \subset \Gamma$ is a subalgebra such that $\Delta \subset A$. Prove that A is a subfield of Γ.

2. Show that if $\Delta \subset \Gamma$ is a subfield, and Γ has finite dimension over Δ, then there are no non-trivial derivations in the Δ-algebra Γ.

3. A complex number z is called *algebraic* if it satisfies an equation of the form

$$\sum_{\nu=1}^{n} \alpha_\nu z^\nu = 0 \qquad (\alpha_\nu \text{ rational})$$

where not all the coefficients α_ν are zero. Prove that the algebraic numbers are a subfield, A, of $\mathbb{C}$ and that A has infinite dimension over the rationals. Prove that there are no non-trivial derivations in A.

Chapter VI

Gradations and homology

In this chapter all vector spaces are defined over a fixed, but arbitrarily chosen field Γ of characteristic 0.

§ 1. G-graded vector spaces

6.1. Definition. Let E be a vector space and G be an abelian group. Suppose that a direct decomposition

$$E = \sum_{\alpha \in I} E_\alpha \tag{6.1}$$

is given and that to every subspace E_α an element $k(\alpha)$ of G is assigned such that the mapping $\alpha \to k(\alpha)$ is injective. Then E is called a *G-graded vector space*. G is called the *group of degrees* for E. The vectors of E_α are called *homogeneous of degree* $k(\alpha)$ and we shall write

$$\deg x = k(\alpha), \qquad x \in E_\alpha .$$

In particular, the zero vector is homogeneous of every degree. If the mapping $\alpha \to k(\alpha)$ is bijective we may use the group G as index set in the decomposition (6.1). Then formula (6.1) reads

$$E = \sum_{k \in G} E_k$$

where E_k denotes the subspace of the homogeneous elements of degree k.

If $G = \mathbb{Z}$, E will be called simply a graded vector space. Suppose that E is a vector space with direct decomposition

$$E = \sum_{k=0}^{\infty} E_k \qquad (k \in \mathbb{Z}).$$

Then by setting $E_k = 0 (k \leqq -1)$ we make E into a *graded space*, and whenever we refer to the graded space $E = \sum_{k=0}^{\infty} E_k$, we shall mean this particular gradation. A gradation of E such that $E_k = 0$, $k \leqq -1$, is called a *positive gradation.*

Now let E be a G-graded space, $E=\sum_{k\in G} E_k$, and consider a subspace $F\subset E$ such that

$$F = \sum_{k\in G} F \cap E_k .$$

Then a G-gradation is induced in F by assigning the degree k to the vectors of $F\cap E_k$. F together with its induced gradation is called a *G-graded subspace* of E.

Suppose next that E^λ is a family of G-graded spaces indexed by a set I and let E be the direct sum of the E^λ. Then a G-gradation is induced in E by

$$E = \sum_{k\in G} E_k \quad \text{where} \quad E_k = \sum_{\lambda\in I} E_k^\lambda .$$

This follows from the relation

$$E = \sum_{\lambda\in I} E^\lambda = \sum_{\lambda\in I}\sum_{k\in G} E_k^\lambda = \sum_{k\in G}\sum_{\lambda\in I} E_k^\lambda = \sum_{k\in G} E_k .$$

6.2. Linear mappings of G-graded spaces. Let E and F be two G-graded spaces and let $\varphi: E\to F$ be a linear map. The map φ is called *homogeneous* if there exists a fixed element $k\in G$ such that

$$\varphi E_j \subset F_{j+k} \qquad j\in G \tag{6.2}$$

k is called the *degree* of the homogeneous mapping φ. The kernel of a homogeneous mapping is a graded subspace of E. In fact, if $\varrho_j: E\to E_j$ and $\sigma_j: F\to F_j$ denote the projection operators in E and F induced by the gradations of E and F it follows from (6.2) that

$$\sigma_{k+j}\circ\varphi = \varphi\circ\varrho_j . \tag{6.3}$$

Relation (6.3) implies that $\ker\varphi$ is stable under the projection operators ϱ_j and hence $\ker\varphi$ is a G-graded subspace of E. Similarly, the image of φ is a G-graded subspace of F.

Now let E be a G-graded vector space, F be an arbitrary vector space (without gradation) and suppose that $\varphi: E\to F$ is a linear map of E *onto* F such that $\ker\varphi$ is a graded subspace of E. Then there is a uniquely determined G-gradation in F such that φ is homogeneous of degree zero. The G-gradation of F is given explicitly by

$$F = \sum_{j\in G} F_j \tag{6.4}$$

where

$$F_j = \varphi(E_j) .$$

To show that (6.4) defines a G-gradation in F, we notice first that since φ is onto,

$$F = \varphi E = \varphi \sum_j E_j = \Sigma_j F_j .$$

To prove that the decomposition (6.4) is direct assume that

$$\sum_j y_j = 0 \quad \text{where} \quad y_j \in F_j .$$

Since $F_j = \varphi E_j$ every y_j can be written in the form $y_j = \varphi x_j$, $x_j \in E_j$. It follows that $\varphi \sum_j x_j = 0$ whence

$$\sum_j x_j \in \ker \varphi .$$

Since $\ker \varphi$ is a graded subspace of E we obtain

$$x_j \in \ker \varphi \quad \text{for each } j$$

whence $y_j = \varphi x_j = 0$. Thus the decomposition (6.4) is direct and hence it defines a G-gradation in F. Clearly, the mapping φ is homogeneous of degree zero with respect to the induced gradation.

Finally, it is clear that any G-gradation of F such that φ is homogeneous of degree zero must assign to the elements of F_j the degree j. In view of the decomposition (6.4) it follows that this G-gradation is uniquely determined by the requirement that φ be homogeneous of degree zero.

This result implies in particular that there is a unique G-gradation determined in the factor space of E with respect to a G-graded subspace such that the canonical projection is homogeneous of degree zero. Such a factor space, together with its G-gradation, is called a *G-graded factor space of E*.

Now let $E = \sum_{k \in G} E_k$ and $F = \sum_{k \in G} F_k$ be two G-graded spaces, and suppose that

$$\varphi : E \to F$$

is a linear mapping homogeneous of degree l. Denote by φ_k the restriction of φ to E_k,

$$\varphi_k : E_k \to F_{k+l} .$$

Then clearly

$$\varphi = \sum_{k \in G} \varphi_k .$$

It follows that φ is injective (surjective, bijective) if and only if each φ_k is injective (surjective, bijective).

6.3. Gradations with respect to several groups. Suppose that $\tau: G \to G'$ is a homomorphism of G into another abelian group G'. Then a G-gradation of E induces a G'-gradation of E by

$$E = \sum_{\beta} E'_{\beta} \quad \text{where} \quad E'_{\beta} = \sum_{\tau(\alpha)=\beta} E_{\alpha}. \tag{6.5}$$

To prove this we note first that

$$E = \sum_{\beta} E'_{\beta}$$

since $E_{\alpha} \subset E'_{\tau\alpha}$. The directness of the decomposition (6.5) follows from the fact that every space E'_{β} is a sum of certain subspaces E_{α} and that the decomposition $E = \sum_{\alpha} E_{\alpha}$ is direct.

A *G p-gradation* is a gradation with respect to the group $G_p = \underbrace{G \oplus \ldots \oplus G}_{p}$.

If $G = \mathbb{Z}$, we refer simply to a p-gradation of E. Given a G p-gradation in E consider the homomorphism

$$\tau: G_p \to G$$

given by

$$\tau(k_1, \ldots k_p) = k_1 + \cdots + k_p.$$

The induced (simple) G-gradation of E is given by

$$E = \sum_{j} F_j, \quad F_j = \sum_{k_1 + \cdots + k_p = j} E_{k_1} \oplus \cdots \oplus E_{k_p}. \tag{6.6}$$

The G-gradation (6.6) of E is called the (simple) G-gradation induced by the given G p-gradation.

Finally, suppose E is a vector space, and assume that

$$E = \sum_{j \in G} E_j, \quad E = \sum_{k \in H} F_k \tag{6.7}$$

define G and H-gradations in E. Then the gradations will be called *compatible* if

$$E = \sum_{j,k} E_j \cap F_k.$$

If the two gradations given by (6.7) are compatible, they determine a $(G \oplus H)$-gradation in E by the assignment

$$(j, k) \to E_j \cap F_k.$$

Conversely, suppose a $(G \oplus H)$-gradation in E is given

$$E = \sum_{j,k} E_{j,k}.$$

Then compatible G and H-gradations of E are defined by

$$E = \sum_{j \in G} E_j, \quad E_j = \sum_{k \in H} E_{j,k}$$

and

$$E = \sum_{k \in H} F_k, \quad F_k = \sum_{j \in G} E_{j,k}.$$

Moreover, the $(G \oplus H)$-gradation of E determined by these G and H-gradations is given by

$$E_{j,k} = E_j \cap F_k \qquad j \in G, k \in H.$$

6.4. The Poincaré series. A gradation $E = \sum_k E_k$ of a vector space E is called *almost finite* if the dimension of every space E_k is finite. To every almost finite positive gradation we assign the formal series

$$P_E(t) = \sum_k \dim E_k \cdot t^k.$$

$P_E(t)$ is called the *Poincaré series* of the graded space E. If the dimension of E is finite, then $P_E(t)$ is a polynomial and

$$P_E(1) = \dim E.$$

The direct sum of two almost finite positively graded spaces E and F is again an almost finite positively graded space, and its Poincaré series is given by

$$P_{E \oplus F}(t) = P_E(t) + P_F(t).$$

Two almost finite positively graded spaces E and F are connected by a homogeneous linear isomorphism of degree 0 if and only if $P_E = P_F$. In fact, suppose

$$\varphi: E \to F$$

is such a homogeneous linear isomorphism of degree 0. Writing

$$\varphi = \sum_{k=0}^{\infty} \varphi_k$$

(cf. sec. 6.2) we obtain that each φ_k is a linear isomorphism,

$$\varphi_k: E_k \xrightarrow{\cong} F_k.$$

Hence

$$\dim E_k = \dim F_k \qquad (k = 0, 1, \ldots) \tag{6.8}$$

and so

$$P_E = P_F.$$

Conversely, assume that $P_E = P_F$. Then (6.8) must hold, and thus there are linear isomorphisms

$$\varphi_k : E_k \xrightarrow{\cong} F_k. \tag{6.9}$$

Since $E = \sum_{k=0}^{\infty} E_k$ we can construct the linear mapping

$$\varphi = \sum_{k=0}^{\infty} \varphi_k : E \to F$$

which is clearly homogeneous of degree zero. Moreover, in view of sec. (6.2) it follows from (6.9) that φ is a linear isomorphism.

6.5. Dual G-graded spaces. Suppose $E = \sum_{k \in G} E_k$ and $F = \sum_{k \in G} F_k$ are two G-graded vector spaces, and assume that

$$\varphi : E \times F \to \Gamma$$

is a bilinear function. Then we say that φ *respects the G-gradations of E and F* if

$$\varphi(E_k \times F_j) = 0 \tag{6.10}$$

for each pair of distinct degrees, $k \neq j$.

Every bilinear function $\varphi : E \times F \to \Gamma$ which respects G-gradations determines bilinear functions $\varphi_k : E_k \times F_k \to \Gamma$ $(k \in G)$ by

$$\varphi \mid E_k \times F_k = \varphi_k, \qquad k \in G. \tag{6.11}$$

Conversely, if any bilinear functions $\varphi_k : E_k \times F_k \to \Gamma$ are given, then a unique bilinear function $\varphi : E \times F \to \Gamma$ which respects G-gradations is determined by (6.10) and (6.11).

In particular, it follows that φ is non-degenerate if and only if each φ_k is non-degenerate. Thus a scalar product which respects G-gradations determines a scalar product between each pair (E_k, F_k), and conversely if a scalar product is defined between each pair (E_k, F_k) then the given scalar product can be extended in a unique way to a G-gradation-respecting scalar product between E and F. E and F, together with a G-gradation-respecting scalar product, will be called *dual G-graded spaces.*

Now suppose that E and F are dual almost finite G-graded spaces.

Then E_k and F_k are dual, and so

$$\dim E_k = \dim F_k \qquad k \in G.$$

In particular, if $G = \mathbb{Z}$ and the gradations of E and F are positive, we have

$$P_E = P_F.$$

Problems

1. Let $\varphi: E \to F$ be an injective linear mapping. Assume that F is a G-graded vector space and that $\operatorname{Im} \varphi$ is a G-graded subspace of F. Prove that there is a unique G-gradation in E so that φ becomes homogeneous of degree zero.

2. Prove that every G-graded subspace of a G-graded vector space has a complementary G-graded subspace.

3. Let E, F be G-graded vector spaces and suppose that $E_1 \subset E$, $F_1 \subset F$ are G-graded subspaces. Let $\varphi: E \to F$ be a linear mapping homogeneous of degree k. Assume that φ can be restricted to E_1, F_1 to obtain a linear mapping $\varphi_1: E_1 \to F_1$ and an induced mapping

$$\bar{\varphi}: E/E_1 \to F/F_1.$$

Prove that φ_1 and $\bar{\varphi}$ are homogeneous of degree k.

4. If φ, E, F are as in problem 3, prove that if φ has a left (right) inverse, then a left (right) *homogeneous* inverse of φ must exist. What are the possible degrees of such a homogeneous left (right) inverse mapping?

5. Let E_1, E_2, E_3 be G-graded vector spaces. Suppose that $\varphi: E_1 \to E_2$ and $\psi: E_1 \to E_3$ are linear mappings, homogeneous of degree k and l respectively. Assume that ψ can be factored over φ. Prove that ψ can be factored over φ with a *homogeneous* linear mapping $\chi: E_2 \to E_3$ and determine the degree of χ.
Hint: See problem 5, chap. II, § 1.

6. Let E, E^* and F, F^* be two pairs of dual G-graded vector spaces. Assume that

$$\varphi: E \to F \quad \text{and} \quad \varphi^*: E^* \leftarrow F^*$$

are dual linear mappings. If φ is homogeneous of degree k, prove that φ^* is homogeneous of degree k.

7. Let E be an almost finite graded space. Suppose that E_1^* and E_2^* are G-graded spaces each of which is dual to the G-graded space E. Construct

a homogeneous linear isomorphism of degree zero

$$\varphi: E_1^* \xrightarrow{\cong} E_2^*$$

such that

$$\langle \varphi y^*, x \rangle = \langle y^*, x \rangle \quad y^* \in E_1^*, \quad x \in E.$$

8. Let E, E^* be a pair of almost finite dual G-graded spaces. Let F be a G-graded subspace of E. Prove that $F^\perp$ is a G-graded subspace of E^* and that $(F^\perp)^\perp = F$.

9. Suppose E, E^*, F are as in problem 8. Let F_1 be a complementary G-graded subspace for F in E (cf. problem 2). Prove that

$$E^* = F^\perp \oplus F_1^\perp$$

and that F, $F_1^\perp$ and F_1, $F^\perp$ are two pairs of dual G-graded spaces.

10. Suppose E, E^* and F, F^* are two pairs of almost finite dual G-graded vector spaces, and let $\varphi: E \to F$ be a linear mapping homogeneous of degree k. Prove that φ^* exists.

11. Suppose E, E^* is a pair of almost finite dual G-graded vector spaces. Let $\{x_\alpha\}$ be a basis of E consisting of the set union of bases for the homogeneous subspaces of E. Prove that a dual basis $\{x^{*\alpha}\}$ in E^* exists.

12. Let E and F be two G-graded vector spaces and $\varphi: E \to F$ be a homogeneous linear mapping of degree k. Assume further that a homomorphism $\omega: G \to H$ is given. Prove that φ is homogeneous of degree $\omega(k)$ with erspect to the induced H-gradation.

§ 2. G-graded algebras

6.6. *G*-graded algebras. Let A be an algebra and suppose that a G-gradation $A = \sum_{k \in G} A_k$ is defined in the vector space A. Then A is called a *G-graded algebra* if for every two homogeneous elements x and y, xy is homogeneous, and

$$\deg(xy) = \deg x + \deg y. \tag{6.12}$$

Suppose that $A = \sum_{k \in G} A_k$ is a graded algebra with identity element e. Then e is homogeneous of degree 0. In fact, writing

$$e = \sum_{k \in G} e_k \qquad e_k \in A_k$$

we obtain for each $x \in A$ that

$$x = x e = \sum_{k \in G} x e_k .$$

Hence if x is homogeneous of degree l

$$\sum_{k \in G} x e_k \in A_{l+k}$$

whence

$$x e_k = 0 \quad \text{for} \quad k \neq 0$$

and so

$$x e_0 = x . \tag{6.13}$$

Since (6.13) holds for each homogeneous vector x it follows that e_0 is a right identity for A, whence

$$e = e e_0 = e_0 .$$

Thus $e \in A_0$ and so it is homogeneous of degree 0.

It is clear that every subalgebra of A that is simultaneously a G-graded subspace, is a G-graded algebra. Such subalgebras are called *G-graded subalgebras*.

Now suppose that $I \subset A$ is a G-graded ideal. Then the factor algebra A/I has a natural G-gradation as a linear space (cf. sec. 6.2) such that the canonical projection $\pi: A \to A/I$ is homogeneous of degree zero. Hence if $\bar{x}$ and $\bar{y}$ are any two homogeneous elements in A/I we have

$$\bar{x} \cdot \bar{y} = \overline{x \cdot y} = \pi(x y)$$

and so $\bar{x}\bar{y}$ is homogeneous. Moreover,

$$\deg(\bar{x}\bar{y}) = \deg(x y) = \deg x + \deg y = \deg \bar{x} + \deg \bar{y} .$$

Consequently, A/I is a G-graded algebra.

More generally, if B is a second algebra without gradation, and $\varphi: A \to B$ is an epimorphism whose kernel is a G-graded ideal in A, then the induced G-gradation (cf. sec. 6.2) makes B into a G-graded algebra.

Now let A and B be G-graded algebras, and assume that $\varphi: A \to B$ is a homogeneous homomorphism of degree k. Then $\ker \varphi$ is a G-graded ideal in A and $\operatorname{Im} \varphi$ is a G-graded subalgebra of B.

Suppose next that A is a G-graded algebra, and $\tau: G \to G'$ is a homomorphism, G' being a second abelian group. Then it is easily checked that the induced G'-gradation of A makes A into a G'-graded algebra.

The reader should also verify that if E is simultaneously a G- and an H-graded algebra such that the gradations are compatible, then the induced $(G \oplus H)$-gradation of A makes A into a $(G \oplus H)$-graded algebra.

A graded algebra A is called *anticommutative* if for every two homogeneous elements x and y

$$x y = (-1)^{\deg x \deg y} y x .$$

If x and y are two homogeneous elements in an associative anticommutative graded algebra such that $\deg x \cdot \deg y$ is even, then x and y commute, and so we obtain the *binomial formula*

$$(x+y)^n = \sum_{i=0}^{n} \binom{n}{i} x^i y^{n-i} .$$

In every graded algebra $A = \sum_k A_k$ an involution ω is defined by

$$\omega x = (-1)^k x , \qquad x \in A_k . \tag{6.14}$$

In fact, if $x \in A_k$ and $y \in A_l$ are two homogeneous elements we have

$$\omega(x y) = (-1)^{k+l} x y = (-1)^k x (-1)^l y = \omega x \cdot \omega y$$

and so ω preserves products. It follows immediately from (6.14) that $\omega^2 = \iota$ and so ω is an involution. ω will be called the *canonical involution* of the graded algebra A.

A homogeneous antiderivation with respect to the canonical involution (6.14) will simply be called an antiderivation in the graded algebra A. It satisfies the relation

$$\Omega(x y) = \Omega x \cdot y + (-1)^k x \cdot \Omega y , \qquad x \in A_k, y \in A .$$

If Ω_1 and Ω_2 are antiderivations of odd degree then $\Omega_1 \Omega_2 + \Omega_2 \Omega_1$ is a derivation. If Ω is an antiderivation of odd degree and θ is a derivation then $\Omega\theta - \theta\Omega$ is an antiderivation (cf. sec. 5.8).

Now assume that A is an associative anticommutative graded algebra and let $h \in A$ be a fixed element of odd (even) degree. Then, if Ω is a homogeneous antiderivation, $\mu(h)\Omega$ is a homogeneous derivation (antiderivation) and if θ is a homogeneous derivation, $\mu(h)\theta$ is a homogeneous antiderivation (derivation) as is easily checked.

Problems

1. Let A be a G-graded algebra and suppose x is an invertible element homogeneous of degree k (cf. problem 7, chap. V, § 1). Prove that x^{-1}

is homogeneous and calculate the degree. Conclude that if A is a positively graded algebra, then $k=0$.

2. Suppose that A is a graded algebra without zero divisors. Prove that every invertible element is homogeneous of degree zero.

3. Let E, F be G-graded vector spaces. Show that the vector space $L_G(E;F)$ generated by the homogeneous linear mappings $\varphi: E \to F$ is a subspace of $L(E;F)$. Define a natural G-gradation in this subspace such that an element $\varphi \in L_G(E;F)$ is homogeneous if and only if it is a homogeneous linear mapping.

4. Prove that the G-graded space $L_G(E;E)$ (E is a G-graded vector space) is a subalgebra of $A(E;E)$. Prove that the G-gradation makes $L_G(E;E)$ into a G-graded algebra (which is denoted by $A_G(E;E)$).

5. Let E be a positively graded vector space. Show that an injective (surjective) linear mapping $\varphi \in L_{\mathbb{Z}}(E;E)$ has degree $\leqq 0$ ($\geqq 0$). Conclude that a homogeneous linear automorphism of E has degree zero.

6. Let A be a positively graded algebra. Show that the subset A_k of A consisting of the linear combinations of homogeneous elements of degree $\geqq k$ is an ideal.

7. Let E, E^* be a pair of almost finite dual G-graded vector spaces. Construct an isomorphism of algebras:

$$\Phi: A_G(E;E) \xrightarrow{\cong} A_G(E^*;E^*)^{\text{opp}}.$$

Hint: See problem 12, chap. V, § 1.
Show that there is a natural G-gradation in $A_G(E^*;E^*)^{\text{opp}}$ such that Φ is homogeneous of degree zero.

8. Consider the G-graded space $L_G(E;E)$ (E is a G-graded vector space). Assign a new gradation to $L_G(E;E)$ by setting

$$\overleftarrow{\deg}\,\varphi = -\deg\varphi$$

whenever $\varphi \in L_G(E;E)$ is a homogeneous element. Show that with this new gradation $L_G(E;E)$ is again a G-graded space and $A_G(E;E)$ is a G-graded algebra. To avoid confusion, we denote these objects by $\tilde{L}_G(E;E)$ and $\tilde{A}_G(E;E)$.

Prove that the scalar product between $L_G(E;E)$ and $\tilde{L}_G(E;E)$ defined by

$$\langle \varphi, \psi \rangle = \operatorname{tr}(\varphi \circ \psi)$$

makes these spaces into dual G-graded vector spaces.

9. Let $A=\sum_p A_p (G=\mathbb{Z})$ be a graded algebra and consider the linear mapping $\theta: A \to A$ defined by

$$\theta x = p x \qquad x \in A_p.$$

Show that θ is a derivation.

§ 3.* Differential spaces and differential algebras

6.7. Differential spaces. A *differential operator* ∂ in a vector space E is a linear mapping $\partial: E \to E$ such that $\partial^2 = 0$. The vectors of $\ker \partial = Z(E)$ are called *cycles* and the vectors of $\operatorname{Im} \partial = B(E)$ are called *boundaries*. It follows from $\partial^2 = 0$ that $B(E) \subset Z(E)$. The factor space

$$H(E) = Z(E)/B(E)$$

is called the *homology space* of E with respect to the differential operator ∂. A vector space E together with a *fixed* differential operator ∂_E, is called a *differential space*.

A linear mapping of a differential space (E, ∂_E) into a differential space (F, ∂_F) is called a *homomorphism* (of differential spaces) if

$$\partial_F \circ \varphi = \varphi \circ \partial_E. \tag{6.15}$$

It follows from (6.15) that φ maps $Z(E)$ into $Z(F)$ and $B(E)$ into $B(F)$. Hence a linear mapping $\varphi_{\#}: H(E) \to H(F)$ is induced by φ. If φ is an isomorphism of differential spaces and φ^{-1} is the linear inverse isomorphism, then by applying φ^{-1} on the left and right of (6.15) we obtain

$$\varphi^{-1} \circ \partial_F = \partial_E \circ \varphi^{-1}$$

and so φ^{-1} is an isomorphism of differential spaces as well.

If ψ is a homomorphism of (F, ∂_F) into a third differential space (G, ∂_G) we have clearly

$$(\psi \circ \varphi)_{\#} = \psi_{\#} \circ \varphi_{\#}.$$

In particular, if φ is an isomorphism of E onto F and φ^{-1} is the inverse isomorphism we have

$$(\varphi^{-1})_{\#} \circ \varphi_{\#} = \iota_{\#} = \iota$$

and

$$\varphi_{\#} \circ (\varphi^{-1})_{\#} = \iota_{\#} = \iota.$$

Consequently, $\varphi_{\#}$ is an isomorphism of $H(E)$ onto $H(F)$.

6.8. The exact triangle. An *exact sequence of differential spaces* is an exact sequence

$$0 \to F \xrightarrow{\varphi} E \xrightarrow{\psi} G \to 0 \tag{6.16}$$

where (F, ∂_F), (E, ∂_E) and (G, ∂_G) are differential spaces, and φ, ψ are homomorphisms.

Suppose we are given an exact sequence of differential spaces then the sequence

$$H(F) \xrightarrow{\varphi_\#} H(E) \xrightarrow{\psi_\#} H(G)$$

is exact at $H(E)$. In fact, clearly, $\psi_\# \circ \varphi_\# = 0$ and so $\operatorname{Im} \varphi_\# \subset \ker \psi_\#$. Conversely, let $\beta \in \ker \psi_\#$ and choose an element $y \in Z(E)$ which represents β. Then, since $\psi_\# \beta = 0$,

$$\psi y = \partial_G z_1, \quad z_1 \in G.$$

Since ψ is surjective, there is an element $y_1 \in E$ such that $\psi y_1 = z_1$. It follows that

$$\psi(y - \partial_E y_1) = \psi y - \partial_G \psi y_1 = \psi y - \partial_G z_1 = 0.$$

Hence, by exactness at E,

$$y - \partial_E y_1 = \varphi x \qquad \text{for some } x \in F.$$

Applying ∂_E we obtain

$$\partial_E \varphi x = \partial_E y = 0$$

whence $\varphi(\partial_F x) = 0$.

Since φ is injective, this implies that $\partial_F x = 0$; i.e., $x \in Z(F)$. Thus x represents an element $\alpha \in H(F)$. It follows from the definitions that

$$\varphi_\# \alpha = \beta$$

and so $\beta \in \operatorname{Im} \varphi_\#$.

It should be observed that the induced homology sequence is *not* short exact. However, there is a linear map, $\chi\colon H(G) \to H(F)$ which makes the triangle

$$\begin{array}{ccc} H(F) & \xrightarrow{\varphi_\#} & H(E) \\ {\scriptstyle\chi}\nwarrow & & \swarrow{\scriptstyle\psi_\#} \\ & H(G) & \end{array} \tag{6.17}$$

exact. It is defined as follows: Let $\gamma \in H(G)$ and let $z \in Z(G)$ be a representative of γ. Choose $y \in E$ such that $\psi y = z$. Then

$$\psi(\partial_E y) = \partial_G \psi y = \partial_G z = 0$$

and so there is an element $x \in F$ such that

$$\varphi x = \partial_E y .$$

Since

$$\varphi(\partial_F x) = \partial_E \varphi x = \partial_E^2 y = 0$$

and since φ is injective, it follows that $\partial_F x = 0$ and so x represents an element α of $H(F)$. It is straightforward to check that α is independent of all choices and so a linear map $\chi: H(G) \to H(F)$ is defined by $\chi(\gamma) = \alpha$. It is not difficult to verify that this linear map makes the triangle (6.17) exact at $H(G)$ and $H(F)$. χ is called the *connecting homomorphism* for the short exact sequence (6.16).

6.9. Dual differential spaces. Suppose (E, ∂) is a differential space and consider the dual space $E^* = L(E)$. Let ∂^* be the dual map of ∂. Then for all $x \in E$, $x^* \in E^*$ we have

$$\langle \partial^* \partial^* x^*, x \rangle = \langle x^*, \partial \partial x \rangle = \langle x^*, 0 \rangle = 0$$

whence $\partial^* \partial^* x^* = 0$ i.e.,

$$(\partial^*)^2 = 0 .$$

Thus (E^*, ∂^*) is again a differential space. It is called the *dual differential space*.

The vectors of $\ker \partial^* = Z(E^*)$ are called *cocycles* (for E) and the vectors of $B(E^*)$ are called *coboundaries* (for E). The factor space

$$H(E^*) = Z(E^*)/B(E^*)$$

is called the *cohomology space* for E.

It will now be shown that the scalar product between E and E^* determines a scalar product between the homology and cohomology spaces. In view of sec. 2.26 and 2.28 we have the relations

$$Z(E^*) = B(E)^{\perp}, \quad Z(E) = B(E^*)^{\perp} \tag{6.18}$$

$$B(E^*) = Z(E)^{\perp}, \quad B(E) = Z(E^*)^{\perp}. \tag{6.19}$$

We can now construct a scalar product between $H(E)$ and $H(E^*)$. Consider the restriction of the scalar product between E and E^* to $Z(E) \times Z(E^*)$,

$$Z(E) \times Z(E^*) \to \Gamma .$$

Then since, in view of (6.18) and (6.19),

$$Z(E^*)^{\perp} \cap Z(E) = B(E) \cap Z(E) = B(E)$$

and

$$Z(E)^{\perp} \cap Z(E^*) = B(E^*) \cap Z(E^*) = B(E^*),$$

the equation

$$\langle \bar{x}^*, \bar{x} \rangle = \langle x^*, x \rangle \qquad \begin{matrix} x^* \in \bar{x}^* \\ x \in \bar{x} \end{matrix}$$

defines a scalar product between $H(E)$ and $H(E^*)$ (cf. sec. 2.23).

Finally, suppose (E, ∂_E), (E^*, ∂_{E^*}) and (F, ∂_F), (F^*, ∂_{F^*}) are two pairs of dual differential spaces. Let

$$\varphi: E \to F$$

be a homomorphism of differential spaces, with dual map $\varphi^*: E^* \leftarrow F^*$.

Then dualizing (6.15) we obtain

$$\varphi^* \circ \partial_F^* = \partial_E^* \circ \varphi^*$$

and so φ^* is a homomorphism of differential spaces. It is clear that the induced mappings

$$\varphi_{\#}: H(E) \to H(F)$$

and

$$(\varphi^*)_{\#}: H(E^*) \leftarrow H(F^*)$$

are again dual with respect to the induced scalar products; i.e.,

$$(\varphi^*)_{\#} = (\varphi_{\#})^*. \tag{6.20}$$

6.10. G-graded differential spaces. Let E be a G-graded space, $E = \sum_{p \in G} E_p$ and consider a differential operator ∂ in E that is homogeneous of some degree k. Then a gradation is induced in $Z(E)$ and $B(E)$ by

$$Z(E) = \sum_{p \in G} Z_p(E) \quad \text{and} \quad B(E) = \sum_{p \in G} B_p(E)$$

where $Z_p(E) = Z(E) \cap E_p$ and $B_p(E) = B(E) \cap E_p$ (cf. sec. 6.2).

Now consider the canonical projection

$$\pi: Z(E) \to H(E).$$

Since π is an onto map and the kernel of π is a graded subspace of $Z(E)$, a G-gradation is induced in the homology space $H(E)$ by

$$H(E) = \sum_{p \in G} H_p(E) \quad \text{where} \quad H_p(E) = \pi Z_p(E).$$

Now consider the subspaces $Z_p(E) \subset Z(E)$ and $B_p(E) \subset B(E)$. The factor space $Z_p(E)/B_p(E)$ is called the *p-th homology space* of the graded differential space E. It is canonically isomorphic to the space $H_p(E)$. In fact, if π_p denotes the restriction of π to the spaces $Z_p(E)$, $H_p(E)$, then

$$\pi_p: Z_p(E) \to H_p(E)$$

is an onto map and the kernel of π_p is given by

$$\ker \pi_p = Z_p(E) \cap \ker \pi = Z_p(E) \cap B(E) = B_p(E).$$

Hence, π_p induces a linear isomorphism of $Z_p(E)/B_p(E)$ onto $H_p(E)$.

If E is a graded space and $\dim H_p(E)$ is finite we write

$$\dim H_p(E) = b_p;$$

b_p is called the *p-th Betti number* of the graded differential space (E, ∂).

If E is an almost finite positively graded space, then clearly, so is $H(E)$. The Poincaré series for $H(E)$ is given by

$$P_{H(E)} = \sum_{p=0}^{\infty} b_p t^p. \tag{6.21}$$

6.11. Dual G-graded differential spaces. Suppose $(E = \sum_{k \in G} E_k, \partial_E)$ and $(E^* = \sum_{k \in G} E_k^*, \partial_E^*)$ is a pair of dual G-graded differential spaces. Then if ∂_E is homogeneous of degree l, we have that $\partial_E E_j \subset E_{j+l}$ and hence

$$\langle \partial_E^* y_i^*, x_j \rangle = \langle y_i^*, \partial_E x_j \rangle = 0 \qquad y_i^* \in E_i^*, x_j \in E_j$$

unless $y_i^* \in E_{j+l}^*$. It follows that

$$\partial_E^* y_i^* \in \bigcap_{j \neq i-l} E_j^{\perp} = E_{i-l}^*$$

and so ∂_E^* is homogeneous of degree $-l$.

Now consider the induced G-gradations in the homology spaces

$$H(E) = \sum_k H_k(E), \quad H(E^*) = \sum_k H_k(E^*).$$

The induced scalar product is given by

$$\langle \pi_{E^*} z^*, \pi_E z \rangle = \langle z^*, z \rangle \qquad \begin{array}{l} z^* \in Z(E^*) \\ z \in Z(E). \end{array} \tag{6.22}$$

Since

$$\pi_{E^*}: Z(E^*) \to H(E^*) \quad \text{and} \quad \pi_E: Z(E) \to H(E)$$

are homogeneous of degree zero, it follows that the scalar product (6.22) respects the gradations. Hence $H(E)$ and $H(E^*)$ are again dual G-graded vector spaces. In particular, if $G=\mathbb{Z}$, the p-th homology and cohomology spaces of E are dual. If $H_p(E)$ has finite dimension we obtain that

$$\dim H_p(E^*) = \dim H_p(E) = b_p .$$

6.12. Differential algebras. Suppose that A is an algebra and that ∂ is a differential operator in the vector space A. Assume further that an involution ω of the algebra A is given such that $\partial\omega+\omega\partial=0$, and that ∂ is an antiderivation with respect to ω; i.e., that

$$\partial(x\,y) = \partial x \cdot y + \omega x \cdot \partial y . \tag{6.23}$$

Then (A, ω, ∂) is called a *differential algebra.*

It follows from (6.23) that the subspace $Z(A)$ is a subalgebra of A. Further, the subspace $B(A)$ is an ideal in the algebra $Z(A)$. In fact, if and $\partial x \in B(A)$ we have

$$\partial(x\,y) = \partial x \cdot y \qquad y \in Z(A)$$

and

$$\partial(\omega y \cdot x) = \omega^2 y \cdot \partial x + \partial(\omega y) \cdot x = y \cdot \partial x \qquad y \in Z(A)$$

whence $\partial x \cdot y \in B(A)$ and $y \cdot \partial x \in B(A)$.

Hence, a multiplication is induced in the homology space $H(A)$. The space $H(A)$ together with this multiplication is called the *homology algebra* of the differential algebra (A, ∂).

The multiplication in $H(A)$ is given by

$$\pi z_1 \, \pi z_2 = \pi(z_1 z_2) \qquad z_1, z_2 \in Z(A)$$

where $\pi: Z(A) \to H(A)$ denotes the canonical projection. If A is associative (commutative) then so is $H(A)$.

Let (A, ∂_A) and (B, ∂_B) be differential algebras. Then a homomorphism $\varphi: A \to B$ is called a *homomorphism of differential algebras* if

$$\varphi\,\omega_A = \omega_B\,\varphi \quad \text{and} \quad \varphi\,\partial_A = \partial_B\,\varphi .$$

It follows easily that the induced mapping $\varphi_\# : H(A) \to H(B)$ is a homomorphism of homology algebras.

Suppose now A is a graded algebra and that ∂ and ω are both homogeneous, ω of degree zero. The A is called a *graded differential algebra.* Consider the induced gradation in $H(A)$. Since the canonical projection $\pi: Z(A) \to H(A)$ is a homogeneous map of degree zero it follows that $H(A)$ is a graded algebra.

If A is an anticommutative graded algebra, then so is $H(A)$ as follows from the fact that π is a homogeneous epimorphism of degree zero.

Problems

1. Let (E, ∂_1), (F, ∂_2) be two differential spaces and define the differential operator ∂ in $E \oplus F$ by

$$\partial = \partial_1 \oplus \partial_2 .$$

Prove that

$$H(E \oplus F) \cong H(E) \oplus H(F).$$

2. Given a differential space (E, ∂), consider a differential subspace; i.e., a subspace E_1 that is stable under ∂. Assume that $\varrho: E \to E_1$ is a linear mapping such that

i) $\varrho \partial = \partial \varrho$
ii) $\varrho y = y \qquad y \in E_1$
iii) $\varrho x - x \in B(E) \qquad x \in Z(E)$.

Prove that the induced mapping

$$\varrho_{\#}: H(E) \to H(E_1)$$

is a linear isomorphism.

3. Let (E, ∂) be a differential space. A *homotopy operator* in E is a linear transformation $h: E \to E$ such that

$$h\partial + \partial h = \iota .$$

Show that a homotopy operator exists in E if and only if $H(E) = 0$.

4. Let (E, ∂_E) and (F, ∂_F) be two differential spaces and let φ, ψ be homomorphisms of differential spaces. Prove that $\varphi_{\#} = \psi_{\#}$ if and only if there exists a linear mapping $h: E \to F$ such that

$$h\partial_E + \partial_F h = \varphi - \psi ;$$

h is called a *homotopy operator connecting* φ and ψ. Show that problem 3 is a special case of problem 4.

5. Let ∂_1, ∂_2 be differential operators in E which commute, $\partial_1 \partial_2 = \partial_2 \partial_1$.

a) Prove that $\partial_1 \partial_2$ is a differential operator in E.

b) Let B_1, B_2, B be the boundaries with respect to ∂_1, ∂_2 and $\partial_1 \partial_2$. Prove that

$$\partial_2(B_1) = \partial_1(B_2) = B .$$

c) Let Z_1, Z_2, Z be the cycles with respect to ∂_1, ∂_2 and $\partial_1\partial_2$. Show that $Z_1+Z_2 \subset Z$. Establish natural linear isomorphisms

$$Z/Z_1 \xrightarrow{\cong} B_1 \cap Z_2 \quad \text{and} \quad Z/Z_2 \xrightarrow{\cong} B_2 \cap Z_1.$$

d) Establish a natural linear isomorphism

$$(B_1 \cap Z_2)/\partial_1(Z_2) \xrightarrow{\cong} (B_2 \cap Z_1)/\partial_2(Z_1)$$

and then show that each of these spaces is linearly isomorphic to $Z/(Z_1 + Z_2)$.

e) Show that ∂_1 induces a differential operator in Z_2. Let $\tilde{H}_1$ denote the corresponding homology space. Assume now that $Z=Z_1+Z_2$ and prove that $\tilde{H}_1$ can be identified with a subspace of the homology space $H_1=Z_1/B_1$. State and prove a similar result for ∂_2.

f) Show that the results a) to e) remain true if $\partial_1\partial_2 = -\partial_2\partial_1$.

6. Let ∂_1, ∂_2 be differential operators in E such that $\partial_1\partial_2 = -\partial_2\partial_1$.

a) Prove that $\partial_1+\partial_2$ and $\partial_1-\partial_2$ are differential operators in E.

b) With the notation of problem 5, assume that

$$B = B_1 \cap B_2 \quad \text{and} \quad Z = Z_1 + Z_2.$$

Prove that the homology space of each of the differential operators in a) is linearly isomorphic to

$$Z_1 \cap Z_2/(Z_1 \cap B_2 + Z_2 \cap B_1).$$

(This is essentially the Künneth theorem of sec. 2.10 volume II.)

7. *Lefschetz formula.* Let $E=\sum_{i=0}^{n} E_i$ be a finite dimensional graded differential space and assume that ∂ is homogeneous of degree -1 ($\partial=0$ in E_0). Let $\varphi\colon E\to E$ be a homomorphism of differential spaces, homogeneous of degree zero. Denote the restrictions of φ (respectively $\varphi_\#$) to E_p (respectively $H_p(E)$) by φ_p (respectively $(\varphi_\#)_p$). Prove the *Lefschetz formula*

$$\sum_{p=0}^{n} (-1)^p \operatorname{tr}(\varphi_\#)_p = \sum_{p=0}^{n} (-1)^p \operatorname{tr}\varphi_p.$$

Conclude that

$$\sum_{p=0}^{n} (-1)^p \dim H_p(E) = \sum_{p=0}^{n} (-1)^p \dim E_p \qquad \text{(EP)}$$

(Euler-Poincaré formula). Express this formula in terms of P_E and $P_{H(E)}$. The number given in (EP), is called the *Euler-Poincaré characteristic* of E.

Chapter VII

Inner product spaces

In this chapter all vector spaces are assumed to be real vector spaces

§ 1. The inner product

7.1. Definition. An *inner product* in a real vector space E is a bilinear function $(\,,)$ having the following properties:

1. Symmetry: $(x, y)=(y, x)$.
2. Positive definiteness: $(x, x)\geqq 0$, and $(x, x)=0$ only for the vector $x=0$.

A vector space in which an inner product is defined is called an *inner product space*. An inner product space of finite dimension is also called a *Euclidean space*.

The *norm* $|x|$ of a vector $x\in E$ is defined as the positive square-root

$$|x| = \sqrt{(x, x)}.$$

A *unit vector* is a vector with the norm 1. The set of all unit vectors is called the *unit-sphere*.

It follows from the bilinearity and symmetry of the inner product that

$$|x + y|^2 = |x|^2 + 2(x, y) + |y|^2$$

whence

$$(x, y) = \tfrac{1}{2}(|x + y|^2 - |x|^2 - |y|^2).$$

This equation shows that the inner product can be expressed in terms of the norm.

The restriction of the bilinear function $(\,,)$ to a subspace $E_1\subset E$ has again properties 1 and 2 and hence every subspace of an inner product space is itself an inner product space.

The bilinear function $(\,,)$ is non-degenerate. In fact, assume that $(a, y)=0$ for a fixed vector $a\in E$ and every vector $y\in E$. Setting $y=a$ we obtain $(a, a)=0$ whence $a=0$. It follows that an inner product space is dual to itself.

7.2. Examples. 1. In the real number-space $\mathbb{R}^n$ the *standard inner product is defined by*

$$(x, y) = \sum_\nu \xi^\nu \eta^\nu,$$

where

$$x = (\xi^1 \ldots \xi^n) \quad \text{and} \quad y = (\eta^1 \ldots \eta^n).$$

2. Let E be an n-dimensional real vector space and $x_\nu (\nu = 1 \ldots n)$ be a basis of E. Then an inner product can be defined by

$$(x, y) = \sum_\nu \xi^\nu \eta^\nu,$$

where

$$x = \sum_\nu \xi^\nu x_\nu, \quad y = \sum_\nu \eta^\nu x_\nu.$$

3. Consider the space C of all continuous functions f in the interval $0 \leqq t \leqq 1$ and define the inner product by

$$(f, g) = \int_0^1 f(t) g(t)\, dt.$$

7.3. Orthogonality. Two vectors $x \in E$ and $y \in E$ are said to be *orthogonal* if $(x, y) = 0$. The definiteness implies that only the zero-vector is orthogonal to itself. A system of p vectors $x_\nu \neq 0$ in which any two vectors x_ν and $x_\mu (\nu \neq \mu)$ are orthogonal, is linearly independent. In fact, the relation

$$\sum_\nu \lambda^\nu x_\nu = 0$$

yields

$$\lambda^\mu (x_\mu, x_\mu) = 0 \qquad (\mu = 1 \ldots p)$$

whence

$$\lambda^\mu = 0 \qquad (\mu = 1 \ldots p).$$

Two subspaces $E_1 \subset E$ and $E_2 \subset E$ are called *orthogonal*, denoted as $E_1 \perp E_2$, if any two vectors $x_1 \in E_1$ and $x_2 \in E_2$ are orthogonal.

7.4. The Schwarz-inequality. Let x and y be two arbitrary vectors of the inner product space E. Then the *Schwarz-inequality* asserts that

$$(x, y)^2 \leqq |x|^2 |y|^2 \tag{7.1}$$

and that equality holds if and only if the vectors are linearly dependent. To prove this consider the function

$$|x + \lambda y|^2$$

of the real variable λ. The definiteness of the inner product implies that

$$|x+\lambda y|^2 \geqq 0 \quad (-\infty<\lambda<\infty).$$

Expanding the norm we obtain

$$\lambda^2 |y|^2 + 2\lambda (x, y) + |x|^2 \geqq 0.$$

Hence the discriminant of the above quadratic expression must be negative or zero,*)

$$(x, y)^2 \leqq |x|^2 |y|^2.$$

Now assume that equality holds in (7.1). Then the discriminant of the quadratic equation

$$\lambda^2 |y|^2 + 2\lambda (x, y) + |x|^2 = 0 \tag{7.2}$$

is zero. Hence equation (7.2) has a real solution λ_0. It follows that

$$|\lambda_0 y + x|^2 = 0,$$

whence

$$\lambda_0 y + x = 0.$$

Thus, the vectors x and y are linearly dependent.

7.5. Angles. Given two vectors $x \neq 0$ and $y \neq 0$, the Schwarz-inequality implies that

$$-1 \leqq \frac{(x, y)}{|x|\,|y|} \leqq 1.$$

Consequently, there exists exactly one real number $\omega\,(0 \leqq \omega \leqq \pi)$ such that

$$\cos \omega = \frac{(x, y)}{|x|\,|y|}. \tag{7.3}$$

The number ω is called the *angle* between the vectors x and y. The symmetry of the inner product implies that the angle is symmetric with respect to x and y. If the vectors x and y are orthogonal, it follows that $\cos \omega = 0$, whence $\omega = \frac{\pi}{2}$

Now assume that the vectors x and y are linearly dependent, $y = \lambda x$, Then

$$\cos \omega = \frac{\lambda}{|\lambda|} = \begin{cases} +1 & \text{if } \lambda > 0 \\ -1 & \text{if } \lambda < 0 \end{cases}$$

*) Without loss of generality we may assume that $y \neq 0$.

and hence

$$\omega = \begin{cases} 0 & \text{if } \lambda > 0 \\ \pi & \text{if } \lambda < 0. \end{cases}$$

With the help of (7.3) the equation

$$|x - y|^2 = |x|^2 - 2(x, y) + |y|^2$$

can be written in the form

$$|x - y|^2 = |x|^2 + |y|^2 - 2|x|\,|y| \cos\omega .$$

This formula is known as the *cosine-theorem*. If the vectors x and y are orthogonal, the cosine-theorem reduces to the *Pythagorean theorem*

$$|x - y|^2 = |x|^2 + |y|^2 .$$

7.6. The triangle-inequality. It follows from the Schwarz-inequality that

$$|x + y|^2 = |x|^2 + 2(x, y) + |y|^2 \leqq |x|^2 + 2|x|\,|y| + |y|^2 = (|x| + |y|)^2 ,$$

whence

$$|x + y| \leqq |x| + |y| . \tag{7.4}$$

Relation (7.4) is called the *triangle-inequality*. To discuss the equality-sign we may exclude the trivial case $y=0$. It will be shown that equality holds in (7.4) if and only if

$$x = \lambda y, \quad \lambda > 0 .$$

The equation

$$|x + y| = |x| + |y|$$

implies that

$$|x|^2 + 2(x, y) + |y|^2 = |x|^2 + 2|x|\,|y| + |y|^2 ,$$

whence

$$(x, y) = |x|\,|y| . \tag{7.5}$$

Thus, the vectors x and y must be linearly dependent,

$$x = \lambda y . \tag{7.6}$$

Equations (7.5) and (7.6) yield $\lambda = |\lambda|$, whence $\lambda \geqq 0$.

Conversely, assume that $x = \lambda y$, where $\lambda \geqq 0$. Then

$$|x + y| = |(\lambda + 1)y| = (\lambda + 1)|y| = \lambda|y| + |y| = |x| + |y| .$$

Given three vectors x, y, z, the triangle-inequality can be written in the form

$$|x - y| \leqq |x - z| + |z - y|. \tag{7.7}$$

As a generalization of (7.7), we prove the *Ptolemy-inequality*

$$|x - y|\,|z| \leqq |y - z|\,|x| + |z - x|\,|y|. \tag{7.8}$$

Relation (7.8) is trivial if one of the three vectors is zero. Hence we may assume that $x \neq 0$, $y \neq 0$ and $z \neq 0$. Define the vectors x', y' and z' by

$$x' = \frac{x}{|x|^2}, \quad y' = \frac{y}{|y|^2}, \quad z' = \frac{z}{|z|^2}.$$

Then

$$|x' - y'|^2 = \frac{1}{|x|^2} - \frac{2(x, y)}{|x|^2\,|y|^2} + \frac{1}{|y|^2} = \frac{|x - y|^2}{|x|^2\,|y|^2},$$

Applying the inequality (7.7) to the vectors x', y' and z' we obtain

$$\frac{|x - y|}{|x|\,|y|} \leqq \frac{|y - z|}{|y|\,|z|} + \frac{|z - x|}{|z|\,|x|},$$

whence (7.8).

7.7. The Riesz theorem. Let E be an inner product space of dimension n and consider the space $L(E)$ of linear functions. Then the spaces $L(E)$ and E are dual with respect to the bilinear function defined by

$$(f, x) \to f(x).$$

On the other hand, E is dual to itself with respect to the inner product. Hence Corollary II to Proposition I sec. 2.33 implies that there is a linear isomorphism $a \to f_a$ of E onto $L(E)$ such that

$$f_a(y) = (a, y).$$

In other words, every linear function f in E can be written in the form

$$f(y) = (a, y)$$

and the vector $a \in E$ is uniquely determined by f (Riesz theorem).

Problems

1. For $x = (\xi^1, \xi^2)$ and $y = (\eta^1, \eta^2)$ in $\mathbb{R}^2$ show that the bilinear function

$$(x, y) = \xi^1 \eta^1 - \xi^2 \eta^1 - \xi^1 \eta^2 + 4\xi^2 \eta^2$$

satisfies the properties listed in sec. 7.1.

2. Consider the space S of all infinite sequences $x=(\xi_1, \xi_2, \ldots)$ such that

$$\sum_\nu \xi_\nu^2 < \infty .$$

Show that $\sum_\nu \xi_\nu \eta_\nu$ converges and that the bilinear function $(x, y)=\sum_\nu \xi_\nu \eta_\nu$ is an inner product.

3. Consider three distinct vectors $x \neq 0$, $y \neq 0$ and $z \neq 0$. Prove that the equation

$$|x-y|\,|z| = |y-z|\,|x| + |z-x|\,|y|$$

holds if and only if the four points $x, y, z, 0$ are contained on a circle such that the pairs x, y and $z, 0$ separate each other.

4. Consider two inner product spaces E_1 and E_2. Prove that an inner product is defined in the direct sum $E_1 \oplus E_2$ by

$$((x_1, x_2), (y_1, y_2)) = (x_1, y_1) + (x_2, y_2) \qquad x_1, y_1 \in E_1, \quad x_2, y_2 \in E_2 .$$

5. Given a subspace E_1 of a finite dimensional inner product space E, consider the factor space E/E_1. Prove that every equivalence class contains exactly one vector which is orthogonal to E_1.

§ 2. Orthonormal bases

7.8. Definition. Let E be an n-dimensional inner product space and $x_\nu\ (\nu=1 \ldots n)$ be a basis of E. Then the bilinear function $(,)$ determines a symmetric matrix

$$g_{\nu\mu} = (x_\nu, x_\mu) \qquad (\nu, \mu = 1 \ldots n). \tag{7.9}$$

The inner product of two vectors

$$x = \sum_\nu \xi^\nu x_\nu \quad \text{and} \quad y = \sum_\nu \eta^\nu x_\nu$$

can be written as

$$(x, y) = \sum_{\nu,\mu} \xi^\nu \eta^\mu (x_\nu, x_\mu) = \sum_{\nu,\mu} g_{\nu\mu} \xi^\nu \eta^\mu \tag{7.10}$$

and hence it appears as a bilinear form with the coefficient-matrix $g_{\nu\mu}$.

The basis $x_\nu\ (\nu=1 \ldots n)$ is called *orthonormal*, if the vectors $x_\nu\ (\nu=1 \ldots n)$ are mutually orthogonal and have the norm 1,

$$(x_\nu, x_\mu) = \delta_{\nu\mu} . \tag{7.11}$$

Then formula (7.10) reduces to

$$(x, y) = \sum_{\nu} \xi^{\nu} \eta^{\nu} \tag{7.12}$$

and in the case $y = x$

$$|x|^2 = \sum_{\nu} \xi^{\nu} \xi^{\nu}.$$

The substitution $y = x_\mu$ in (7.12) yields

$$(x, x_\mu) = \xi^{\mu} \qquad (\mu = 1 \dots n). \tag{7.13}$$

Now assume that $x \neq 0$, and denote by θ_μ the angle between the vectors x and $x_\mu (\mu = 1 \dots n)$. Formulas (7.3) and (7.13) imply that

$$\cos \theta_\mu = \frac{\xi^{\mu}}{|x|} \qquad (\mu = 1 \dots n). \tag{7.14}$$

If x is a unit-vector (7.14) reduces to

$$\cos \theta_\mu = \xi^{\mu} \qquad (\mu = 1 \dots n). \tag{7.15}$$

These equations show that the components of a unit-vector x relative to an orthonormal basis are equal to the cosines of the angles between x and the basisvectors x_μ.

7.9. The Schmidt-orthogonalization. In this section it will be shown that an orthonormal basis can be constructed in every inner product space of finite dimension. Let $a_\nu (\nu = 1 \dots n)$ be an arbitrary basis of E. Starting out from this basis a new basis $b_\nu (\nu = 1 \dots n)$ will be constructed whose vectors are mutually orthogonal. Let

$$b_1 = a_1 .$$

Then put

$$b_2 = a_2 + \lambda b_1$$

and determine the scalar λ such that $(b_1, b_2) = 0$. This yields

$$(a_2, b_1) + \lambda (b_1, b_1) = 0 .$$

Since $b_1 \neq 0$, this equation can be solved with respect to λ. The vector b_2 thus obtained is different from zero because otherwise a_1 and a_2 would be linearly dependent.

To obtain b_3, set

$$b_3 = a_3 + \mu b_1 + \nu b_2$$

and determine the scalars μ and ν such that

$$(b_1, b_3) = 0 \quad \text{and} \quad (b_2, b_3) = 0 .$$

This yields

$$(a_3, b_1) + \mu(b_1, b_1) = 0$$

and

$$(a_3, b_2) + \nu(b_2, b_2) = 0.$$

Since $b_1 \neq 0$ and $b_2 \neq 0$, these equations can be solved with respect to μ and ν. The linear independence of the vectors a_1, a_2, a_3 implies that $b_3 \neq 0$. Continuing this way we finally obtain a system of n vectors $b_\nu \neq 0\, (\nu = 1 \dots n)$ such that

$$(b_\nu, b_\mu) = 0 \qquad (\nu \neq \mu).$$

It follows from the criterion in sec. 7.3, that the vectors b_ν are linearly independent and hence they form a basis of E. Consequently the vectors

$$e_\nu = \frac{b_\nu}{|b_\nu|} \qquad (\nu = 1 \dots n)$$

form an orthonormal basis.

7.10. Orthogonal transformations. Consider two orthogonal bases x_ν and $\bar{x}_\nu\, (\nu = 1 \dots n)$ of E. Denote by α_ν^μ the matrix of the basis-transformation $x_\nu \to \bar{x}_\nu$,

$$\bar{x}_\nu = \sum_\mu \alpha_\nu^\mu x_\mu. \tag{7.16}$$

The relations

$$(x_\nu, x_\mu) = \delta_{\nu\mu} \quad \text{and} \quad (\bar{x}_\nu, \bar{x}_\mu) = \delta_{\nu\mu}$$

imply that

$$\sum_\lambda \alpha_\nu^\lambda \alpha_\mu^\lambda = \delta_{\nu\mu}. \tag{7.17}$$

This equation shows that the product of the matrix (α_ν^μ) and the transposed matrix is equal to the unit-matrix. In other words, the transposed matrix coincides with the inverse matrix. A matrix of this kind is called *orthogonal.*

Hence, two orthonormal bases are related by an orthogonal matrix. Conversely, given an orthonormal basis $x_\nu\, (\nu = 1 \dots n)$ and an orthogonal $n \times n$-matrix (α_ν^μ), the basis $\bar{x}_\nu$ defined by (7.16) is again orthonormal.

7.11. Orthogonal complement. Let E be an inner product space (of finite or infinite dimension) and E_1 be a subspace of E. Denote by $E_1^\perp$ the set of all vectors which are orthogonal to E_1. Obviously, $E_1^\perp$ is again a subspace of E and the intersection $E_1 \cap E_1^\perp$ consists of the zero-vector only. $E_1^\perp$ is called the *orthogonal complement* of E_1. If E has finite dimension, then we have that

$$\dim E_1 + \dim E_1^\perp = \dim E$$

and hence $E_1 \cap E_1^\perp = 0$ implies that

$$E = E_1 \oplus E_1^\perp. \tag{7.18}$$

Select an orthonormal basis $y_\mu\,(\mu = 1 \dots m)$ of E_1. Given a vector $x \in E$ and a vector

$$y = \sum_\mu \eta^\mu y_\mu$$

of E_1 consider the difference

$$z = x - y.$$

Then

$$(z, y_\mu) = (x, y_\mu) - (y, y_\mu) = (x, y_\mu) - \eta^\mu.$$

This equation shows that z is contained in $E_1^\perp$ if and only if

$$\eta^\mu = (x, y_\mu) \qquad (\mu = 1 \dots m).$$

We thus obtain the decomposition

$$x = p + h \tag{7.19}$$

where

$$p = \sum_\mu (x, y_\mu)\, y_\mu \quad \text{and} \quad h = x - p.$$

The vector p is called the *orthogonal projection* of x onto E_1.

Passing over to the norm in the decomposition (7.19) we obtain the relation

$$|x|^2 = |p|^2 + |h|^2. \tag{7.20}$$

Formula (7.20) yields *Bessel's-inequality*

$$|x| \geqq |p|$$

showing that the norm of the projection never exceeds the norm of x. The equality holds if and only if $h = 0$, i.e. if and only if $x \in E_1$. The number $|h|$ is called the *distance* of x from the subspace E_1.

Problems

1. Starting from the basis

$$a_1 = (1, 0, 1) \quad a_2 = (2, 1, -3) \quad a_3 = (-1, 1, 0)$$

of the number-space $\mathbb{R}^3$ construct an orthonormal basis by the Schmidt-orthogonalization process.

2. Let E be an inner product space and consider E as dual to itself. Prove that the orthonormal bases are precisely the bases which are dual to themselves.

3. Given an inner product space E and a subspace E_1 of finite dimension consider a decomposition

$$x = x_1 + x_2 \qquad x_1 \in E_1$$

and the projection

$$x = p + h \qquad p \in E_1, h \in E_1^{\perp}.$$

Prove that

$$|x_2| \geqq |h|$$

and that equality is assumed only if $x_1 = p$ and $x_2 = h$.

4. Let C be the space of all continuous functions in the interval $0 \leqq t \leqq 1$ with the inner product defined as in sec. 7.2. If C^1 denotes the subspace of all continuously differentiable functions, show that $(C^1)^{\perp} = 0$.

5. Consider a subspace E_1 of E. Assume an orthogonal decomposition

$$E_1 = F_1 \oplus G_1 \quad F_1 \perp G_1 .$$

Establish the relations

$$F_1^{\perp} = E_1^{\perp} \oplus G_1, E_1^{\perp} \perp G_1 \quad \text{and} \quad G_1^{\perp} = E_1^{\perp} \oplus F_1, E_1^{\perp} \perp F_1 .$$

6. Let F^3 be the space of all polynomials of degree $\leqq 2$. Define the inner product of two polynomials as follows:

$$(P, Q) = \int_{-1}^{1} P(t)\, Q(t)\, dt .$$

The vectors $1, t, t^2$ form a basis in F^3. Orthogonalize and orthonormalize this basis. Generalize the result for the case of the space F^n of polynomials of degree $\leqq n-1$.

§ 3. Normed determinant functions

7.12. Definition. Let E be an n-dimensional inner product space and $\Delta_0 \neq 0$ be a determinant function in E. Since E is dual to itself we have in view of (4.21)

$$\Delta_0(x_1, \ldots x_n)\, \Delta_0(y_1 \ldots y_n) = \alpha \det(x_i, y_j) \qquad x_i \in E, y_i \in E$$

where α is a real constant. Setting $x_i = y_i = e_i$ where e_i is an orthonormal basis we obtain

$$\alpha = \Delta_0(e_1 \dots e_n)^2 \tag{7.21}$$

and so the constant α is positive. Now define a determinant function Δ by

$$\Delta = \pm \frac{\Delta_0}{\sqrt{\alpha}}. \tag{7.22}$$

Then we have

$$\Delta(x_1 \dots x_n)\Delta(y_1 \dots y_n) = \det(x_i, y_j). \tag{7.23}$$

A determinant function in an inner product space which satisfies (7.23) is called a *normed determinant function*. It follows from (7.22) that there are precisely two normed determinant functions Δ and $-\Delta$ in E.

Now assume that an orientation is defined in E. Then one of the functions Δ and $-\Delta$ represents the orientation. Consequently, *in an oriented inner product space there exists exactly one normed determinant function representing the given orientation.*

7.13. Angles in an oriented plane. With the help of a normed determinant-function it is possible to attach a sign to the angle between two vectors of a 2-dimensional oriented inner product space. Consider the normed determinant function Δ which represents the given orientation. Then the identity (7.23) yields

$$|x|^2 |y|^2 - (x, y)^2 = \Delta(x, y)^2. \tag{7.24}$$

Now assume that $x \neq 0$ and $y \neq 0$. Dividing (7.24) by $|x|^2 |y|^2$ we obtain the relation

$$\frac{(x, y)^2}{|x|^2 |y|^2} + \frac{\Delta(x, y)^2}{|x|^2 |y|^2} = 1.$$

Consequently, there exists exactly one real number θ mod 2π such that

$$\cos\theta = \frac{(x, y)}{|x|\,|y|} \quad \text{and} \quad \sin\theta = \frac{\Delta(x, y)}{|x|\,|y|}. \tag{7.25}$$

This number is called the *oriented angle* between x and y.

If the orientation is changed, Δ has to be replaced by $-\Delta$, and hence θ changes into $-\theta$.

Furthermore it follows from (7.25) that θ changes sign if the vectors x and y are interchanged and that

$$\theta(x, -y) = \theta(x, y) + \pi \mod 2\pi.$$

7.14. The Gram determinant. Given p vectors $x_\nu(\nu=1\dots p)$ in an inner product space E, the *Gram determinant* $G(x_1\dots x_p)$ is defined by

$$G(x_1\dots x_p)=\det\begin{pmatrix}(x_1,x_1)\dots(x_1,x_p)\\ \vdots \qquad\qquad \vdots\\ (x_p,x_1)\dots(x_p,x_p)\end{pmatrix}. \tag{7.26}$$

It will be shown that

$$G(x_1\dots x_p)\geqq 0 \tag{7.27}$$

and that equality holds if and only if the vectors $(x_1\dots x_p)$ are linearly dependent. In the case $p=2$ (7.27) reduces to the Schwarz-inequality.

To prove (7.27), assume first that the vectors $x_\nu(\nu=1\dots p)$ are linearly dependent. Then the rows of the matrix (7.26) are also linearly dependent whence

$$G(x_1\dots x_p)=0.$$

If the vectors $x_\nu(\nu=1\dots p)$ are linearly independent, they generate a p-dimensional subspace E_1 of E. E_1 is again an inner product space. Denote by Δ_1 a normed determinant function in E_1. Then it follows from (7.23) that

$$G(x_1\dots x_p)=\Delta_1(x_1\dots x_p)^2.$$

The linear independence of the vectors $x_\nu(\nu=1\dots p)$ implies that $\Delta_1(x_1\dots x_p)\neq 0$, whence

$$G(x_1\dots x_p)>0.$$

7.15. The volume of a parallelepiped. Let p linearly independent vectors $a_\nu(\nu=1\dots p)$ be given in E. The set

$$x=\sum_\nu \lambda^\nu a_\nu \qquad 0\leqq\lambda^\nu\leqq 1 \quad (\nu=1\dots p) \tag{7.28}$$

is called the *p-dimensional parallelepiped* spanned by the vectors a_ν $(\nu=1\dots p)$. The volume $V(a_1\dots a_p)$ of the parallelepiped is defined by

$$V(a_1\dots a_p)=|\Delta_1(a_1\dots a_p)|, \tag{7.29}$$

where Δ_1 is a normed determinant function in the subspace generated by the vectors $a_\nu(\nu=1\dots p)$.

In view of the identity (7.23) formula (7.29) can be written as

$$V(a_1\dots a_p)^2=\det\begin{pmatrix}(a_1,a_1)\dots(a_1,a_p)\\ \vdots \qquad\qquad \vdots\\ (a_p,a_1)\dots(a_p,a_p)\end{pmatrix}. \tag{7.30}$$

In the case $p=2$ the above formula yields

$$V(a_1,a_2)^2 = |a_1|^2\,|a_2|^2 - (a_1,a_2)^2 = |a_1|^2\,|a_2|^2 \sin^2\theta\,, \tag{7.31}$$

where θ denotes the angle between a_1 and a_2. Taking the square-root on both sides of (7.31), we obtain the well-known formula for the area of a parallelogram:

$$V(a_1,a_2) = |a_1|\,|a_2|\,|\sin\theta|\,.$$

Going back to the general case, select an integer i ($1 \leqq i \leqq p$) and decompose a_i in the form

$$a_i = \sum_{\nu \neq i} \xi^\nu a_\nu + h_i\,, \quad \text{where} \quad (h_i, a_\nu) = 0 \quad (\nu \neq i)\,. \tag{7.32}$$

Then (7.29) can be written as

$$V(a_1 \dots a_p) = |\Delta_1(a_1 \dots a_{i-1}, h_i, a_{i+1} \dots a_p)|\,.$$

Employing the identity (7.23) and observing that $(h_i, a_\nu) = 0\,(\nu \neq i)$ we obtain*)

$$V(a_1 \dots a_p)^2 = \det\begin{pmatrix} (a_1,a_1)\dots(\hat{a}_1,\hat{a}_i)\dots(a_1,a_p) \\ \vdots \qquad\qquad\qquad \vdots \\ (\hat{a}_i,\hat{a}_1)\;\dots(\hat{a}_i,\hat{a}_i)\;\dots(\hat{a}_i,\hat{a}_p) \\ \vdots \qquad\qquad\qquad \vdots \\ (a_p,a_1)\dots(\hat{a}_p,\hat{a}_i)\dots(a_p,a_p) \end{pmatrix}(h_i,h_i)\,. \tag{7.33}$$

The determinant in this equation represents the square of the volume of the $(p-1)$-dimensional parallelepiped generated by the vectors $(a_1 \dots \hat{a}_i \dots a_p)$. We thus obtain the formula

$$V(a_1 \dots a_p) = V(a_1 \dots \hat{a}_i \dots a_p)\cdot|h_i| \qquad (1 \leqq i \leqq p)$$

showing that the volume $V(a_1 \dots a_p)$ is the product of the volume $V(a_1 \dots \hat{a}_i \dots a_p)$ of the ith "base" and the corresponding height.

7.16. The cross product. Let E be an oriented 3-dimensional Euclidean space and Δ be the normed determinant function which represents the orientation. Given two vectors $x \in E$ and $y \in E$ consider the linear function f defined by

$$f(z) = \Delta(x, y, z)\,. \tag{7.34}$$

In view of the Riesz-theorem there exists precisely one vector $u \in E$ such that

$$f(z) = (u, z)\,. \tag{7.35}$$

*) The symbol $\hat{a}_i$ indicates that the vector a_i is deleted.

The vector u is called the *cross product* of x and y and is denoted by $x \times y$. Relations (7.34) and (7.35) yield

$$(x \times y, z) = \Delta(x, y, z). \tag{7.36}$$

It follows from the linearity of Δ in x and y that the cross product is distributive

$$\begin{aligned}(\lambda x_1 + \mu x_2) \times y &= \lambda x_1 \times y + \mu x_2 \times y \\ x \times (\lambda y_1 + \mu y_2) &= \lambda x \times y_1 + \mu x \times y_2\end{aligned}$$

and hence it defines an algebra in E. The reader should observe that the cross product depends on the orientation of E. If the orientation is reversed then the cross product changes its sign.

From the skew symmetry of Δ we obtain that

$$x \times y = - y \times x.$$

Setting $z = x$ in (7.36) we obtain that

$$(x \times y, x) = 0.$$

Similarly it follows that

$$(x \times y, y) = 0$$

and so the cross product is orthogonal to both factors.

It will now be shown that $x \times y \neq 0$ if and only if x and y are linearly independent. In fact, if $y = \lambda x$, it follows immediately from the skew symmetry that $x \times y = 0$. Conversely, assume that the vectors x and y are linearly independent. Then choose a vector $z \in E$ such that the vectors x, y, z form a basis of E. It follows from (7.36) that

$$(x \times y, z) = \Delta(x, y, z) \neq 0$$

whence $x \times y \neq 0$.

Formula (7.36) yields for $z = x \times y$

$$\Delta(x, y, x \times y) = |x \times y|^2. \tag{7.37}$$

If x and y are linearly independent it follows from (7.37) that $\Delta(x, y, x \times y) > 0$ and so the basis x, y, $x \times y$ is positive with respect to the given orientation.

Finally the identity

$$(x_1 \times x_2, y_1 \times y_2) = (x_1, y_1)(x_2, y_2) - (x_1, y_2)(x_2, y_1) \tag{7.38}$$

will be proved. We may assume that the vectors x_1, x_2 are linearly independent because otherwise both sides of (7.38) are zero. Multiplying the relations

$$\Delta(x_1, x_2, x_3) = (x_1 \times x_2, x_3)$$

and

$$\Delta(y_1, y_2, y_3) = (y_1 \times y_2, y_3)$$

we obtain in view of (7.23)

$$(x_1 \times x_2, x_3)(y_1 \times y_2, y_3) = \det(x_i, y_j).$$

Setting $y_3 = x_1 \times x_2$ and expanding the determinant on the right hand side by the last row we obtain that

$$\begin{aligned}(x_1 \times x_2, x_3)(y_1 \times y_2, x_1 \times x_2) = \\ (x_1 \times x_2, x_3)[(x_1, y_1)(x_2, y_2) - (x_1, y_2)(x_2, y_1)].\end{aligned} \tag{7.39}$$

Since x_1 and x_2 are linearly independent we have that $x_1 \times x_2 \neq 0$ and hence x_3 can be chosen such that $(x_1 \times x_2, x_3) \neq 0$. Hence formula (7.39) implies (7.38).

Formula (7.38) yields for $x_1 = y_1 = x$ and $x_2 = y_2 = y$

$$|x \times y|^2 = |x|^2 |y|^2 - (x, y)^2. \tag{7.40}$$

If $\theta\,(0 \leqq \theta \leqq \pi)$ denotes the angle between x and y we can rewrite (7.40) in the form

$$|x \times y| = |x|\,|y| \sin\theta \qquad x \neq 0, y \neq 0.$$

Now we establish the *triple identity*

$$x \times (y \times z) = (x, z)\, y - (x, y)\, z. \tag{7.41}$$

In fact, let $u \in E$ be arbitrary. Then formulae (7.36) and (7.38) yield

$$\begin{aligned}(x \times (y \times z), u) &= \Delta(x, y \times z, u) = -\Delta(y \times z, x, u) \\ &= -(y \times z, x \times u) = -(y, x)(z, u) + (y, u)(z, x) \\ &= (-(y, x)\, z + (z, x)\, y, u).\end{aligned}$$

Since $u \in E$ is arbitrary, (7.41) follows. From (7.41) we obtain the *Jacobi identity*

$$x \times (y \times z) + y \times (z \times x) + z \times (x \times y) = 0.$$

Finally note that if e_1, e_2, e_3 is a positive orthonormal basis of E we have the relations

$$e_1 \times e_2 = e_3, \quad e_2 \times e_3 = e_1, \quad e_3 \times e_1 = e_2.$$

Problems

1. Given a vector $a \neq 0$ determine the locus of all vectors x such that $x-a$ is orthogonal to $x+a$.

2. Prove that the cross product defines a Lie-algebra in a 3-dimensional inner product space (cf. problem 8, Chap. V, § 1).

3. Let e be a given unit vector of an n-dimensional inner product space E and E_1 be the orthogonal complement of e. Show that the distance of a vector $x \in E$ from the subspace E_1 is given by

$$d = |(x, e)|.$$

4. Prove that the area of the parallelogram generated by the vectors x_1 and x_2 is given by

$$A = 2\sqrt{s(s-a)(s-b)(s-c)},$$

where

$$a = |x_1|, \quad b = |x_2|, \quad c = |x_2 - x_1|, \quad s = \tfrac{1}{2}(a+b+c).$$

5. Let $a \neq 0$ and b be two given vectors of an oriented 3-space. Prove that the equation $x \times a = b$ has a solution if and only if $(a, b) = 0$. If this condition is satisfied and x_0 is a particular solution, show that the general solution is $x_0 + \lambda a$.

6. Consider an oriented inner product space of dimension 2. Given two positive orthonormal bases (e_1, e_2) and $(\bar{e}_1, \bar{e}_2)$, prove that

$$\begin{aligned} \bar{e}_1 &= e_1 \cos\omega - e_2 \sin\omega \\ \bar{e}_2 &= e_1 \sin\omega + e_2 \cos\omega . \end{aligned}$$

where ω is the oriented angle between e_1 and $\bar{e}_1$.

7. Let a_1 and a_2 be two linearly independent vectors of an oriented Euclidean 3-space and F be the plane generated by a_1 and a_2. Introduce an orientation in F such that the basis a_1, a_2 is positive. Prove that the angle between two vectors

$$x = \xi^1 a_1 + \xi^2 a_2 \quad \text{and} \quad y = \eta^1 a_1 + \eta^2 a_2$$

is determined by the equations

$$\cos\theta = \frac{\sum\limits_{\nu,\mu} (a_\nu, a_\mu)\, \xi^\nu \eta^\mu}{|x|\,|y|} \quad \text{and} \quad \sin\theta = \frac{\xi^1 \eta^2 - \xi^2 \eta^1}{|x|\,|y|} |a_1 \times a_2| (-\pi < \theta \leqq \pi).$$

8. Given an orthonormal basis $e_\nu (\nu=1, 2, 3)$ in the 3-space, define linear transformations φ_ν by

$$\varphi_\nu x = e_\nu \times x \qquad (\nu = 1,2,3).$$

Prove that

$$\sum_\nu \varphi_\nu^2 = -2\iota.$$

9. Let Δ be a determinant function in the plane. Prove the identity

$$(x, x_1)\Delta(x_2, x_3) + (x, x_2)\Delta(x_3, x_1) + (x, x_3)\Delta(x_1, x_2) = 0 \qquad x_i \in E,\ x \in E.$$

10. Let e_i $(i=1, 2, 3)$ be unit vectors in an oriented plane and denote by Θ_{ij} the oriented angle determined by e_i and e_j $(i<j)$. Prove the formulae

$$\cos\Theta_{13} = \cos\Theta_{12}\cos\Theta_{23} - \sin\Theta_{12}\sin\Theta_{23}$$

$$\sin\Theta_{13} = \sin\Theta_{12}\cos\Theta_{23} + \cos\Theta_{12}\sin\Theta_{23}.$$

11. Let x, y, z be three vectors of a plane such that x and y are linearly independent and that $x+y+z=0$.

a) Prove that the ordered pairs x, y; y, z and z, x represent the same orientation. Then show that

$$\theta(x, y) + \theta(y, z) + \theta(z, x) = 2\pi$$

where the angles refer to the above orientation.

b) Prove that

$$\theta(y, -x) + \theta(z, -y) + \theta(x, -z) = \pi.$$

What is the geometric significance of the two above relations?

12. Given p vectors $x_1, \ldots, x_p$ prove the inequality

$$G(x_1, \ldots, x_p) \leqq |x_1|^2 |x_2|^2 \ldots |x_p|^2$$

Then derive *Hadamard's inequality* for a determinant

$$\det\begin{pmatrix} a_{11} \cdots a_{1n} \\ \vdots \quad\ \vdots \\ a_{n1} \cdots a_{nn} \end{pmatrix}^2 \leqq \sum_{k=1}^n |a_{1k}|^2 \cdot \sum_{k=1}^n |a_{2k}|^2 \ldots \sum_{k=1}^n |a_{nk}|^2.$$

§ 4. Duality in an inner product space

7.18. The isomorphism τ. Let E be an inner product space of dimension n and let E^* be a vector space which is dual to E with respect to a scalar product $\langle , \rangle$. Since E^* and E are both dual to E it follows from sec. 2.33 that there is a linear isomorphism $\tau: E \to E^*$ such that

$$\langle \tau x, y \rangle = (x, y) \quad x, y \in E. \tag{7.42}$$

With the aid of this isomorphism we can introduce a positive definite inner product in E^* given by

$$(x^*, y^*) = (\tau^{-1} x^*, \tau^{-1} y^*). \tag{7.43}$$

Now introduce a scalar product in $E \times E^*$ by

$$\langle x, x^* \rangle = \langle x^*, x \rangle. \tag{7.44}$$

Then it follows from (7.42) and (7.44) that

$$\langle \tau x, y \rangle = (x, y) = (y, x) = \langle \tau y, x \rangle = \langle x, \tau y \rangle.$$

This relation shows that the dual mapping $\tau^*: E^* \leftarrow E$ coincides with τ and so τ is dual to itself.

Let e_ν, $e^{*\nu}$ $(\nu = 1 \ldots n)$ be a pair of dual bases of E and E^* and consider the matrices

$$g_{\nu\lambda} = (e_\nu, e_\lambda) \quad \text{and} \quad g^{\nu\lambda} = (e^{*\nu}, e^{*\lambda}). \tag{7.45}$$

It follows from the symmetry of the inner product that the matrices (7.45) are symmetric. On the other hand, the linear isomorphism $\tau: E \to E^*$ determines an $n \times n$-matrix $\alpha_{\lambda\nu}$ by

$$\tau e_\lambda = \sum_\nu \alpha_{\lambda\nu} e^{*\nu}.$$

Taking the inner product with e_μ yields

$$\alpha_{\lambda\mu} = \langle \tau e_\lambda, e_\mu \rangle = (e_\lambda, e_\mu) = g_{\lambda\mu}$$

and hence we can write

$$\tau e_\lambda = \sum_\mu g_{\lambda\mu} e^{*\mu}. \tag{7.46}$$

A similar argument shows that

$$\tau^{-1} e^{*\lambda} = \sum_\mu g^{\lambda\mu} e_\mu. \tag{7.47}$$

From (7.46) and (7.47) we obtain

$$\sum_\mu g_{\lambda\mu} g^{\mu\kappa} = \delta_\lambda^\kappa$$

and hence the matrices (7.45) are inverse to each other.

If

$$x=\sum_{\lambda}\xi^{\lambda}e_{\lambda} \tag{7.48}$$

is an arbitrary vector of E we can write

$$\tau x=\sum_{\lambda}\xi_{\lambda}e^{*\lambda}. \tag{7.49}$$

The numbers ξ_{λ} are called the *covariant components* of x with respect to the basis $e_{\lambda}(\lambda=1\dots n)$. It follows from (7.48) that

$$\xi_{\lambda}=\langle\tau x,e_{\lambda}\rangle=\langle\tau e_{\lambda},x\rangle=\sum_{\nu}\langle\tau e_{\lambda},e_{\nu}\rangle\xi^{\nu}=\sum_{\nu}g_{\lambda\nu}\xi^{\nu}$$

whence

$$\xi_{\lambda}=\sum_{\nu}g_{\lambda\nu}\xi^{\nu}. \tag{7.50}$$

We finally note that the covariant components of a vector $x\in E$ are its inner products with the basis vectors. In fact, from (7.48) and (7.50) we obtain that

$$(x,e_{\nu})=\sum_{\lambda}\xi^{\lambda}(e_{\lambda},e_{\nu})=\sum_{\lambda}g_{\lambda\nu}\xi^{\lambda}=\sum_{\nu}g_{\nu\lambda}\xi^{\lambda}=\xi_{\nu}.$$

If the basis $e_{\nu}(\nu=1\dots n)$ is orthonormal we have that $g_{\lambda\nu}=\delta_{\lambda\nu}$ and hence formulae (7.46) simplify to

$$\tau e_{\lambda}=e^{*\lambda}.$$

It follows that τ maps every orthonormal basis of E into the dual basis. Moreover, the equations (7.50) reduce in the case of an orthonormal basis to

$$\xi_{\lambda}=\xi^{\lambda}.$$

Problems

1. Let $e_i(i=1\dots n)$ be a basis of E consisting of unit vectors. Given a vector $x\in E$ write

$$x=p_i+h_i$$

where p_i is the orthogonal projection of x onto the subspace defined by $(x,e_i)=0$. Show that

$$|h_i|=|\xi_i| \qquad i=1\dots n$$

where the ξ_i are the covariant components of x with respect to the basis e_i.

2. Let E, E^* be a dual pair of finite dimensional vector spaces and consider a linear isomorphism $\tau:E\to E^*$. Find necessary and sufficient con-

ditions such that the bilinear function defined by

$$(x, y) = \langle \tau x, y \rangle \qquad x, y \in E$$

be a positive definite inner product.

§ 5. Normed vector spaces

7.19. Norm-functions. Let E be a real linear space of finite or infinite dimension. A *norm-function* in E is a real-valued function $\| \quad \|$ having the following properties:

N_1: $\|x\| \geqq 0$ for every $x \in E$, and $\|x\| = 0$ only if $x = 0$.
N_2: $\|x + y\| \leqq \|x\| + \|y\|$.
N_3: $\|\lambda x\| = |\lambda| \cdot \|x\|$.

A linear space in which a norm-function is defined is called a *normed linear space*. The *distance* of two vectors x and y of a normed linear space is defined by

$$\varrho(x, y) = \|x - y\|.$$

N_1, N_2 and N_3 imply respectively

$$\varrho(x, y) > 0 \quad \text{if} \quad x \neq y$$
$$\varrho(x, y) \leqq \varrho(x, z) + \varrho(z, y) \quad \text{(triangle inequality)}$$
$$\varrho(x, y) = \varrho(y, x).$$

Hence ϱ is a metric in E and so it defines a topology in E, called the *norm topology*. It follows from N_2 and N_3 that the linear operations are continuous in this topology and so E becomes a topological vector space.

7.20. Examples. 1. Every inner product space is a normed linear space with the norm defined by

$$\|x\| = \sqrt{(x, x)}.$$

2. Let C be the linear space of all continuous functions f in the interval $0 \leqq t \leqq 1$. Then a norm is defined in C by

$$\|f\| = \max_{0 \leqq t \leqq 1} |f(t)|.$$

Conditions N_1 and N_3 are obviously satisfied. To prove N_2 observe that

$$|f(t) + g(t)| \leqq |f(t)| + |g(t)| \leqq \|f\| + \|g\|, \qquad (0 \leqq t \leqq 1)$$

whence

$$\|f + g\| \leqq \|f\| + \|g\|.$$

3. Consider an n-dimensional (real or complex) linear space E and let $x_\nu\,(\nu=1\dots n)$ be a basis of E. Define the norm of a vector

$$x=\sum_\nu \xi^\nu x_\nu$$

by

$$\|x\|=\sum_\nu |\xi^\nu|\,.$$

7.21. Bounded linear transformations. A linear transformation $\varphi: E\to E$ of a normed space is called *bounded* if there exists a number M such that

$$\|\varphi x\| \leqq M\,\|x\| \qquad x\in E\,. \tag{7.51}$$

It is easily verified that a linear transformation is bounded if and only if it is continuous. It follows from N_2 and N_3 that a linear combination of bounded transformations is again bounded. Hence, the set $B(E;E)$ of all bounded linear transformations is a subspace of $L(E;E)$.

Let $\varphi: E\to E$ be a bounded linear transformation. Then the set $\|\varphi x\|$, $\|x\|=1$ is bounded. Its least upper bound will be denoted by $\|\varphi\|$,

$$\|\varphi\| = \sup_{\|x\|=1} \|\varphi x\|\,. \tag{7.52}$$

It follows from (7.52) that

$$\|\varphi x\| \leqq \|\varphi\|\cdot\|x\| \qquad x\in E\,.$$

Now it will be shown that the function $\varphi\to\|\varphi\|$ thus obtained is indeed a norm-function in $B(E;E)$. Conditions N_1 and N_3 are obviously satisfied. To prove N_2 let φ and ψ be two bounded linear transformations. Then

$$\|(\varphi+\psi)x\| = \|\varphi x+\psi x\| \leqq \|\varphi x\| + \|\psi x\| \leqq (\|\varphi\|+\|\psi\|)\cdot\|x\| \qquad x\in E$$

and consequently,

$$\|\varphi+\psi\| \leqq \|\varphi\|+\|\psi\|\,.$$

The norm-function $\|\varphi\|$ has the following additional property:

$$\|\psi\circ\varphi\| \leqq \|\psi\|\cdot\|\varphi\|\,. \tag{7.53}$$

In fact,

$$\|(\psi\circ\varphi)x\| \leqq \|\psi\|\cdot\|\varphi x\| \leqq \|\psi\|\cdot\|\varphi\|\cdot\|x\| \qquad x\in E$$

whence (7.53).

7.22. Normed spaces of finite dimension. Suppose now that E is a normed vector space of finite dimension. Then it will be shown that the norm topology of E coincides with the natural topology (cf. sec. 1.22). Since the linear operations are continuous it has only to be shown that a linear function is continuous in the norm topology. Let $e_\nu\,(\nu=1,\dots,n)$ be a basis of E. Then we have in view of N_2 and N_3 that

$$\|x\| = \|\sum_\nu \xi^\nu e_\nu\| \leqq \sum_\nu |\xi^\nu| \|e_\nu\|.$$

This relation implies that the function $x \to \|x\|$ is continuous in the natural topology.

Now consider the set $Q \subset E$ defined by

$$Q = \{x = \sum_\nu \xi^\nu e_\nu \mid \sum_\nu |\xi^\nu| = 1\}.$$

Since Q is compact in the natural topology and $\|x\| \neq 0$ for $x \in Q$ it follows that there exists a positive constant m such that

$$\|x\| \geq m \qquad x \in Q.$$

Now N_3 yields

$$\|x\| \geq m \sum_\nu |\xi^\nu| \qquad x \in E$$

whence

$$|\xi^\nu| \leq \frac{\|x\|}{m} \qquad \nu = 1, \ldots, n. \tag{7.54}$$

Let f be a linear function in E. Then we have in view of (7.54) that

$$|f(x)| = |\sum_\nu \xi^\nu f(e_\nu)| \leq \frac{\|x\|}{m} \sum_\nu |f(e_\nu)| \leq M \|x\|$$

and so f is continuous. This completes the proof.

Since every linear transformation φ of E is continuous (cf. sec. 1.22) it follows that φ is bounded and hence $B(E; E) = L(E; E)$. Thus $L(E; E)$ becomes a normed space, the norm of a transformation φ being given by

$$\|\varphi\| = \max_{\|x\| = 1} \|\varphi x\|.$$

Problems

1. Let E be a normed linear space and E_1 be a subspace of E. Show that a norm-function is defined in the factor-space E/E_1 by

$$\|\bar{x}\| = \inf_{x \in \bar{x}} \|x\| \qquad \bar{x} \in E/E_1.$$

2. An infinite sequence of vectors $x_\nu (\nu = 1, 2 \ldots)$ of a normed linear space E is called *convergent* towards x if the following condition holds: To every positive number ε there exists an integer N such that

$$\|x_n - x\| < \varepsilon \quad \text{if} \quad n > N.$$

a) Prove that every convergent sequence satisfies the following *Cauchy-criterion:* To every positive number ε there exists an integer N such that

$$\|x_n - x_m\| < \varepsilon \quad \text{if} \quad n > N \quad \text{and} \quad m > N .$$

b) Prove that every Cauchy-sequence*) in a normed linear space of finite dimension is convergent.

c) Give an example showing that the assertion b) is not necessarily correct if the dimension of E is infinite.

3. A normed linear space is called *complete* if every Cauchy-sequence is convergent. Let E be a complete normed linear space and φ be a linear transformation of E such that $\|\varphi\| < 1$. Prove that the series $\sum_{\nu=0}^{\infty} \varphi^\nu$ is convergent and that the linear transformation

$$\psi = \sum_{\nu=0}^{\infty} \varphi^\nu$$

has the following properties:

a) $(\iota - \varphi) \circ \psi = \psi \circ (\iota - \varphi) = \iota$.

b) $\|\psi\| \leqq \dfrac{1}{1 - \|\varphi\|}$.

§ 6. The algebra of quaternions

7.23. Definition. Let E be an oriented Euclidean space of dimension 4. Choose a unit vector e, and let E_1 denote the orthogonal complement of e. Let E_1 have the orientation induced by the orientation of E and by e (cf. sec. 4.29). Observe that every vector $x \in E$ can be uniquely decomposed in the form

$$x = \lambda e + x_1 \qquad \lambda \in \mathbb{R},\ x_1 \in E_1 .$$

Now consider the bilinear map $E \times E \to E$ defined by

$$e \cdot x = x, \quad x \cdot e = x \qquad x \in E \tag{7.55}$$

$$x \cdot y = -(x, y)\, e + x \times y \qquad x, y \in E_1 \tag{7.56}$$

where $\times$ denotes the cross product in the oriented 3-space E, cf. sec. 7.16.

It is easily checked (by means of formula (7.41)) that this bilinear map makes E into an associative algebra. This algebra is called the *algebra*

*) i.e. a sequence satisfying the Cauchy-criterion.

of quaternions and is denoted by $\mathbb{H}$. In view of (7.55), e is the unit element of $\mathbb{H}$ while (7.56) shows that

$$x \cdot y + y \cdot x = -2(x, y)\, e \qquad x, y \in E_1$$

and so the algebra $\mathbb{H}$ is *not* commutative.

The elements of E are called *quaternions*, and the elements of E_1 are called *pure quaternions*. The *conjugate* of a quaternion $x = \lambda e + x_1$, $(\lambda \in \mathbb{R},\ x_1 \in E_1)$ is defined by

$$\bar{x} = \lambda e - x_1$$

It is easily verified that

$$\overline{x + y} = \bar{x} + \bar{y}, \quad \overline{x \cdot y} = \bar{y} \cdot \bar{x} \qquad x, y \in E$$

and

$$\bar{\bar{x}} = x \qquad x \in E.$$

Note that $x \in E_1$ if and only if $\bar{x} = -x$. Next, let $x \in \mathbb{H}$ be arbitrary and write

$$x = \lambda e + x_1 \qquad \lambda \in \mathbb{R},\ x_1 \in E_1.$$

Then

$$x \cdot \bar{x} = \lambda^2 e + (x_1, x_1)\, e = (x, x)\, e = |x|^2 \cdot e.$$

Similarly,

$$\bar{x} \cdot x = |x|^2 \cdot e.$$

Thus we have the relations

$$x \cdot \bar{x} = \bar{x} \cdot x = |x|^2\, e.$$

Now assume that $x \neq 0$ and define x^{-1} by

$$x^{-1} = \frac{1}{|x|^2}\, \bar{x}.$$

Then the relations above yield

$$x \cdot x^{-1} = x^{-1} x = e \qquad (x \neq 0).$$

Thus every non-zero element in $\mathbb{H}$ has a left and right inverse and so $\mathbb{H}$ is a division algebra.

Finally note that the multiplication in $\mathbb{H}$ satisfies the relations

$$(x \cdot y, x \cdot z) = |x|^2 (y, z) \qquad x, y, z \in \mathbb{H} \tag{7.57}$$

and

$$(y \cdot x, z \cdot x) = (y, z)\, |x|^2, \tag{7.58}$$

which are easily verified using (7.36) and (7.38). In particular, we have

$$|x \cdot y| = |x| \cdot |y| \qquad x, y \in \mathbb{H}.$$

Proposition I: The 3-linear function in E_1 given by

$$\Delta(x_1, x_2, x_3) = (x_1 \cdot x_2, x_3) \qquad x_1, x_2, x_3 \in E_1$$

is a normed determinant function in E_1 and represents the orientation of E_1.

Proof: First we show that Δ is skew symmetric. In fact, formulae (7.58) and (7.57) imply that

$$\Delta(x_1, x, x) = (x_1 \cdot x, x) = (x_1, e)\,|x|^2 = 0$$

and

$$\Delta(x, x_2, x) = (x \cdot x_2, x) = (x_2, e)\,|x|^2 = 0.$$

Thus Δ is skew symmetric.

Next observe that

$$\begin{aligned}\Delta(x_1, x_2, x_3) &= \tfrac{1}{2}\{\Delta(x_1, x_2, x_3) - \Delta(x_2, x_1, x_3)\}\\ &= \tfrac{1}{2}(x_1 \cdot x_2 - x_2 \cdot x_1, x_3) = (x_1 \times x_2, x_3).\end{aligned}$$

This relation shows that Δ is a normed determinant function in E_1 and represents the orientation of E_1.

7.24. Associative division algebras. Let A be an associative division algebra (cf. sec. 5.1) with unit element e. Observe that the real numbers, the complex numbers and the quaternions form associative division algebras over $\mathbb{R}$. The dimensions of these algebras are respectively 1, 2 and 4. We shall show that these are the *only* finite dimensional associative division algebras over $\mathbb{R}$.

Let A be any such algebra with unit element e. Let $a \in A$ be arbitrary and consider the $n+1$ powers

$$a^\nu \,(\nu = 0, \ldots, n,\ a^0 = e) \qquad n = \dim A.$$

Since these elements are linearly dependent we have a non-trivial relation

$$\sum_{\nu=0}^{n} \lambda_\nu a^\nu = 0 \qquad \lambda_\nu \in \mathbb{R}.$$

This relation can be written in the form

$$f(a) = 0 \tag{7.59}$$

where f is the polynomial given by

$$f(t) = \sum_{\nu=0}^{n} \lambda_\nu t^\nu \qquad \lambda_\nu \in \mathbb{R}.$$

By the fundamental theorem of algebra, f is a product of irreducible polynomials of degree 1 and 2. Since A is a division algebra, it follows from (7.59) that a satisfies an equation of the form

$$p(a) = 0$$

where p is of degree two or one. Equivalently, every element $a \in A$ satisfies an equation of the form

$$(a + \alpha e)^2 = -\beta^2 e \qquad \alpha, \beta \in \mathbb{R}.$$

Lemma I: If x and y are elements of A satisfying $x^2 = -e$ and $y^2 = -e$, then

$$xy + yx = 2\lambda e \qquad -1 \leqq \lambda \leqq 1.$$

Proof: Without loss of generality we may assume that $y \neq \pm x$. Then the elements e, x, y are linearly independent as is easily checked.

Now observe that $x+y$ and $x-y$ satisfy quadratic equations

$$(x+y)^2 + \alpha(x+y) + 2\beta e = 0 \qquad \alpha, \beta \in \mathbb{R} \tag{7.60}$$

and

$$(x-y)^2 + \gamma(x-y) + 2\delta e = 0 \qquad \gamma, \delta \in \mathbb{R}. \tag{7.61}$$

Adding these equations and observing that $x^2 = y^2 = -e$ we obtain

$$(\alpha+\gamma)x + (\alpha-\gamma)y + 2(\beta+\delta-2)e = 0.$$

Since the vectors e, x and y are linearly independent, the equation above yields

$$\alpha + \gamma = 0, \qquad \alpha - \gamma = 0$$
$$\beta + \delta = 2$$

and therefore

$$\alpha = 0, \qquad \gamma = 0.$$

Now equations (7.60) and (7.61) reduce to

$$(x+y)^2 = -2\beta e \tag{7.62}$$

and

$$(x-y)^2 = -2\delta e. \tag{7.63}$$

Since the vectors e, x, y are linearly independent the polynomial $t^2 + 2\beta$ must be irreducible and so $\beta > 0$. Similarly, $\delta > 0$. Now the equation $\beta + \delta = 2$ implies that $0 < \beta < 2$. Finally, relation (7.62) yields

$$xy + yx = 2(1-\beta)e.$$

Since $0 < \beta < 2$, the lemma follows.

Lemma II: Let F be the subset of A consisting of those elements x which satisfy

$$x^2 = -\gamma^2 e \qquad \text{for some } \gamma \in \mathbb{R}.$$

Then

(i) F is a vector space.

(ii) $A = (e) \oplus F$ where (e) denotes the 1-dimensional subspace generated by e.

(iii) If $x \in F$ and $y \in F$, then $xy + yx \in (e)$ and $xy - yx \in F$.

Proof: (i) If $x \in F$, then clearly, $\lambda x \in F$ for $\lambda \in \mathbb{R}$. Now let $x \in F$ and $y \in F$. We may assume that $x^2 = -e$ and $y^2 = -e$. Then Lemma I yields

$$(x+y)^2 = x^2 + y^2 + xy + yx = 2(\lambda - 1)e, \qquad -1 \leqq \lambda \leqq 1.$$

It follows that $x + y \in F$. Thus F is a vector space.

(ii) Clearly, $(e) \cap F = 0$. Finally, let $a \in A$ be arbitrary. Then a satisfies an equation of the form

$$(a + \alpha e)^2 = -\beta^2 e \qquad \alpha, \beta \in \mathbb{R}.$$

Set

$$x_1 = -\alpha e \quad \text{and} \quad x_2 = a + \alpha e.$$

Then $x_1 \in (e)$, $x_2 \in F$ and $x_1 + x_2 = a$.

(iii) Let $x \in F$ and $y \in F$. Again we may assume that $x^2 = -e$ and $y^2 = -e$. Then, by Lemma I, $xy + yx \in (e)$.

To show that $xy - yx \in F$ observe that

$$xy - yx = -(x+y)(x-y) = (x-y)(x+y).$$

Thus

$$(xy - yx)^2 = -(x+y)(x-y)^2(x+y).$$

Since, by (i), $x + y \in F$ and $x - y \in F$ we have

$$(x+y)^2 = -\alpha^2 e \qquad \alpha \in \mathbb{R}$$
$$(x-y)^2 = -\beta^2 e \qquad \beta \in \mathbb{R}$$

whence

$$(xy - yx)^2 = -\alpha^2 \beta^2 e$$

and so $xy - yx \in F$.

7.25. The inner product in F. Let $x \in F$ and $y \in F$. Then, by Lemma II, $xy + yx \in (e)$. Thus a symmetric bilinear function, $(\,,)$, is defined in F by

$$xy + yx = -2(x, y)e \qquad x, y \in F. \tag{7.64}$$

Since

$$x^2 \equiv -(x, x)e$$

it follows that $(\,,)$ is positive definite and so it makes F into a Euclidean space.

On the other hand, again by Lemma II sec. 7.24, $xy-yx\in F$. Hence a skew symmetric bilinear map $\Psi\colon F\times F\to F$ is defined by

$$xy-yx=2\Psi(x,y) \qquad x,y\in F. \tag{7.65}$$

Formulae (7.64) and (7.65) imply that

$$xy=-(x,y)\,e+\Psi(x,y) \qquad x,y\in F. \tag{7.66}$$

Finally, observe that

$$(xy-yx,y)=0 \qquad x,y\in F \tag{7.67}$$

since

$$(xy-yx)\,y+y(xy-yx)=xy^2-y^2x=\beta^2(xe-ex)=0.$$

7.26. Theorem: *Let A be a finite dimensional associative division algebra over $\mathbb{R}$. Then $A\cong\mathbb{R}$, $A\cong\mathbb{C}$ or $A\cong\mathbb{H}$.*

Proof: We may assume that $n\geqq 2$ ($n=\dim A$). If $n=2$, then F has dimension 1. Choose a vector $j\in F$ such that $j^2=-e$. Then an isomorphism $A\xrightarrow{\cong}\mathbb{C}$ is given by

$$\alpha e+\beta j\mapsto\alpha+i\beta \qquad \alpha,\beta\in\mathbb{R}.$$

Now consider the case $n>2$. Then $\dim F\geqq 2$. Hence there are unit vectors $e_1\in F$, $e_2\in F$ such that

$$(e_1,e_2)=0.$$

It follows that

$$e_1e_2+e_2e_1=-2(e_1,e_2)\,e=0.$$

Now set

$$e_3=e_1e_2.$$

Then we have

$$e_3^2=e_1e_2e_1e_2=-e_1^2e_2^2=-e$$

whence $e_3\in F$. Moreover,

$$e_1e_3=-e_2, \qquad e_2e_3=e_1.$$

These relations imply that

$$(e_3,e_3)=1$$

and

$$(e_1,e_3)=(e_2,e_3)=0.$$

Thus the vectors e_1,e_2,e_3 are orthonormal.

In particular, it follows that $\dim F \geqq 3$.

Now we show that the vectors e_1, e_2, e_3 span F. In fact, let $z \in F$. We may assume that $z^2 = -e$. Then

$$\begin{aligned} z e_3 - e_3 z &= z e_1 e_2 - e_1 e_2 z \\ &= [-2(z, e_1) e - e_1 z] e_2 - e_1 [-2(z, e_2) - z e_2] \\ &= -2(z, e_1) e_2 + 2(z, e_2) e_1 . \end{aligned}$$

On the other hand,

$$z e_3 + e_3 z = -2(z, e_3) e .$$

Adding these equations we obtain

$$z e_3 = -(z, e_1) e_2 + (z, e_2) e_1 - (z, e_3) e$$

whence

$$z = (z, e_1) e_1 + (z, e_2) e_2 + (z, e_3) e_3 .$$

This shows that z is a linear combination of e_1, e_2 and e_3. Hence $\dim F = 3$ and so $\dim A = 4$.

Finally, we show that A is the quaternion algebra. Let Δ be the trilinear function in F defined by

$$\Delta(x, y, z) = (\Psi(x, y), z) \qquad x, y, z \in F . \tag{7.68}$$

We show that Δ is skew symmetric. Clearly, $\Delta(x, x, z) = 0$. On the other hand, formula (7.67) yields

$$\Delta(x, y, y) = (\Psi(x, y), y) = \tfrac{1}{2}(x y - y x, y) \equiv 0 .$$

Thus Δ is a determinant function in F. Since

$$\Delta(e_1, e_2, e_3) = \tfrac{1}{2}(e_1 e_2 - e_2 e_1, e_3) = (e_3, e_3) = 1 ,$$

Δ is a *normed* determinant function in the Euclidean space F. In particular, Δ specifies an orientation and so the cross product is defined in F. Now relation (7.68) implies that

$$\Psi(x, y) = x \times y \qquad x, y \in F . \tag{7.69}$$

Combining (7.66) and (7.69) yields

$$x y = -(x, y) e + x \times y \qquad x, y \in F .$$

Since on the other hand,

$$x e = e x = x \qquad x \in A$$

and $A = (e) \oplus F$, it follows that A is the algebra of quaternions.

Remark: It is a deep theorem of Adams that if one does not require associativity there is only *one* additional finite dimensional division algebra over $\mathbb{R}$. It is the algebra of *Cayley numbers* defined in Chap. XI, § 2, problem 6.

Problems

1. Let $a \neq 0$ be a quaternion which is not a negative multiple of e. Show that the equation $x^2 = a$ has exactly two solutions.

2. If y_i $(i = 1, 2, 3)$ are three vectors in E_1 prove the identity

$$y_1 y_2 y_3 = -\Delta(y_1, y_2, y_3)\, e - (y_1, y_2)\, y_3 + (y_1, y_3)\, y_2 - (y_2, y_3)\, y_1.$$

3. Let p be a fixed quaternion and consider the linear transformation φ given by $\varphi x = px$. Show that the characteristic polynomial of φ reads

$$f(t) = (t^2 - 2(p, e)\, t + |p|^2)^2.$$

Conclude that φ has no real eigenvalues unless p is a multiple of e.

4. Let y_1, y_2 and y_3 be quaternions orthogonal to e. Prove the identity

$$\begin{aligned}&(y_1 y_2, y_2 y_3) + (y_2 y_3, y_3 y_1) + (y_3 y_1, y_1 y_2)\\ &= (y_1, y_2)(y_2, y_3) + (y_2, y_3)(y_3, y_1) + (y_3, y_1)(y_1, y_2) + \Delta(y_1, y_2, y_3)^2.\end{aligned}$$

5. Let p and q be quaternions orthogonal to e and consider the linear transformations of E given by

$$\sigma(x) = px + xq \quad \text{and} \quad \tau x = px - xq.$$

Show that

$$\mathrm{ad}(\sigma) = -(|p|^2 - |q|^2) \cdot \tau$$

where ad denotes the classical adjoint (cf. sec. 4.6).

Chapter VIII

Linear mappings of inner product spaces

In this chapter all linear spaces are assumed to be real and to have finite dimension

§ 1. The adjoint mapping

8.1. Definition. Consider two inner product spaces E and F and assume that a linear mapping $\varphi: E \to F$ is given. If E^* and F^* are two linear spaces dual to E and F respectively, the mapping φ induces a dual mapping $\varphi^*: F^* \to E^*$. The mappings φ and φ^* are related by

$$\langle y^*, \varphi x \rangle = \langle \varphi^* y^*, x \rangle \qquad x \in E, y^* \in F^*. \tag{8.1}$$

Since inner products are defined in E and in F, these linear spaces can be considered as dual to themselves. Then the dual mapping is a linear mapping of F into E. This mapping is called the *adjoint mapping* of φ and will be denoted by $\tilde{\varphi}$. Replacing the scalar product by the inner product in (8.1) we obtain the relation

$$(\varphi x, y) = (x, \tilde{\varphi} y) \quad x \in E, y \in F. \tag{8.2}$$

In this way every linear mapping φ of an inner product space E into an inner product space F determines a linear mapping $\tilde{\varphi}$ of F into E.

The adjoint mapping $\tilde{\tilde{\varphi}}$ of $\tilde{\varphi}$ is again φ. In fact, the mappings $\tilde{\varphi}$ and $\tilde{\tilde{\varphi}}$ are related by

$$(\tilde{\varphi} y, x) = (y, \tilde{\tilde{\varphi}} x). \tag{8.3}$$

Equations (8.2) and (8.3) yield

$$(\varphi x, y) = (\tilde{\tilde{\varphi}} x, y) \qquad x \in E, y \in F$$

whence $\tilde{\tilde{\varphi}} = \varphi$. Hence, the relation between a linear mapping and the adjoint mapping is symmetric.

As it has been shown in sec. 2.26 the subspaces $\operatorname{Im} \varphi$ and $\ker \tilde{\varphi}$ are orthogonal complements. We thus obtain the orthogonal decomposition

$$F = \operatorname{Im}\varphi \oplus \ker\tilde{\varphi}. \tag{8.4}$$

8.2. The relation between the matrices. Employing two bases x_ν $(\nu=1\dots n)$ and $y_\mu\,(\mu=1\dots m)$ of E and of F, we obtain from the mappings φ and $\tilde{\varphi}$ two matrices α_ν^μ and $\tilde{\alpha}_\mu^\nu$ *) defined by the equations

$$\varphi x_\nu = \sum_\kappa \alpha_\nu^\kappa y_\kappa$$

and

$$\tilde{\varphi} y_\mu = \sum_\lambda \tilde{\alpha}_\mu^\lambda x_\lambda .$$

Substituting $x=x_\nu$ and $y=y_\mu$ in (8.2) we obtain the relation

$$\sum_\kappa \alpha_\nu^\kappa (y_\kappa, y_\mu) = \sum_\lambda \tilde{\alpha}_\mu^\lambda (x_\nu, x_\lambda). \tag{8.5}$$

Introducing the components

$$g_{\nu\lambda} = (x_\nu, x_\lambda) \quad \text{and} \quad h_{\mu\kappa} = (y_\mu, y_\kappa)$$

of the metric tensors we can write the relation (8.5) as

$$\sum_\kappa \alpha_\nu^\kappa h_{\kappa\mu} = \sum_\nu \tilde{\alpha}_\mu^\lambda g_{\nu\lambda} .$$

Multiplication by the inverse matrix $g^{\nu\varrho}$ yields the formula

$$\tilde{\alpha}_\mu^\varrho = \sum_{\kappa,\,\nu} \alpha_\nu^\kappa h_{\kappa\mu} g^{\nu\varrho} . \tag{8.6}$$

Now assume that the bases $x_\nu (\nu=1\dots n)$ and $y_\mu\,(\mu=1\dots m)$ are orthonormal,

$$g_{\nu\lambda} = \delta_{\nu\lambda}, \quad h_{\kappa\mu} = \delta_{\kappa\mu} .$$

Then formula (8.6) reduces to

$$\tilde{\alpha}_\mu^\varrho = \alpha_\varrho^\mu .$$

This relation shows that with respect to orthonormal bases, the matrices of adjoint mappings are transposed to each other.

8.3. The adjoint linear transformation. Let us now consider the case that $F=E$. Then to every linear transformation φ of E corresponds an adjoint transformation $\tilde{\varphi}$. Since $\tilde{\varphi}$ is dual to φ relative to the inner pro-

*) The subscript indicates the row.

duct, it follows that

$$\det \tilde{\varphi} = \det \varphi \quad \text{and} \quad \operatorname{tr} \tilde{\varphi} = \operatorname{tr} \varphi .$$

The adjoint mapping of the product $\psi \circ \varphi$ is given by

$$\widetilde{\psi \circ \varphi} = \tilde{\varphi} \circ \tilde{\psi} .$$

The matrices of $\tilde{\varphi}$ and φ relative to an orthonormal basis are transposes of each other.

Suppose now that e and $\tilde{e}$ are eigenvectors of φ and $\tilde{\varphi}$ respectively. Then we have that

$$\varphi e = \lambda e \quad \text{and} \quad \tilde{\varphi} \tilde{e} = \tilde{\lambda} \tilde{e}$$

whence in view of (8.2)

$$(\tilde{\lambda} - \lambda)(e, \tilde{e}) = 0 .$$

It follows that $(e, \tilde{e}) = 0$ whenever $\tilde{\lambda} \neq \lambda$; that is, any two eigenvectors of φ and $\tilde{\varphi}$ whose eigenvalues are different are orthogonal.

8.4. The relation between linear transformations and bilinear functions. Given a linear transformation $\varphi: E \to E$ consider the bilinear function

$$\Phi(x, y) = (\varphi x, y) . \tag{8.7}$$

The correspondence $\varphi \to \Phi$ defines a linear mapping

$$\varrho: L(E; E) \to B(E, E) , \tag{8.8}$$

where $B(E, E)$ denotes the space of bilinear functions in $E \times E$. It will be shown that this linear mapping is a linear isomorphism of $L(E; E)$ onto $B(E, E)$. To prove that ϱ is regular, assume that a certain φ determines the zero-function. Then $(\varphi x, y) = 0$ for every $x \in E$ and every $y \in E$, whence $\varphi = 0$.

It remains to be shown that ϱ is a mapping onto $B(E, E)$. Given a bilinear function Φ, choose a fixed vector $x \in E$ and consider the linear function f_x defined by

$$f_x(y) = \Phi(x, y) .$$

By the Riesz-theorem (cf. sec. 7.7) this function can be written in the form

$$f_x(y) = (x', y)$$

where the vector $x' \in E$ is uniquely determined by x.

Define a linear transformation $\varphi: E \to E$ by

$$\varphi x = x'.$$

Then

$$\Phi(x, y) = (\varphi x, y) \qquad x \in E, y \in E.$$

Thus, there is a one-to-one correspondence between the linear transformations of E and the bilinear functions in E. In particular, the identity-map corresponds to the bilinear function defined by the inner product.

Let $\tilde{\Phi}$ be the bilinear function which corresponds to the adjoint transformation. Then

$$\tilde{\Phi}(x, y) = (\tilde{\varphi} x, y) = (x, \varphi y) = (\varphi y, x) = \Phi(y, x).$$

This equation shows that the bilinear functions $\tilde{\Phi}$ and Φ are obtained from each other by interchanging the arguments.

8.5. Normal transformations. A linear transformation $\varphi: E \to E$ is called *normal*, if

$$\tilde{\varphi} \circ \varphi = \varphi \circ \tilde{\varphi}. \tag{8.9}$$

The above condition is equivalent to

$$(\varphi x, \varphi y) = (\tilde{\varphi} x, \tilde{\varphi} y) \qquad x, y \in E. \tag{8.10}$$

In fact, assume that φ is normal. Then

$$(\varphi x, \varphi y) = (x, \tilde{\varphi} \varphi y) = (x, \varphi \tilde{\varphi} y) = (\tilde{\varphi} x, \tilde{\varphi} y).$$

Conversely, condition (8.10) implies that

$$(y, \tilde{\varphi} \varphi x) = (\varphi y, \varphi x) = (\tilde{\varphi} y, \tilde{\varphi} x) = (y, \varphi \tilde{\varphi} x)$$

whence (8.9).

Formula (8.10) is equivalent to

$$|\varphi x|^2 = |\tilde{\varphi} x|^2 \qquad x \in E.$$

This relation implies that the kernels of φ and $\tilde{\varphi}$ coincide,

$$\ker \varphi = \ker \tilde{\varphi}.$$

Hence, the orthogonal decomposition (8.4) can be written in the form

$$E = \ker \varphi \oplus \operatorname{Im} \varphi. \tag{8.11}$$

Relation (8.11) implies that the restriction of φ to $\operatorname{Im} \varphi$ is regular. Hence, φ^2 has the same rank as φ. The same argument shows that all the transformations φ^k $(k=2, 3\ldots)$ have the same rank as φ.

It is easy to verify that if φ is a normal transformation then so is $\varphi-\lambda\iota$, $\lambda\in\mathbb{R}$. Hence it follows that

$$\ker(\varphi-\lambda\iota)=\ker(\tilde{\varphi}-\lambda\iota).$$

In other words, φ and $\tilde{\varphi}$ have the same eigenvectors. Now the result at the end of sec. 8.3. implies that every two eigenvectors of a normal transformation whose eigenvalues are different must be orthogonal.

Let $\varphi: E\to E$ be a linear transformation and assume that an orthogonal decomposition

$$E=E_1\oplus\cdots\oplus E_r$$

is given such that the subspaces E_i are stable. Denote by φ_i the restriction of φ to E_i. Then φ is normal if and only if the subspaces E_i are stable under $\tilde{\varphi}_i$ and the transformations φ_i are normal.

In fact, assume that φ is normal and let $x_i\in E_i$ be arbitrary. Then we have for every $x_j\in E_j$, $j\neq i$

$$(\tilde{\varphi}x_i, x_j)=(x_i,\varphi x_j)=0.$$

This implies that $\tilde{\varphi}x_i\in E_j^{\perp}$, $j\neq i$ whence $\tilde{\varphi}x_i\in E_i$. Thus E_i is stable under $\tilde{\varphi}$. The normality of φ_i follows immediately from the relation

$$|\varphi_i x|^2=|\varphi x|^2=|\tilde{\varphi}x|^2=|\tilde{\varphi}_i x|^2 \quad x\in E_i.$$

Conversely, assume that E_i is stable under $\tilde{\varphi}_i$ and that φ_i is normal. Then we have for every vector

$$x=\sum_i x_i \qquad x_i\in E_i$$

that

$$|\varphi x|^2=\sum_i|\varphi x_i|^2=\sum_i|\varphi_i x_i|^2=\sum_i|\tilde{\varphi}_i x_i|^2=\sum_i|\tilde{\varphi}x_i|^2=|\tilde{\varphi}x|^2$$

and so φ is normal.

Problems

1. Consider two inner product spaces E and F. Prove that an inner product is defined in the space $L(E;F)$ by

$$(\varphi,\psi)=\operatorname{tr}(\tilde{\psi}\circ\varphi) \qquad \varphi,\psi\in L(E;F).$$

Derive the inequality

$$(\operatorname{tr}(\tilde{\psi}\varphi))^2\leqq\operatorname{tr}(\tilde{\psi}\psi)\operatorname{tr}(\tilde{\varphi}\varphi)$$

and show that equality holds only if φ, ψ are linearly dependent.

2. Let $\varphi: E \to E$ be a linear transformation and $\tilde{\varphi}$ be the adjoint transformation. Prove that if $F \subset E$ is stable under φ, then $F^{\perp}$ is stable under $\tilde{\varphi}$.

3. Prove that the matrix of a normal transformation of a 2-dimensional space with respect to an orthonormal basis has the form

$$\begin{pmatrix} \alpha & \beta \\ -\beta & \alpha \end{pmatrix} \quad \text{or} \quad \begin{pmatrix} \alpha & \beta \\ \beta & \delta \end{pmatrix}.$$

§ 2. Selfadjoint mappings

8.6. Eigenvalue problem. A linear transformation $\varphi: E \to E$ is called *selfadjoint* if $\tilde{\varphi} = \varphi$ or equivalently

$$(\varphi x, y) = (x, \varphi y) \qquad x, y \in E.$$

The above equation implies that the matrix of a selfadjoint transformation relative to an orthonormal basis is symmetric.

If $\varphi: E \to E$ is a selfadjoint transformation and $F \subset E$ is a stable subspace then the orthogonal complement $F^{\perp}$ is stable as well. In fact, let $z \in F^{\perp}$ be any vector. Then we have for every $y \in F$

$$(\varphi z, y) = (z, \varphi y) = 0$$

whence $\varphi z \in F^{\perp}$.

It is the aim of this paragraph to show that a selfadjoint transformation of an n-dimensional inner product space E has n eigenvectors which are mutually orthogonal.

Define the function F by

$$F(x) = \frac{(x, \varphi x)}{(x, x)} \qquad x \neq 0. \tag{8.12}$$

This function is defined for all vectors $x \neq 0$. As a quotient of continuous functions, F is also continuous. Moreover, F is homogeneous of degree zero, i.e.

$$F(\lambda x) = F(x) \qquad (\lambda \neq 0). \tag{8.13}$$

Consider the function F on the unit sphere $|x| = 1$. Since the unit sphere is a bounded and closed subset of E, F assumes a minimum on the sphere $|x| = 1$. Let e_1 be a unit vector such that

$$F(e_1) \leqq F(x) \tag{8.14}$$

for all vectors $|x|=1$. Relations (8.13) and (8.14) imply that

$$F(e_1) \leqq F(x) \tag{8.15}$$

for all vectors $x \neq 0$. In fact, if $x \neq 0$ is an arbitrary vector, consider the corresponding unit-vector e. Then $x=|x|e$, whence in view of (8.13)

$$F(x) = F(e) \geqq F(e_1).$$

Now it will be shown that e_1 is an eigenvector of φ. Let y be an arbitrary vector and define the function f by

$$f(t) = F(e_1 + ty). \tag{8.16}$$

Then it follows from (8.15) that f assumes a minimum at $t=0$, whence $f'(0)=0$. Inserting the expression (8.12) into (8.16) we can write

$$f(t) = \frac{(e_1 + ty, \varphi e_1 + t\varphi y)}{(e_1 + ty, e_1 + ty)}.$$

Differentiating this function at $t=0$ we obtain

$$f'(0) = (e_1, \varphi y) + (y, \varphi e_1) - 2(e_1, \varphi e_1)(e_1, y). \tag{8.17}$$

Since φ is selfadjoint,

$$(e_1, \varphi y) = (\varphi e_1, y)$$

and hence equation (8.17) can be written as

$$f'(0) = 2(\varphi e_1, y) - 2(e_1, \varphi e_1)(e_1, y). \tag{8.18}$$

We thus obtain

$$(\varphi e_1 - (e_1, \varphi e_1) e_1, y) = 0 \tag{8.19}$$

for every vector $y \in E$. This implies that

$$\varphi e_1 = (e_1, \varphi e_1) e_1,$$

i.e. e_1 is an eigenvector of φ and the corresponding eigenvalue is

$$\lambda_1 = (e_1, \varphi e_1).$$

8.7. Representation in diagonal form. Once an eigenvector of φ has been constructed it is easy to find a system of n orthogonal eigenvectors. In fact, consider the 1-dimensional subspace (e_1) generated by e_1. Then (e_1) is stable under φ and hence so is the orthogonal complement E_1 of

(e_1). Clearly the induced linear transformation is again selfadjoint and hence the above construction can be applied to E_1. Hence, there exists an eigenvector e_2 such that $(e_1, e_2)=0$.

Continuing this way we finally obtain a system of n eigenvectors e_ν $(\nu=1\ldots n)$ such that

$$(e_\nu, e_\mu) = \delta_{\nu\mu}.$$

The eigenvectors e_ν form an orthonormal basis of E. In this basis the mapping φ has the form

$$\varphi e_\nu = \lambda_\nu e_\nu \tag{8.20}$$

where λ_ν denotes the eigenvalue of e_ν. These equations show that the matrix of a selfadjoint mapping has diagonal form if the eigenvectors are used as a basis.

8.8. The eigenvector-spaces. If λ is an eigenvalue of φ, the corresponding *eigen-space* $E(\lambda)$ is the set of all vectors x satisfying the equation $\varphi x=\lambda x$. Two eigen-spaces $E(\lambda)$ and $E(\lambda')$ corresponding to different eigenvalues are orthogonal. In fact, assume that

$$\varphi e = \lambda e \quad \text{and} \quad \varphi e' = \lambda' e'.$$

Then

$$(e', \varphi e) = \lambda (e, e') \quad \text{and} \quad (e, \varphi e') = \lambda' (e, e').$$

Subtracting these equations we obtain

$$(\lambda' - \lambda)(e, e') = 0,$$

whence $(e, e')=0$ if $\lambda' \neq \lambda$.

Denote by $\lambda_\nu (\nu=1\ldots r)$ the different eigenvalues of φ. Then every two eigenspaces $E(\lambda_i)$ and $E(\lambda_j) (i\neq j)$ are orthogonal. Since every vector $x\in E$ can be written as a linear combination of eigenvectors it follows that the direct sum of the spaces $E(\lambda_i)$ is E. We thus obtain the orthogonal decomposition

$$E = E(\lambda_1) \oplus \cdots \oplus E(\lambda_r). \tag{8.21}$$

Let φ_i be the transformation induced by φ in $E(\lambda_i)$. Then

$$\varphi_i x = \lambda_i x \qquad x \in E(\lambda_i).$$

This implies that the characteristic polynomial of φ_i is given by

$$\det(\varphi_i - \lambda \iota) = (\lambda_i - \lambda)^{k_i} \qquad (i = 1 \ldots r) \tag{8.22}$$

where k_i is the dimension of $E(\lambda_i)$. It follows from (8.21), and (8.22) that the characteristic polynomial of φ is equal to the product

$$\det(\varphi - \lambda\iota) = (\lambda_1 - \lambda)^{k_1} \dots (\lambda_r - \lambda)^{k_r}. \tag{8.23}$$

The representation 8.23 shows that the characteristic polynomial of a selfadjoint transformation has n real zeros, if every zero is counted with its multiplicity. As another consequence of (8.23) we note that the dimension of the eigen-space $E(\lambda_i)$ is equal to the multiplicity of the zero λ_i in the characteristic polynomial.

8.9. The characteristic polynomial of a symmetric matrix. The above result implies that a symmetric $n \times n$-matrix $A = (\alpha_\nu^\mu)$ has n real eigenvalues. In fact, consider the transformation

$$\varphi x_\nu = \sum_\mu \alpha_\nu^\mu x_\mu \qquad (\nu = 1 \dots n)$$

where $x_\nu (\nu = 1 \dots n)$ is an orthonormal basis of E. Then φ is selfadjoint and hence the characteristic polynomial of φ has the form (8.23). At the same time we know that

$$\det(\varphi - \lambda\iota) = \det(A - \lambda J). \tag{8.24}$$

Equations (8.23) and (8.24) yield

$$\det(A - \lambda J) = (\lambda_1 - \lambda)^{k_1} \dots (\lambda_r - \lambda)^{k_r}.$$

8.10. Eigenvectors of bilinear functions. In sec. 8.4 a one-to-one correspondence between all the bilinear functions Φ in E and all the linear transformations $\varphi: E \to E$ has been established. A bilinear function Φ and the corresponding transformation φ are related by the equation

$$\Phi(x, y) = (\varphi x, y) \qquad x, y \in E.$$

Using this relation, we define eigenvectors and eigenvalues of a bilinear function to be the eigenvectors and eigenvalues of the corresponding transformation. Let e be an eigenvector of Φ and λ be the corresponding eigenvalue. Then

$$\Phi(e, y) = (\varphi e, y) = \lambda(e, y) \tag{8.25}$$

for every vector $y \in E$.

Now assume that the bilinear function Φ is symmetric,

$$\Phi(x, y) = \Phi(y, x).$$

Then the corresponding transformation φ is selfadjoint. Consequently, there exists an orthonormal system of n eigenvectors e_ν

$$\varphi e_\nu = \lambda_\nu e_\nu \qquad (\nu = 1 \dots n). \tag{8.26}$$

This implies that

$$\Phi(e_\nu, e_\mu) = \lambda_\nu (e_\nu, e_\mu) = \lambda_\nu \delta_{\nu\mu}.$$

Hence, to every symmetric bilinear function Φ in E there exists an orthonormal basis of E in which the matrix of Φ has diagonal-form.

Problems

1. Prove by direct computation that a symmetric 2×2-matrix has only real eigenvalues.

2. Compute the eigenvalues of the matrix

$$\begin{pmatrix} 4 & -1 & 2 \\ -1 & -2 & -\frac{5}{2} \\ 2 & -\frac{5}{2} & 1 \end{pmatrix}.$$

3. Find the eigenvalues of the bilinear function

$$\Phi(x, y) = \sum_{\nu \neq \mu} \xi^\nu \eta^\mu.$$

4. Prove that the product of two selfadjoint transformations φ and ψ is selfadjoint if and only if $\psi \circ \varphi = \varphi \circ \psi$.

5. A selfadjoint transformation φ is called *positive*, if

$$(x, \varphi x) \geqq 0$$

for every $x \in E$. Given a positive selfadjoint transformation φ, prove that there exists exactly one positive selfadjoint transformation ψ such that $\psi^2 = \varphi$.

6. Given a selfadjoint mapping φ, consider a vector $b \in (\ker \varphi)^\perp$. Prove that there exists exactly one vector $a \in \ker \varphi^\perp$ such that $\varphi a = b$.

7. Let φ be a selfadjoint mapping and let $e_\nu (\nu = 1 \dots n)$ be a system of n orthonormal eigenvectors with eigenvalues λ_ν. Define the mapping φ_λ by

$$\varphi_\lambda = \varphi - \lambda \iota$$

where λ is a real parameter. Prove that

$$\varphi_\lambda^{-1} x = \sum_\nu \frac{(x, e_\nu)}{\lambda_\nu - \lambda} e_\nu \qquad x \in E.$$

provided that λ is not an eigenvalue of φ.

8. Let φ be a linear transformation of a real n-dimensional linear space E. Show that an inner product can be introduced in E such that φ becomes a selfadjoint mapping if and only if φ has n linearly independent eigenvectors.

9. Let φ be a linear transformation of E and $\tilde{\varphi}$ the adjoint map. Denote by $|\varphi|$ the norm of φ which is induced by the Euclidean norm of E (cf. sec. (7.21)). Prove that

$$|\varphi|^2 = \lambda$$

where λ is the largest eigenvalue of the selfadjoint mapping $\tilde{\varphi} \circ \varphi$.

10. Let φ be any linear transformation of an inner product space E. Prove that $\varphi \tilde{\varphi}$ is a positive self-adjoint mapping. Prove that

$$(x, \varphi \tilde{\varphi} x) \geqq 0$$

with equality only if $x \in \ker \varphi$.

11. Prove that a regular linear transformation φ of a Euclidean space can be uniquely written in the form

$$\varphi = \sigma \circ \tau$$

where σ is a positive selfadjoint transformation and τ is a rotation (cf. sec. 8.19).

Hint: Use problems 5 and 10. (This is essentially the *unitary trick* of Weyl).

§ 3. Orthogonal projections

8.11. Definition. A linear transformation $\pi: E \to E$ of an inner product space is called an *orthogonal projection* if it is selfadjoint and satisfies the condition $\pi^2 = \pi$. For every orthogonal projection we have the orthogonal decomposition

$$E = \ker \pi \oplus \operatorname{Im} \pi$$

and the restriction of π to $\operatorname{Im} \pi$ is the identity. Clearly every orthogonal projection is normal. Conversely, a normal transformation φ which satisfies the relation $\varphi^2 = \varphi$ is an orthogonal projection. In fact, since $\varphi^2 = \varphi$ we can write

$$x = \varphi x + x_1 \qquad x_1 \in \ker \varphi .$$

Since φ is normal we have that $\ker \varphi = \ker \tilde{\varphi}$ and so it follows that $x_1 \in \ker \tilde{\varphi}$.

Hence we obtain for an arbitrary vector $y \in E$

$$(x, \varphi y) = (\varphi x, \varphi y) + (x_1, \varphi y) = (\varphi x, \varphi y) + (\tilde{\varphi} x_1, y) = (\varphi x, \varphi y)$$

whence

$$(x, \varphi y) = (y, \varphi x).$$

It follows that φ is selfadjoint.

To every subspace $E_1 \subset E$ there exists precisely one orthogonal projection π such that $\operatorname{Im} \pi = E_1$. It is clear that π is uniquely determined by E_1. To obtain π consider the orthogonal complement $E_1^\perp$ and define π by

$$\pi y = y, y \in E_1; \quad \pi z = 0, z \in E_1^\perp.$$

Then it is easy to verify that $\pi^2 = \pi$ and $\tilde{\pi} = \pi$.

Consider two subspaces E_1 and E_2 of E and the corresponding orthogonal projections $\pi_1 : E \to E$ and $\pi_2 : E \to E$. It will be shown that $\pi_2 \circ \pi_1 = 0$ if and only if E_1 and E_2 are orthogonal to each other. Assume first that $E_1 \perp E_2$. Then $\pi_1 x \in E_2^\perp$ for every vector $x \in E$, whence $\pi_2 \circ \pi_1 = 0$. Conversely, the equation $\pi_2 \circ \pi_1 = 0$ implies that $\pi_1 x \in E_2^\perp$ for every vector $x \in E$, whence $E_1 \in E_2^\perp$.

8.12. Sum of two projections. The sum of two projections $\pi_1 : E \to E$ and $\pi_2 : E \to E$ is again a projection if and only if the subspaces E_1 and E_2 are orthogonal. Assume first that $E_1 \perp E_2$ and consider the transformation $\pi = \pi_1 + \pi_2$. Then

$$\pi(x_1 + x_2) = \pi x_1 + \pi x_2 = x_1 + x_2 \qquad x_1 \in E_1, x_2 \in E_2.$$

Hence, π reduces to the identity-map in the sum $E_1 \oplus E_2$. On the other hand,

$$\pi x = 0 \quad \text{if} \quad x \in E_1^\perp \cap E_2^\perp.$$

But $E_1^\perp \cap E_2^\perp$ is the orthogonal complement of the sum $E_1 \oplus E_2$ and hence π is the projection of E onto $E_1 \oplus E_2$.

Conversely, assume that $\pi_1 + \pi_2$ is a projection. Then

$$(\pi_1 + \pi_2)^2 = \pi_1 + \pi_2,$$

whence

$$\pi_1 \circ \pi_2 + \pi_2 \circ \pi_1 = 0. \tag{8.27}$$

This equation implies that

$$\pi_1 \circ \pi_2 \circ \pi_1 + \pi_2 \circ \pi_1 = 0 \tag{8.28}$$

and

$$\pi_1 \circ \pi_2 + \pi_1 \circ \pi_2 \circ \pi_1 = 0. \tag{8.29}$$

Adding (8.28) and (8.29) and using (8.27) we obtain

$$\pi_1 \circ \pi_2 \circ \pi_1 = 0. \tag{8.30}$$

The equations (8.30) and (8.28) yield

$$\pi_2 \circ \pi_1 = 0.$$

This implies that $E_1 \perp E_2$, as has been shown at the end of sec. 8.11.

8.13. Difference of two projections. The difference $\pi_1 - \pi_2$ of two projections $\pi_1: E \to E$ and $\pi_2: E \to E$ is a projection if and only if E_2 is a subspace of E_1. To prove this, consider the mapping

$$\varphi = \iota - (\pi_1 - \pi_2) = (\iota - \pi_1) + \pi_2 .$$

Since $\iota - \pi_1$ is the projection $E \to E_1^\perp$, it follows that φ is a projection if and only if $E_1^\perp \subset E_2^\perp$, i.e., if and only if $E_1 \supset E_2$. If this condition is fulfilled, φ is the projection onto the subspace $E_1^\perp \oplus E_2$. This implies that $\pi_1 - \pi_2 = \iota - \varphi$ is the projection onto the subspace

$$(E_1^\perp \oplus E_2)^\perp = E_1 \cap E_2^\perp .$$

This subspace is the orthogonal complement of E_2 relative to E_1.

8.14. Product of two projections. The product of two projections $\pi_1: E \to E$ and $\pi_2: E \to E$ is an orthogonal projection if and only if the projections commute. Assume first that $\pi_2 \circ \pi_1 = \pi_1 \circ \pi_2$. Then

$$\pi_2 \pi_1 x = \pi_2 x = x \text{ for every vector } \quad x \in E_1 \cap E_2 . \qquad (8.31)$$

On the other hand, $\pi_2 \circ \pi_1$ reduces to the zero-map in the subspace $(E_1 \cap E_2)^\perp = E_1^\perp + E_2^\perp$. In fact, consider a vector

$$x = x_1^\perp + x_2^\perp \qquad x_1^\perp \in E_1^\perp, x_2^\perp \in E_2^\perp .$$

Then

$$\pi_2 \pi_1 x = \pi_2 \pi_1 x_1^\perp + \pi_2 \pi_1 x_2^\perp = \pi_2 \pi_1 x_1^\perp + \pi_1 \pi_2 x_2^\perp = 0 . \qquad (8.32)$$

Equations (8.31) and (8.32) show that $\pi_2 \circ \pi_1$ is the projection $E \to E_1 \cap E_2$.

Conversely, if $\pi_2 \circ \pi_1$ is a projection, it follows that

$$\pi_2 \circ \pi_1 = \widetilde{\pi_2 \circ \pi_1} = \tilde{\pi}_1 \circ \tilde{\pi}_2 = \pi_1 \circ \pi_2 .$$

Problems

1. Prove that a subspace $J \subset E$ is stable under the projection $\pi: E \to E$ if and only if

$$J = J \cap E_1 \oplus J \cap E_1^\perp .$$

2. Prove that two projections $\pi_1: E \to E$ and $\pi_2: E \to E$ commute if and only if

$$E_1 + E_2 = E_1 \cap E_2 + E_1 \cap E_2^{\perp} + E_1^{\perp} \cap E_2$$

$$E_1 = \operatorname{Im} \pi_1 \qquad E_2 = \operatorname{Im} \pi_2 .$$

3. The *reflection* ϱ of E at a subspace E_1 is defined by

$$\varrho x = p - h$$

where $x = p + h \, (p \in E_1, h \in E_1^{\perp})$. Show that the reflection ϱ and the projection $\pi: E \to E_1$ are related by

$$\varrho = 2\pi - \iota .$$

4. Consider a linear transformation φ of a real linear space E. Prove that an inner product can be introduced in E such that φ becomes an orthogonal projection if and only if $\varphi^2 = \varphi$.

5. Given a selfadjoint mapping φ of E, consider the distinct eigenvalues λ_i and the corresponding eigenspaces $E_i \, (i = 1 \ldots r)$. If π_i denotes the orthogonal projection $E \to E_i$ prove the relations:

a) $\pi_i \circ \pi_j = 0 \quad (i \neq j)$.

b) $\sum_i \pi_i = \iota$.

c) $\sum_i \lambda_i \pi_i = \varphi$.

§ 4. Skew mappings

8.15. Definition. A linear transformation ψ in E is called *skew* if $\tilde{\psi} = -\psi$. The above condition is equivalent to the relation

$$(\psi x, y) + (x, \psi y) = 0 \qquad x, y \in E. \tag{8.33}$$

It follows from (8.33) that the matrix of a skew mapping relative to an orthonormal basis is skew symmetric.

Substitution of $y = x$ in (8.33) yields the equation

$$(x, \psi x) = 0 \qquad x \in E \tag{8.34}$$

showing that every vector is orthogonal to its image-vector. Conversely, a transformation ψ having this property is skew. In fact, replacing x by $x + y$ in (8.34) we obtain

$$(x + y, \psi x + \psi y) = 0 ,$$

whence

$$(y, \psi x) + (x, \psi y) = 0 .$$

It follows from (8.34) that a skew mapping can only have the eigenvalue $\lambda=0$.

The relation $\tilde{\psi}=-\psi$ implies that

$$\operatorname{tr}\psi = 0$$

and

$$\det\psi = (-1)^n \det\psi .$$

The last equation shows that

$$\det\psi = 0$$

if the dimension of E is odd. More generally, it will now be shown that the rank of a skew transformation is always even. Since every skew mapping is normal (see sec. 8.5) the image space is the orthogonal complement of the kernel. Consequently, the induced transformation $\psi_1 : \operatorname{Im}\psi \to \operatorname{Im}\psi$ is regular. Since ψ_1 is again skew, it follows that the dimension of $\operatorname{Im}\psi$ must be even.

It follows from this result that the rank of a skew-symmetric matrix is always even.

8.16. The normal-form of a skew symmetric matrix. In this section it will be shown that to every skew mapping ψ an orthonormal basis a_ν $(\nu=1\ldots n)$ can be constructed in which the matrix of ψ has the form

$$\begin{pmatrix} 0 & \kappa_1 & & & & & & & \\ -\kappa_1 & 0 & & & & & & & \\ & & \cdot & & & & & & \\ & & & \cdot & & & & & \\ & & & & \cdot & & & & \\ & & & & & 0 & \kappa_p & & \\ & & & & & -\kappa_p & 0 & & \\ & & & & & & & 0 & \\ & & & & & & & & \cdot \\ & & & & & & & & & \cdot \\ & & & & & & & & & & \cdot \\ & & & & & & & & & & & 0 \end{pmatrix} \tag{8.35}$$

Consider the mapping $\varphi=\psi^2$. Then $\tilde{\varphi}=\varphi$. According to the result of sec. 8.7, there exists an orthonormal basis $e_\nu\,(\nu=1\ldots n)$ in which φ has the form

$$\varphi e_\nu = \lambda_\nu e_\nu \qquad (\nu = 1 \ldots n).$$

All the eigenvalues λ_ν are negative or zero. In fact, the equation

$$\varphi e = \lambda e \qquad |e| = 1$$

implies that

$$\lambda = (e, \varphi e) = (e, \psi^2 e) = -(\psi e, \psi e) \leqq 0 .$$

Since the rank of ψ is even and ψ^2 has the same rank as ψ, the rank of φ must be even. Consequently, the number of negative eigenvalues is even and we can enumerate the vectors $e_\nu\,(\nu = 1 \ldots n)$ such that

$$\lambda_\nu < 0 \quad (\nu = 1 \ldots 2p) \quad \text{and} \quad \lambda_\nu = 0 \quad (\nu = 2p+1 \ldots n).$$

Define an orthonormal basis $a_\nu (\nu = 1 \ldots n)$ as follows:

$$a_1 = e_1, \qquad a_2 = \frac{1}{\kappa_1}\psi e_1, \quad \text{where } \kappa_1 = \sqrt{-\lambda_1}.$$

Next, choose a vector e_2 which is not a linear combination of a_1 and a_2 and set

$$a_3 = e_2, \qquad a_4 = \frac{1}{\kappa_2}\psi e_2, \qquad \kappa_2 = \sqrt{-\lambda_2}.$$

etc.

In this basis the matrix of ψ has the form (8.35).

Problems

1. Show that every skew mapping φ of a 2-dimensional inner product space satisfies the relation

$$(\varphi x, \varphi y) = \det \varphi \cdot (x, y).$$

2. *Skew transformations of 3-space.* Let E be an oriented Euclidean 3-space.

(i) Show that every vector $a \in E$ determines a skew transformation φ_a of E given by $\varphi_a(x) = a \times x$.

(ii) Show that

$$\varphi_{a \times b} = \varphi_a \circ \varphi_b - \varphi_b \circ \varphi_a \qquad a, b \in E.$$

(iii) Show that every skew map $\varphi\colon E \to E$ determines a unique vector $a \in E$ such that $\varphi = \varphi_a$.

Hint: Consider the skew 3-linear map $\Phi\colon E \times E \times E \to E$ given by $\Phi(x, y, z) = (\varphi x, y) \cdot z + (\varphi y, z) \cdot x + (\varphi z, x) \cdot y$ and choose $a \in E$ to be the vector determined by

$$\Phi(x, y, z) = \Delta(x, y, z) a$$

(cf. sec. 4.3, Proposition III).

(iv) Let e_1, e_2, e_3 be a positive orthonormal basis of E. Show that the vector a of part (iii) is given by

$$a = \alpha_{23}\, e_1 + \alpha_{31}\, e_2 + \alpha_{12}\, e_3$$

where (α_{ij}) is the matrix of φ with respect to this basis.

3. Assume that $\varphi \neq 0$ and ψ are two skew mappings of the 3-space having the same kernel. Prove that $\psi = \lambda\varphi$ where λ is a scalar.

4. Applying the result of problem 3 to the mappings

$$\varphi x = (a_1 \times a_2) \times x$$

and

$$\psi x = a_2(a_1, x) - a_1(a_2, x)$$

prove the formula

$$(a_1 \times a_2) \times a_3 = a_2(a_1, a_3) - a_1(a_2, a_3).$$

5. Prove that a linear transformation $\varphi: E \to E$ satisfies the relation $\tilde{\varphi} = \lambda\varphi$, $\lambda \in \mathbb{R}$ if and only if φ is selfadjoint or skew.

6. Show that every skew symmetric bilinear function Φ in an oriented 3-space E can be represented in the form

$$\Phi(x, y) = (x \times y, a)$$

and that the vector a is uniquely determined by Φ.

7. Prove that the product of a selfadjoint mapping and a skew mapping has trace zero.

8. Prove that the characteristic polynomial of a skew mapping satisfies the equation

$$\chi(-\lambda) = (-1)^n \chi(\lambda).$$

From this relation derive that the coefficient of $\lambda^{n-\nu}$ is zero for every odd ν.

9. Let φ be a linear transformation of a real linear space E. Prove that an inner product can be defined in E such that φ becomes a skew mapping if and only if the following conditions are satisfied: 1. The space E can be decomposed into $\ker \varphi$ and stable planes. 2. The mappings which are induced in these planes have positive determinant and trace zero.

10. Given a skew symmetric 4×4-matrix $A = (\alpha_{\nu\mu})$ verify the identity

$$\det A = (\alpha_{12}\alpha_{34} + \alpha_{13}\alpha_{42} + \alpha_{14}\alpha_{23})^2.$$

§ 5. Isometric mappings

8.17. Definition. Consider two inner product spaces E and F. A linear mapping $\varphi: E \to F$ is called *isometric* or *orthogonal* if the inner product is preserved under φ,

$$(\varphi x_1, \varphi x_2) = (x_1, x_2) \quad x_1, x_2 \in E.$$

Setting $x_1 = x_2 = x$ we find

$$|\varphi x| = |x| \quad x \in E.$$

Conversely, the above relation implies that φ is isometric. In fact,

$$2(\varphi x_1, \varphi x_2) = |\varphi(x_1 + x_2)|^2 - |\varphi x_1|^2 - |\varphi x_2|^2$$
$$= |x_1 + x_2|^2 - |x_1|^2 - |x_2|^2 = 2(x_1, x_2).$$

Since an isometric mapping preserves the norm it is always injective.

We assume in the following that the spaces E and F have the same dimension. Then every isometric mapping $\varphi: E \to F$ is a linear isomorphism of E onto F and hence there exists an inverse isomorphism $\varphi^{-1}: F \to E$. The isometry of φ implies that

$$(\varphi x, y) = (x, \varphi^{-1} y) \qquad x \in E, y \in F,$$

whence

$$\tilde{\varphi} = \varphi^{-1}. \tag{8.36}$$

Conversely every linear isomorphism φ satisfying the equation (8.36) is isometric. In fact,

$$(\varphi x_1, \varphi x_2) = (x_1, \tilde{\varphi} \varphi x_2) = (x_1, \varphi^{-1} \varphi x_2) = (x_1, x_2).$$

The image of an orthonormal basis a_ν $(\nu = 1 \ldots n)$ of E under an isometric mapping is an orthonormal basis of F. Conversely, a linear mapping which sends an orthonormal basis a_ν of E into an orthonormal basis b_ν $(\nu = 1 \ldots n)$ of F is isometric. To prove this, consider two vectors

$$x_1 = \sum_\nu \xi_1^\nu a_\nu \quad \text{and} \quad x_2 = \sum_\nu \xi_2^\nu a_\nu;$$

then

$$\varphi x_1 = \sum_\nu \xi_1^\nu b_\nu \quad \text{and} \quad \varphi x_2 = \sum_\nu \xi_2^\nu b_\nu,$$

whence

$$(\varphi x_1, \varphi x_2) = \sum_{\nu, \mu} \xi_1^\nu \xi_2^\mu (b_\nu, b_\mu) = \sum_{\nu, \mu} \xi_1^\nu \xi_2^\mu \delta_{\nu\mu} = \sum_\nu \xi_1^\nu \xi_2^\nu = (x_1, x_2).$$

It follows from this remark that an isometric mapping can be defined between any two inner product spaces E and F of the same dimension: Select orthonormal bases a_ν and b_ν $(\nu = 1 \ldots n)$ in E and in F respectively and define φ by $\varphi a_\nu = b_\nu$ $(\nu = 1 \ldots n)$.

8.18. The condition for the matrix. Assume that an isometric mapping $\varphi: E \to F$ is given. Employing two bases a_ν and b_ν $(\nu = 1 \ldots n)$ we obtain from φ an $n \times n$-matrix α_ν^μ by the equations

$$\varphi a_\nu = \sum_\mu \alpha_\nu^\mu b_\mu.$$

Then the equations

$$(\varphi a_\nu, \varphi a_\mu) = (a_\nu, a_\mu)$$

can be written as

$$\sum_{\lambda,\kappa} \alpha_\nu^\lambda \alpha_\mu^\kappa (b_\lambda, b_\kappa) = (a_\nu, a_\mu).$$

Introducing the matrices

$$g_{\nu\mu} = (a_\nu, a_\mu) \quad \text{and} \quad h_{\lambda\kappa} = (b_\lambda, b_\kappa)$$

we obtain the relation

$$\sum_{\lambda,\kappa} \alpha_\nu^\lambda \alpha_\mu^\kappa h_{\lambda\kappa} = g_{\nu\mu}. \tag{8.37}$$

Conversely, (8.37) implies that the inner products of the basis vectors are preserved under φ and hence that φ is an isometric mapping.

If the bases a_ν and b_ν are orthonormal,

$$g_{\nu\mu} = \delta_{\nu\mu}, \quad h_{\lambda\kappa} = \delta_{\lambda\kappa},$$

relation (8.37) reduces to

$$\sum_\lambda \alpha_\nu^\lambda \alpha_\mu^\lambda = \delta_{\nu\mu}$$

showing that the matrix of an isometric mapping relative to orthonormal bases is orthogonal.

8.19. Rotations. A *rotation* of an inner product space E is an isometric mapping of E into itself. Formula (8.36) implies that

$$(\det \varphi)^2 = 1$$

showing that the determinant of a rotation is ± 1.

A rotation is called *proper* if $\det \varphi = +1$ and *improper* if $\det \varphi = -1$.

Every eigenvalue of a rotation is ± 1. In fact, the equation $\varphi e = \lambda e$ implies that $|e| = |\lambda|\,|e|$, whence $\lambda = \pm 1$. A rotation need not have eigenvecvectors as can already be seen in the plane (cf. sec. 4, 17).

Suppose now that the dimension of E is odd and let φ be a proper rotation. Then it follows from sec. 4.20 that φ has at least one positive eigenvalue λ. On the other hand we have that $\lambda = \pm 1$ whence $\lambda = 1$. Hence, every proper rotation of an odd-dimensional space has the eigenvalue 1. The corresponding eigenvector e satisfies the equation $\varphi e = e$; that is, e remains invariant under φ. A similar argument shows that to every improper rotation of an odd-dimensional space there exists a vector e such that $\varphi e = -e$. If the dimension of E is even, nothing can be said about the existence of eigenvalues for a proper rotation. However, to every improper rotation, there is at least one invariant vector and at least one vector e such that $\varphi e = -e$ (cf. sec. 4.20).

Let $\varphi: E \to E$ be a rotation and assume that $F \subset E$ is a stable subspace. Then the orthogonal complement $F^{\perp}$ is stable as well. In fact, if $z \in F^{\perp}$ is arbitrary we have for every $y \in F$

$$(\varphi z, y) = (z, \varphi^{-1} y) = 0$$

whence $\varphi z \in F^{\perp}$.

The product of two rotations is obviously again a rotation and the inverse of a rotation is also a rotation. In other words, the set of all rotations of an n-dimensional inner product space forms a group, called the *general orthogonal group*. The relation

$$\det(\varphi_2 \circ \varphi_1) = \det \varphi_2 \cdot \det \varphi_1$$

implies that the set of all proper rotations forms a subgroup, the *special orthogonal group*.

A linear transformation of the form $\lambda\varphi$ where $\lambda \geqq 0$ and φ is a proper rotation is called a *homothety*.

8.20. Decomposition into stable planes and straight lines. With the aid of the results of § 2 it will now be shown that for every rotation φ there exists an orthogonal decomposition of E into stable subspaces of dimension 1 and 2. Denote by E_1 and E_2 the eigenspaces which correspond to the eigenvalues $\lambda = +1$ and $\lambda = -1$ respectively. Then E_1 is orthogonal to E_2. In fact, let $x_1 \in E_1$ and $x_2 \in E_2$ be two arbitrary vectors. Then

$$\varphi x_1 = x_1 \quad \text{and} \quad \varphi x_2 = -x_2 .$$

These equations yield

$$(x_1, x_2) = -(x_1, x_2),$$

whence $(x_1, x_2) = 0$.

It follows from sec. 8.19 that the subspace $F = (E_1 \oplus E_2)^{\perp}$ is again stable under φ. Moreover, F does not contain an eigenvector of φ and hence F has even dimension. Now consider the selfadjoint mapping

$$\psi = \varphi + \tilde{\varphi} = \varphi + \varphi^{-1}$$

of F. The result of sec. 8.6 assures that there exists an eigenvector e of ψ. If λ denotes the corresponding eigenvalue we have the relation

$$\varphi e + \varphi^{-1} e = \lambda e .$$

Applying φ we obtain

$$\varphi^2 e = \lambda \varphi e - e. \tag{8.38}$$

Since there are no eigenvectors of φ in F the vectors e and φe are linearly independent and hence they generate a plane F_1. Equation (8.38) shows that this plane is stable under φ. The induced mapping is a *proper* rotation (otherwise there would be eigenvectors in F_1).

The orthogonal complement $F_1^\perp$ of F_1 with respect to F is again stable under φ and hence the same construction can be applied to $F_1^\perp$. Continuing in this way we finally obtain an orthogonal decomposition of F into mutually orthogonal stable planes.

Now select orthonormal bases in E_1, E_2 and in every stable plane. These bases determine together an orthonormal basis of E. In this basis the matrix of φ has the form

$$\begin{pmatrix} \varepsilon_1 & & & & & & & \\ & \ddots & & & & & & \\ & & \varepsilon_p & & & & & \\ & & & \cos\theta_1 & \sin\theta_1 & & & \\ & & & -\sin\theta_1 & \cos\theta_1 & & & \\ & & & & & \ddots & & \\ & & & & & & \cos\theta_k & \sin\theta_k \\ & & & & & & -\sin\theta_k & \cos\theta_k \end{pmatrix} \qquad \begin{array}{l} \varepsilon_\nu = \pm 1 \quad (\nu = 1 \dots p) \\ 2k = n - p \end{array}$$

where θ_i $(i = 1 \dots k)$ denotes the corresponding rotation angle (cf. sec. 8.21).

Problems

1. Given a skew transformation ψ of E, prove that

$$\varphi = (\psi + \iota) \circ (\psi - \iota)^{-1}$$

is a rotation and that -1 is not an eigenvalue of φ. Conversely, if φ is a rotation, not having the eigenvalue -1 prove that

$$\psi = (\varphi - \iota) \circ (\varphi + \iota)^{-1}$$

is a skew mapping.

2. Let φ be a regular linear transformation of a Euclidean space E such that $(\varphi x, \varphi y) = 0$ whenever $(x, y) = 0$. Prove that φ is of the form $\varphi = \lambda \tau$, $\lambda \neq 0$ where τ is a rotation.

3. Assume two inner products Φ and Ψ in E such that all oriented angles with respect to Φ and Ψ coincide. Prove that $\Psi(x, y)=\lambda\Phi(x, y)$ where $\lambda>0$ is a constant.

4. Prove that every normal non-selfadjoint transformation of a plane is a homothety.

5. Let φ be a mapping of the inner product space E into itself such that $\varphi 0=0$ and

$$|\varphi x-\varphi y|=|x-y| \qquad x, y \in E.$$

Prove that φ is then linear.

6. Prove that to every proper rotation φ there exists a continuous family $\varphi_t\,(0\leqq t\leqq 1)$ of rotations such that $\varphi_0=\varphi$ and $\varphi_1=\iota$.

7. Let φ be a linear automorphism of an n-dimensional real linear space E. Show that an inner product can be defined in E such that φ becomes a rotation if and only if the following conditions are fulfilled:

(i) The space E can be decomposed into stable planes and stable straight lines.

(ii) Every stable straight line remains pointwise fixed or is reflected at the point 0.

(iii) In every irreducible invariant plane a linear automorphism ψ is induced such that

$$\det\psi=1 \quad \text{and} \quad |\operatorname{tr}\psi|<2.$$

8. If φ is a rotation of an n-dimensional Euclidean space, show that $|\operatorname{tr}\varphi|\leqq n$.

9. Prove that the characteristic polynomial of a proper rotation satisfies the relation

$$f(\lambda)=(-\lambda)^n f(\lambda^{-1}).$$

10. Let E be an inner product space of dimension $n>2$. Consider a proper rotation τ which commutes with all proper rotations. Prove that $\tau=\varepsilon\iota$ where $\varepsilon=1$ if n is odd and $\varepsilon=\pm 1$ if n is even.

§ 6. Rotations of Euclidean spaces of dimension 2, 3 and 4

8.21. Proper rotations of the plane. Let E be an oriented Euclidean plane and let Δ denote the normed positive determinant function in E. Then a linear transformation $j\colon E\to E$ is determined by the equation

$$\Delta(x, y)=(jx, y) \qquad x, y \in E.$$

The transformation j has the following properties:

1) $(jx, y) + (x, jy) = 0$
2) $(jx, jy) = (x, y)$
3) $j^2 = -\iota$
4) $\det j = 1$.

In fact, 1) follows directly from the definition. To verify 2) observe that the identity (7.24), sec. 7.13, implies that

$$(x, y)^2 + (jx, y)^2 = |x|^2 \cdot |y|^2 \qquad x, y \in E.$$

Setting $y = jx$ we obtain, in view of 1),

$$(jx, jx)^2 = |jx|^2 \cdot |x|^2.$$

Since j is injective (as follows from the definition) we obtain

$$|jx| = |x|.$$

Now the relations $\tilde{j} = -j$ and $\tilde{j} = j^{-1}$ imply that $j^2 = -\iota$. Finally, we obtain from 1), 2), 3) and from the definition of j

$$\Delta(jx, jy) = (j^2 x, jy) = -(x, jy) = (jx, y) = \Delta(x, y)$$

whence

$$\det j = 1.$$

The transformation j is called the *canonical complex structure* of the oriented Euclidean plane E.

Next, let φ be any proper rotation of E. Fix a non-zero vector x and denote by Θ the oriented angle between x and φx (cf. sec. 7.13). It is determined mod 2π by the equation

$$\varphi x = x \cdot \cos \Theta + j(x) \cdot \sin \Theta. \tag{8.39}$$

We shall show that

$$\cos \Theta = \tfrac{1}{2} \operatorname{tr} \varphi \qquad \text{and} \qquad \sin \Theta = \tfrac{1}{2} \operatorname{tr}(j \circ \varphi). \tag{8.40}$$

In fact, by definition of the trace we have

$$\Delta(\varphi x, y) + \Delta(x, \varphi y) = \Delta(x, y) \cdot \operatorname{tr} \varphi. \tag{8.41}$$

Inserting (8.39) into (8.41) and using properties 2) and 3) of j we obtain

$$2 \cos \Theta \cdot \Delta(x, y) = \Delta(x, y) \cdot \operatorname{tr} \varphi$$

and so the first relation (8.40) follows. The second relation is proved in a similar way.

Relations (8.40) show that the angle Θ is independent of x. It is called the *rotation angle of* φ and is denoted by $\Theta(\varphi)$.

Now (8.39) reads

$$\varphi x = x \cdot \cos\Theta(\varphi) + j(x) \cdot \sin\Theta(\varphi) \qquad x \in E.$$

In particular, we have

$$\Theta(\iota) = 0, \qquad \Theta(-\iota) = \pi, \qquad \Theta(j) = \frac{\pi}{2}.$$

If

$$\psi x = x \cdot \cos\Theta(\psi) + j(x) \cdot \sin\Theta(\psi)$$

is a second proper rotation with rotation angle $\Theta(\psi)$ a simple calculation shows that

$$(\varphi \circ \psi)\, x = (\psi \circ \varphi)\, x = x \cdot \cos\big(\Theta(\varphi) + \Theta(\psi)\big) + j(x) \cdot \sin\big(\Theta(\varphi) + \Theta(\psi)\big).$$

Thus any two proper rotations commute and the rotation angle for their product is given by

$$\Theta(\psi \circ \varphi) = \Theta(\varphi) + \Theta(\psi) \qquad (\text{mod } 2\pi). \tag{8.42}$$

Finally observe that if e_1, e_2 is a positive orthonormal basis of E then we have the relations

$$\varphi e_1 \equiv e_1 \cos\theta(\varphi) + e_2 \sin\theta(\varphi) \qquad \varphi e_2 = -e_1 \sin\theta(\varphi) + e_2 \cos\theta(\varphi).$$

Remark: If E is a non-oriented plane we can still assign a rotation angle to a proper rotation φ. It is defined by the equation

$$\cos\Theta(\varphi) = \tfrac{1}{2}\operatorname{tr}\varphi \qquad (0 \leqq \Theta(\varphi) \leqq \pi). \tag{8.43}$$

Observe that in this case $\Theta(\varphi)$ is always between 0 and π whereas in the oriented case $\Theta(\varphi)$ can be normalized between $-\pi$ and π.

8.22. Proper rotations of 3-space. Consider a proper rotation φ of a 3-dimensional inner product space E. As has been shown in sec. 8.19, there exists a 1-dimensional subspace E_1 of E whose vectors remain fixed. If φ is different from the identity-map there are no other invariant vectors (an invariant vector is a vector $x \in E$ such that $\varphi x = x$).

In fact, assume that a and b are two linearly independent invariant vectors. Let $c\,(c \neq 0)$ be a vector which is orthogonal to a and to b. Then $\varphi c = \lambda c$ where $\lambda = \pm 1$. Now the equation $\det \varphi = 1$ implies that $\lambda = +1$ showing that φ is the identity.

In the following discussion it is assumed that $\varphi \neq \iota$. Then the invariant vectors generate a 1-dimensional subspace E_1 called the *axis* of φ.

To determine the axis of a given rotation φ consider the skew mapping

$$\psi = \tfrac{1}{2}(\varphi - \tilde{\varphi}) \tag{8.44}$$

and introduce an orientation in E. Then ψ can be written in the form

$$\psi x = u \times x \qquad u \in E \tag{8.45}$$

(cf. problem 2, § 4). The vector u which is uniquely determined by φ is called the *rotation-vector*. The rotation vector is contained in the axis of φ. In fact, let $a \neq 0$ be a vector on the axis. Then equations (8.45) and (8.44) yield

$$u \times a = \psi a = \tfrac{1}{2}(\varphi a - \tilde{\varphi} a) = \tfrac{1}{2}(\varphi a - \varphi^{-1} a) = 0 \tag{8.46}$$

showing that u is a multiple of a. Hence (8.45) can be used to find the rotation axis provided that the rotation vector is different from zero.

This exceptional case occurs if and only if $\varphi = \tilde{\varphi}$ i.e. if and only if $\varphi = \varphi^{-1}$. Then φ has the eigenvalues $1, -1$ and -1. In other words, φ is a reflection at the rotation axis.

8.23. The rotation angle. Consider the plane F which is orthogonal to E_1. Then φ transforms F into itself and the induced rotation φ_1 is again proper. Denote by θ the rotation angle of φ_1 (cf. remark at the end of see 8.21). Then, in view of (8.43),

$$\cos\theta = \tfrac{1}{2}\operatorname{tr}\varphi_1 .$$

Observing that

$$\operatorname{tr}\varphi = \operatorname{tr}\varphi_1 + 1$$

we obtain the formula

$$\cos\theta = \tfrac{1}{2}(\operatorname{tr}\varphi - 1).$$

To find a formula for $\sin\theta$ consider the orientation of F which is induced by E and by the vector u (cf. sec. 4.29)*). This orientation is represented by the normed determinant function

$$\Delta_1(y, z) = \frac{1}{|u|}\Delta(u, y, z)$$

where Δ is the normed determinant function representing the orientation of E. Then formula (7.25) yields

$$\sin\theta = \Delta_1(y, \varphi y) = \frac{1}{|u|}\Delta(u, y, \varphi y) \tag{8.47}$$

*) It is assumed that $u \neq 0$.

where y is an arbitrary unit vector of F. Now

$$\begin{aligned}\Delta(u, y, \varphi y) &= \det\varphi\,\Delta(\varphi^{-1}u, \varphi^{-1}y, y)\\ &= \Delta(u, \varphi^{-1}y, y) = -\Delta(u, y, \varphi^{-1}y)\end{aligned}$$

and hence equation (8.47) can be written as

$$\sin\theta = -\frac{1}{|u|}\Delta(u, y, \varphi^{-1}y). \tag{8.48}$$

By adding formulae (8.47) and (8.48) we obtain

$$\sin\theta = \frac{1}{2|u|}\Delta(u, y, \varphi y - \varphi^{-1}y) = \frac{1}{|u|}\Delta(u, y, \psi y). \tag{8.49}$$

Inserting the expression (8.45) in (8.49), we thus obtain

$$\sin\theta = \frac{1}{|u|}\Delta(u, y, u \times y) = \frac{1}{|u|}|u \times y|^2. \tag{8.50}$$

Since y is a unit-vector orthogonal to u, it follows that

$$|u \times y| = |u|\,|y| = |u|$$

and hence (8.50) yields the formula

$$\sin\theta = |u|.$$

This equation shows that $\sin\theta$ is positive and hence that $0<\theta<\pi$ if the above orientation of F is used.

Altogether we thus obtain the following geometric significance of the rotation-vector u:

1. u is contained in the axis of φ.
2. The norm of u is equal to $\sin\theta$.
3. If the orientation induced by u is used in F, then θ is contained in the interval $0<\theta<\pi$.

Let us now compare the rotations φ and φ^{-1}. φ^{-1} has obviously the same axis as φ. Observing that $\varphi^{-1} = \tilde{\varphi}$ we see that the rotation vector of φ^{-1} is $-u$. This implies that the inverse rotation induces the inverse orientation in the plane F.

To obtain an explicit expression for u select a positive orthonormal basis e_1, e_2, e_3 in E and let α_ν^μ be the corresponding matrix of φ. Then ψ has the matrix

$$\beta_\nu^\mu = \tfrac{1}{2}(\alpha_\nu^\mu - \alpha_\mu^\nu)$$

and the components of u are given by

$$u^1 = \tfrac{1}{2}(\alpha_2^3 - \alpha_3^2) \quad u^2 = \tfrac{1}{2}(\alpha_3^1 - \alpha_1^3) \quad u^3 = \tfrac{1}{2}(\alpha_1^2 - \alpha_2^1).$$

It should be observed that a proper rotation is not uniquely determined by its rotation vector. In fact, if φ_1 and φ_2 are two rotations about the same axis whose rotation angles satisfy $\theta_1 + \theta_2 = \pi$, then the rotation vectors of φ_1 and φ_2 coincide. Conversely, if φ_1 and φ_2 have the same rotation vector, then their rotation axes coincide and the rotation angles satisfy either $\theta_1 = \theta_2$ or $\theta_1 + \theta_2 = \pi$.

Since $\cos(\pi - \theta) = -\cos\theta$ it follows from the remark above that a rotation is completely determined by its rotation vector and the cosine of the rotation angle.

8.24. Proper rotations and quaternions. Let E be an oriented 4-dimensional Euclidean space. Make E into the algebra of quaternions as described in sec. 7.23. Fix a unit vector a and consider the linear transformation

$$\varphi x = a x \qquad x \in E.$$

Then φ is a proper rotation. In fact,

$$|\varphi x| = |a x| = |a|\,|x| = |x|.$$

To show that φ is proper choose a continuous map f from the closed unit interval to S^3 such that

$$f(0) = e, \qquad f(1) = a$$

and set

$$\varphi_t(x) = f(t)\,x \qquad x \in E.$$

Then every map φ_t is a rotation whence

$$\det \varphi_t = \pm 1 \qquad 0 \leqq t \leqq 1.$$

Since $\det \varphi_t$ is a continuous function of t it follows that

$$\det \varphi_1 = \det \varphi_0 = \det \iota = +1.$$

This implies that

$$\det \varphi = +1$$

and so φ is a proper rotation. In the same way it follows that the linear transformation

$$\psi x = x a \qquad x \in E$$

is a proper rotation.

Next, let a and b be fixed unit vectors and consider the proper rotation τ of E given by

$$\tau(x) = a\,x\,b^{-1} \qquad x \in E. \tag{8.51}$$

If $a = b = p$ (p a unit vector) we have

$$\tau(x) = p\,x\,p^{-1} \qquad x \in E. \tag{8.52}$$

It follows that $\tau(e) = e$ and so τ induces a proper rotation, τ_1, of the orthogonal complement, E_1, of E. We shall determine the axis and the rotation angle of τ_1.

Let $q \in E_1$ be the vector given by

$$q = p - \lambda e \qquad \lambda = (p, e).$$

Then

$$\tau_1(q) = \tau(p) - \lambda\,\tau(e) = p - \lambda e = q$$

and so q determines the axis of the rotation τ_1. We shall assume that $p \neq \pm e$ so that $q \neq 0$.

To determine the rotation angle of τ_1 consider the 2-dimensional subspace, F of E_1 orthogonal to q. We shall show that, for $z \in F$,

$$\tau z = (2\lambda^2 - 1)\,z + 2\lambda\, q \times z. \tag{8.53}$$

In fact, the equations

$$p = \lambda e + q \quad \text{and} \quad p^{-1} = \lambda e - q$$

yield

$$p\,z\,p^{-1} = (\lambda e + q)\,z(\lambda e - q) = \lambda^2 z + \lambda(q\,z - z\,q) - q\,z\,q.$$

Since

$$q\,z - z\,q = 2(q \times z)$$

and

$$q\,z + z\,q = -2(q, z)e = 0$$

we obtain

$$p\,z\,p^{-1} = (2\lambda^2 - 1)\,z + 2\lambda(q \times z) \qquad z \in F$$

and so (8.53) follows.

Let F have the orientation induced by the orientation of E_1 and the vector q (cf. sec. 4.29). Then the rotation angle Θ is determined by the equations (cf. see 8.21)

$$\cos\Theta = (z, \tau z)$$

and

$$\sin\Theta = \frac{1}{|q|}\,\Delta(q, z, \tau z) \qquad 0 \leqq \Theta \leqq 2\pi$$

where z is any unit vector in F.

Using formula (8.53) we obtain

$$\cos\Theta = 2\lambda^2 - 1$$

and

$$\sin\Theta = \frac{2\lambda}{|q|}\Delta(q, z, q\times z) = \frac{2\lambda}{|q|}|q\times z|^2 = 2\lambda\,|q|.$$

Since

$$\lambda = \cos\omega \quad\text{and}\quad |q| = \sin\omega$$

where ω denotes the angle between e and p $(0<\omega<\pi)$ these relations can be written in the form

$$\cos\Theta = 2\cos^2\omega - 1 = \cos 2\omega$$

and

$$\sin\Theta = 2\cos\omega\sin\omega = \sin 2\omega.$$

These relations imply that

$$\Theta = 2\omega.$$

Thus we have shown that the axis of τ_1 is generated by q and that the rotation angle of τ_1 is twice the angle between p and e.

Proposition I: Two unit quaternions p_1 and p_2 determine the same rotation of E_1 only if $p_2 = \pm p_1$. Moreover, every proper rotation of E_1 can be represented in the form (8.52).

Proof: The first part of the proposition follows directly from the result above. To prove the second part let σ be any proper rotation of E_1. We may assume that $\sigma \neq \iota$. Let a be a unit vector such that $\sigma a = a$ and let $F \subset E_1$ be the plane orthogonal to a. Give F the orientation induced by a and denote the rotation angle of σ by ϑ $(0<\vartheta<2\pi)$. Set

$$p = e\cos\frac{\vartheta}{2} + a\sin\frac{\vartheta}{2}.$$

Then the rotation τ given by

$$\tau x = p\,x\,p^{-1} \qquad x\in E_1$$

coincides with σ. In fact, if q is the vector determined by $p = \lambda e + q$, $q\in E_1$, then

$$q = a\sin\frac{\vartheta}{2}$$

and so τ and σ have the same rotation axis.

Next, observe that since $\sin\frac{\vartheta}{2} > 0$, q is a positive multiple of a and so the vectors q and a induce the same orientation in F. But, as has been shown above, the rotation angle of τ with respect to this orientation is given by

$$\Theta = 2 \cdot \frac{\vartheta}{2} = \vartheta .$$

It follows that $\tau = \sigma$. This completes the proof.

Proposition II: Every proper rotation of E can be written in the form (8.51). Moreover, if a_1, b_1 and a_2, b_2 are two pairs of unit vectors such that the corresponding rotations coincide, then

$$a_2 = \varepsilon a_1 \quad \text{and} \quad b_2 = \varepsilon b,$$

where $\varepsilon = \pm 1$.

Proof: Let σ be a proper rotation of E and define τ by

$$\tau(x) = \sigma(x)\,\sigma(e)^{-1} \qquad x \in E .$$

Then τ is again a proper rotation and satisfies $\tau(e) = e$. Thus τ restricts to a proper rotation of E_1. Hence, by Proposition I, there is a unit vector p such that

$$\tau(x) = p\,x\,p^{-1} \qquad x \in E .$$

It follows that

$$\sigma(x) = \tau(x)\,\sigma(e) = p\,x\,p^{-1}\,\sigma(e) = a\,x\,b^{-1}$$

with $a = p$ and $b = \sigma(e)^{-1}\,p$.

Finally, assume that

$$a_1\,x\,b_1^{-1} = a_2\,x\,b_2^{-1} \qquad x \in E .$$

Setting $x = e$ yields

$$a_1\,b_1^{-1} = a_2\,b_2^{-1} .$$

Now set

$$p = a_2^{-1}\,a_1 = b_2^{-1}\,b_1 .$$

Then we obtain

$$p\,x\,p^{-1} = x \qquad x \in E .$$

Now Proposition I implies that $p = \varepsilon e$, $\varepsilon = \pm 1$ and so

$$a_2 = \varepsilon a_1 , \qquad b_2 = \varepsilon b_1 .$$

This completes the proof.

Problems

1. Show that the rotation vector of a proper rotation φ of 3-space is zero if and only if φ is of the form

$$\varphi = \iota \quad \text{or} \quad \varphi x = -x + 2(x, e)\, e$$

where e is a unit vector.

2. Let φ be a linear automorphism of a real 2-dimensional linear space E. Prove that an inner product can be introduced in E such that φ becomes a proper rotation if and only if

$$\det \varphi = 1 \quad \text{and} \quad |\operatorname{tr} \varphi| \leqq 2 .$$

3. Consider the set H of all homothetic transformations φ of the plane. Prove:

a) If $\varphi_1 \in H$ and $\varphi_2 \in H$, then $\lambda \varphi_1 + \mu \varphi_2 \in H$.

b) If the multiplication is defined in H in the natural way, the set H becomes a commutative field.

c) Choose a fixed unit-vector e. Then, to every vector $x \in E$ there exists exactly one homothetic mapping φ_x such that $\varphi_x e = x$. Define a multiplication in E by $xy = \varphi_x y$. Prove that E becomes a field under this multiplication and that the mapping $x \to \varphi_x$ defines an isomorphism of E onto H.

d) Prove that E is isomorphic to the field of complex numbers.

4. Given an improper rotation φ of the plane construct an orthonormal basis e_1, e_2 such that $\varphi e_1 = e_1$ and $\varphi e_2 = -e_2$.

5. Show that every skew mapping ψ of the plane is homothetic. If $\psi \neq 0$, prove that the angle of the corresponding rotation is equal to $+\frac{\pi}{2}$ if the orientation is defined by the determinant function

$$\Delta(x, y) = (\psi x, y) \qquad x, y \in E .$$

6. Find the axis and the angle of the rotation defined by

$$\begin{aligned}
\varphi e_1 &= \tfrac{1}{3}(-e_1 + 2e_2 - 2e_3) \\
\varphi e_2 &= \tfrac{1}{3}(2e_1 + 2e_2 + e_3) \\
\varphi e_3 &= \tfrac{1}{3}(2e_1 - e_2 - 2e_3)
\end{aligned}$$

where $e_\nu\ (\nu = 1, 2, 3)$ is a positive orthonormal basis.

7. If φ is a proper rotation of the 3-space, prove the relation

$$\det(\varphi + \iota) = 4(1 + \cos\theta)$$

where θ is the rotation angle.

8. Consider an orthogonal 3×3-matrix (α_ν^μ) whose determinant is $+1$. Prove the relation

$$\left(\sum_\nu \alpha_\nu^\nu - 1\right)^2 + \sum_{\nu<\mu} (\alpha_\mu^\nu - \alpha_\nu^\mu)^2 = 4.$$

9. Let e be a unit-vector of an oriented 3-space and $\theta\,(-\pi < \theta \leqq \pi)$ be a given angle. Denote by F the plane orthogonal to e. Consider the proper rotation φ whose axis is generated by e and whose angle is θ if the orientation induced by e is used in F. Prove the formula

$$\varphi x = x \cos\theta + e(e,x)(1 - \cos\theta) + (e \times x)\sin\theta.$$

10. Prove that two proper rotations of the 3-space commute if and only if they have the same axis.

11. Let φ be a proper rotation of the 3-space not having the eigenvalue -1.

Prove that the skew transformations

$$\chi = (\varphi - \iota) \circ (\varphi + \iota)^{-1} \quad \text{and} \quad \psi = \tfrac{1}{2}(\varphi - \tilde{\varphi})$$

are connected by the equation

$$\chi = \frac{1}{1 + \cos\theta}\psi$$

where θ denotes the rotation-angle of φ.

12. Assume that an improper rotation $\varphi \neq -\iota$ of the Euclidean 3-space is given.

a) Prove that the vectors x for which $\varphi x = -x$, form a 1-dimensional subspace E_1.

b) Prove that a proper rotation φ_1 is induced in the plane F orthogonal to E_1. Defining the rotation-vector u as in sec. 8.22, prove that φ_1 is the identity if and only if $u=0$.

c) Show that the rotation-angle of φ_1 is given by

$$\cos\theta = \tfrac{1}{2}(\operatorname{tr}\varphi + 1)$$

and that $0 < \theta < \pi$ if the induced orientation is used in F.

13. Let a be a vector in an oriented Euclidean 3-space such that $|a| \leqq 1$. Consider the linear transformation φ_a given by

$$\varphi_a x = x\cos(\pi \cdot |a|) + \tfrac{1}{2} a \cdot (a,x) f\left(\frac{|a|}{2}\right)^2 + (a \times x) f(|a|)$$

where f is defined by

$$f(t) = \frac{1}{t}\sin \pi t \qquad t \in \mathbb{R}.$$

(i) Show that φ_a is a proper rotation whose axis is generated by a and whose rotation angle is $\Theta = \pi \cdot |a|$.

(ii) Show that $\varphi_0 = \iota$ and that

$$\varphi_a x = -x + 2a(a, x)$$

if $|a| = 1$.

(iii) Suppose that $a \neq b$. Show that $\varphi_a = \varphi_b$ if and only if $|a| = 1$ and $b = -a$. Conclude that there is a $1-1$ correspondence between the proper rotations of $\mathbb{R}^3$ and the straight lines in $\mathbb{R}^4$.

14. Let E be an oriented 3-dimensional inner product space.

a) Consider E together with the cross product as an algebra. Show that the set of non-zero endomorphisms of this algebra is precisely the group of proper rotations of E.

b) Suppose a multiplication is defined in E such that every proper rotation τ is an endomorphism,

$$\tau(xy) = \tau x \cdot \tau y .$$

Show that

$$xy = \lambda(x \times y)$$

where λ is a constant.

15. Let σ be a skew linear transformation of the Euclidean space $\mathbb{H}$.

a) Show that σ can be written in the form

$$\sigma x = px + xq \qquad x \in \mathbb{H}$$

where $p \in E_1$, $q \in E_1$ ($E_1 = e^{\perp}$), and that the vectors p and q are uniquely determined by σ.

b) Show that E_1 is stable under σ if and only if $q = -p$.

c) Establish the formula

$$\det \sigma = (|p|^2 - |q|^2)^2 .$$

16. Let p be a unit vector in E and consider the rotation $\tau x = pxp^{-1}$. Show that the rotation vector (cf. sec. 8.22) of τ is given by

$$u = 2\lambda(p - \lambda e) \qquad \lambda = (p, e) .$$

17. Let $p \neq \pm e$ be a unit quaternion. Denote by F the plane spanned by e and p and let $F^{\perp}$ be its orthogonal complement. Orient F so that the vectors e, p form a positive basis and give $F^{\perp}$ the induced orientation. Consider the rotations

$$\varphi x = px \quad \text{and} \quad \psi x = xp .$$

a) Show that the planes F and $F^{\perp}$ are stable under φ and ψ and that $\varphi=\psi$ in F.

b) Denote by Θ the common rotation angle of φ and ψ in F and by $\Theta_{\varphi}^{\perp}$, $\Theta_{\psi}^{\perp}$ the rotation angles for φ and ψ in $F^{\perp}$. Show that $\Theta_{\varphi}^{\perp}=\Theta$ and $\Theta_{\psi}^{\perp}=-\Theta$.

§ 7. Differentiable families of linear automorphisms

8.25. Differentiation formulae. Let E be an n-dimensional inner product space and let $L(E;E)$ be the space of all linear transformations of E. It has been shown in sec. 7.21 that a norm is defined in the space $L(E;E)$ by the equation

$$\|\varphi\| = \max_{|x|=1} |\varphi x|.$$

A continuous mapping $t\to\varphi(t)$ of a closed interval $t_0\leqq t\leqq t_1$ into the space $L(E;E)$ will be called a *continuous family of linear transformations* or a *continuous curve in* $L(E;E)$. A continuous curve $\varphi(t)$ is called *differentiable* if the limit

$$\lim_{\Delta t\to 0}\frac{\varphi(t+\Delta t)-\varphi(t)}{\Delta t}=\dot{\varphi}(t)$$

exists for every $t\,(t_0\leqq t\leqq t_1)$. The mapping $\dot{\varphi}$ is obviously again linear for every fixed t.

The following formulae are immediate consequences of the above definition:

1. $(\lambda\varphi+\mu\psi)^{\cdot}=\lambda\dot{\varphi}+\mu\dot{\psi}$ (λ, μ constants)
2. $(\psi\circ\varphi)^{\cdot}=\dot{\psi}\circ\varphi+\psi\circ\dot{\varphi}$
3. $\dot{\tilde{\varphi}}=\tilde{\dot{\varphi}}$
4. If $\varphi_\nu(t)\,(\nu=1\ldots p)$ are p differentiable curves in $L(E;E)$ and Φ is a p-linear function in $L(E;E)$, then

$$\frac{d}{dt}\Phi(\varphi_1(t)\ldots\varphi_p(t))=\sum_{\nu=1}^{p}\Phi(\varphi_1(t)\ldots\dot{\varphi}_\nu(t)\ldots\varphi_p(t)).$$

A curve $\varphi(t)\,(t_0\leqq t\leqq t_1)$ is called *continuously differentiable* if the mapping $t\to\dot{\varphi}(t)$ is again continuous. Throughout this paragraph all differentiable curves are assumed to be continuously differentiable.

8.26. Differentiable families of linear automorphisms. Our first aim is to establish a correspondence between all differentiable families of linear automorphisms on the one hand and all continuous families of linear transformations on the other hand. Let a differentiable family

$\varphi(t)(t_0 \leqq t \leqq t_1)$ of linear automorphisms be given such that $\varphi(t_0)=\iota$. Then a continuous family $\psi(t)$ of linear transformations is defined by

$$\psi(t) = \dot{\varphi}(t) \circ \varphi(t)^{-1}.$$

Interpreting t as time we obtain the following physical significance of the mappings $\psi(t)$: Let x be a fixed vector of E and put

$$x(t) = \varphi(t)x.$$

Then the velocity vector $\dot{x}(t)$ is given by

$$\dot{x}(t) = \dot{\varphi}(t)x = \dot{\varphi}(t)\varphi(t)^{-1}x(t) = \psi(t)x(t).$$

Hence, the mapping $\psi(t)$ associates with every vector $x(t)$ its velocity at the instant t.

Now it will be shown that, conversely, to every continuous curve $\psi(t)$ in $L(E; E)$ there exists exactly one differentiable family $\varphi(t)(t_0 \leqq t \leqq t_1)$ of linear automorphisms satisfying the differential equation

$$\dot{\varphi}(t) = \psi(t) \circ \varphi(t) \tag{8.54}$$

and the initial-condition $\varphi(t_0)=\iota$. First of all we notice that the differential equation (8.54) together with the above initial condition is equivalent to the integral equation

$$\varphi(t) = \iota + \int_{t_0}^{t} \psi(s) \circ \varphi(s)\,ds \quad (t_0 \leqq t \leqq t_1). \tag{8.55}$$

In the next section the solution of the integral equation (8.55) will be constructed by the method of successive approximations.

8.27. The Picard iteration process. Define the curves $\varphi_n(t)(n=0,1\ldots)$ by the equations

$$\left.\begin{aligned} &\varphi_0(t) = \iota \\ &\text{and} \\ &\varphi_{n+1}(t) = \iota + \int_{t_0}^{t} \psi(s) \circ \varphi_n(s)\,ds \quad (n = 0, 1 \ldots) \end{aligned}\right\} t_0 \leqq t \leqq t_1. \tag{8.56}$$

Introducing the differences

$$\Delta_n(t) = \varphi_n(t) - \varphi_{n-1}(t) \qquad (n = 1, 2, \ldots) \tag{8.57}$$

we obtain from (8.56) the relations

$$\Delta_n(t) = \int_{t_0}^{t} \psi(s) \circ \Delta_{n-1}(s) ds \qquad (n = 2, 3, \ldots). \tag{8.58}$$

Equation (8.57) yields for $n=1$

$$\Delta_1(t) = \varphi_1(t) - \varphi_0(t) = \int_{t_0}^{t} \psi(s) ds .$$

Define the number M by

$$M = \max_{t_0 \leqq t \leqq t_1} |\psi(t)| .$$

Then

$$|\Delta_1(t)| \leqq M(t - t_0). \tag{8.59}$$

Employing the equation (8.58) for $n=2$ we obtain in view of (8.59)

$$|\Delta_2(t)| \leqq M^2 \int_{t_0}^{t} (s - t_0) ds \leqq \frac{M^2}{2} (t - t_0)^2$$

and in general

$$|\Delta_n(t)| \leqq \frac{M^n}{n!} (t - t_0)^n \qquad (n = 1, 2 \ldots).$$

Now relations (8.57) imply that

$$\varphi_{n+p}(t) - \varphi_n(t) = \sum_{\nu=n+1}^{n+p} \Delta_\nu(t),$$

whence

$$\begin{aligned} |\varphi_{n+p}(t) - \varphi_n(t)| &\leqq \sum_{\nu=n+1}^{n+p} |\Delta_\nu(t)| \leqq \sum_{\nu=n+1}^{n+p} \frac{M^\nu}{\nu!} (t - t_0)^\nu \leqq \\ &\leqq \sum_{\nu=n+1}^{n+p} \frac{M^\nu}{\nu!} (t_1 - t_0)^\nu . \end{aligned} \tag{8.60}$$

Let $\varepsilon > 0$ be an arbitrary number. It follows from the convergence of the series $\sum_\nu \frac{M^\nu}{\nu!} (t_1 - t_0)^\nu$ that there exists an integer N such that

$$\sum_{\nu=n+1}^{n+p} \frac{M^\nu}{\nu!} (t_1 - t_0)^\nu < \varepsilon \quad \text{for} \quad n > N \quad \text{and} \quad p \geqq 1 . \tag{8.61}$$

The inequalities (8.60) and (8.61) yield

$$|\varphi_{n+p}(t) - \varphi_n(t)| < \varepsilon \quad \text{for} \quad n > N \quad \text{and} \quad p \geqq 1 .$$

These relations show that the sequence $\varphi_n(t)$ is uniformly convergent in the interval $t_0 \leqq t \leqq t_1$,

$$\lim_{n\to\infty} \varphi_n(t) = \varphi(t).$$

In view of the uniform convergence, equation (8.56) implies that

$$\varphi(t) = \iota + \int_{t_0}^{t} \psi(s) \circ \varphi(s)\, ds \quad (t_0 \leqq t \leqq t_1). \tag{8.62}$$

As a uniform limit of continuous curves the curve $\varphi(t)$ is itself continuous. Hence, the right hand-side of (8.62) is differentiable and so $\varphi(t)$ must be differentiable. Differentiating (8.62) we obtain the relation

$$\dot{\varphi}(t) = \psi(t) \circ \varphi(t)$$

showing that $\varphi(t)$ satisfies the differential equation (8.54). The equation $\varphi(t_0) = \iota$ is an immediate consequence of (8.62).

8.28. The determinant of $\varphi(t)$. It remains to be shown that the mappings $\varphi(t)$ are linear automorphisms. This will be done by proving the formula

$$\det \varphi(t) = e^{\int_{t_0}^{t} \operatorname{tr} \psi(t)\, dt} \tag{8.63}$$

Let $\Delta \neq 0$ be a determinant function in E. Then

$$\Delta(\varphi(t)x_1 \ldots \varphi(t)x_n) = \det \varphi(t)\, \Delta(x_1 \ldots x_n) \qquad x_\nu \in E.$$

Differentiating this equation and using the differential equation (8.54) we obtain

$$\begin{aligned} &\sum_\nu \Delta(\varphi(t)x_1 \ldots \psi(t)\varphi(t)x_\nu \ldots \varphi(t)x_n) \\ &= \frac{d}{dt} \det \varphi(t) \cdot \Delta(x_1 \ldots x_n). \end{aligned} \tag{8.64}$$

Observing that

$$\begin{aligned} &\sum_\nu \Delta(\varphi(t)x_1 \ldots \psi(t)\varphi(t)x_\nu \ldots \varphi(t)x_n) \\ &= \operatorname{tr} \psi(t)\, \Delta(\varphi(t)x_1 \ldots \varphi(t)x_n) \\ &= \operatorname{tr} \psi(t) \det \varphi(t)\, \Delta(x_1 \ldots x_n) \end{aligned}$$

we obtain from (8.64) the differential equation

$$\frac{d}{dt} \det \varphi(t) = \operatorname{tr} \psi(t) \cdot \det \varphi(t) \tag{8.65}$$

for the function $\det \varphi(t)$. Integrating this differential equation and observing the initial condition

$$\det \varphi(t_0) = \det \iota = 1$$

we find (8.63).

8.29. Uniqueness of the solution. Assume that $\varphi_1(t)$ and $\varphi_2(t)$ are two solutions of the differential equation (8.54) with the initial condition $\varphi(t_0) = \iota$. Consider the difference

$$\varphi(t) = \varphi_2(t) - \varphi_1(t).$$

The curve $\varphi(t)$ is again a solution of the differential equation (8.54) and it satisfies the initial condition $\varphi(t_0) = 0$. This implies the inequality

$$|\varphi(t)| = \left| \int_{t_0}^{t} \dot{\varphi}(s)\, ds \right| \leqq \int_{t_0}^{t} |\dot{\varphi}(s)|\, ds \leqq M \int_{t_0}^{t} |\varphi(s)|\, ds. \tag{8.66}$$

Now define the function F by

$$F(t) = \int_{t_0}^{t} |\varphi(s)|\, ds. \tag{8.67}$$

Then (8.66) implies the relation

$$\dot{F}(t) \leqq M F(t).$$

Multiplying by e^{-tM} we obtain

$$\dot{F}(t) e^{-tM} - M e^{-tM} F(t) \leqq 0,$$

whence

$$\frac{d}{dt}(F(t) e^{-tM}) \leqq 0.$$

Integrating this inequality and observing that $F(t_0) = 0$, we obtain

$$F(t) e^{-tM} \leqq 0$$

and consequently

$$F(t) \leqq 0 \qquad (t_0 \leqq t \leqq t_1). \tag{8.68}$$

On the other hand it follows from (8.67) that

$$F(t) \geqq 0 \quad (t_0 \leqq t \leqq t_1). \tag{8.69}$$

Relations (8.68) and (8.69) imply that $F(t) \equiv 0$ whence $\varphi(t) \equiv 0$. Consequently, the two solutions $\varphi_1(t)$ and $\varphi_2(t)$ coincide.

8.30. 1-parameter groups of linear automorphisms. A differentiable family of linear automorphisms $\varphi(t)(-\infty<t<\infty)$ is called a *1-parameter group*, if

$$\varphi(t+\tau)=\varphi(t)\circ\varphi(\tau) \qquad t,\tau\in\mathbb{R}. \tag{8.70}$$

Equation (8.70) implies indeed that the automorphisms $\varphi(t)$ form an (abelian) group. Inserting $t=0$ we find $\varphi(0)=\iota$. Now equation (8.70) yields

$$\varphi(t)\circ\varphi(-t)=\iota$$

showing that for every automorphism $\varphi(t)$ the inverse automorphism $\varphi(t)^{-1}$ is contained in the family $\varphi(t)(-\infty<t<\infty)$. In addition it follows from (8.70) that the group $\varphi(t)$ is commutative.

Differentiation of (8.70) with respect to t yields

$$\dot\varphi(t+\tau)=\dot\varphi(t)\circ\varphi(\tau).$$

Inserting $t=0$ we obtain the differential equation

$$\dot\varphi(\tau)=\psi\circ\varphi(\tau) \qquad (-\infty<\tau<\infty) \tag{8.71}$$

where $\psi=\dot\varphi(0)$. Conversely, consider the differential equation (8.71) where ψ is a given transformation of E. It will be shown that the solution $\varphi(\tau)$ of this differential equation with the initial condition $\varphi(0)=\iota$ is a 1-parameter group of automorphisms. To prove this let τ be fixed and consider the curves

$$\varphi_1(t)=\varphi(t+\tau) \tag{8.72}$$

and

$$\varphi_2(t)=\varphi(t)\circ\varphi(\tau). \tag{8.73}$$

Differentiating the equations (8.72) and (8.73) we obtain

$$\dot\varphi_1(t)=\dot\varphi(t+\tau)=\psi\circ\varphi(t+\tau)=\psi\circ\varphi_1(t) \tag{8.74}$$

and

$$\dot\varphi_2(t)=\dot\varphi(t)\circ\varphi(\tau)=\psi\circ\varphi(t)\circ\varphi(\tau)=\psi\circ\varphi_2(t). \tag{8.75}$$

Relations (8.74) and (8.75) show that the two curves $\varphi_1(t)$ and $\varphi_2(t)$ satisfy the same differential equation. Moreover,

$$\varphi_1(0)=\varphi_2(0)=\varphi(\tau).$$

Thus, it follows from the uniqueness theorem of sec. 8.29 that $\varphi_1(t)\equiv\varphi_2(t)$ whence (8.70).

8.31. Differentiable families of rotations. Let $\varphi(t)(t_0 \leqq t \leqq t_1)$ be a differentiable family of *rotations* such that $\varphi(t_0)=\iota$. Since $\det \varphi(t)=\pm 1$ for every t and $\det \varphi(t_0)=+1$ it follows from the continuity that $\det \varphi(t)=+1$, i.e. all rotations $\varphi(t)$ are proper.

Now it will be shown that the linear transformations

$$\psi(t)=\dot{\varphi}(t)\circ\varphi(t)^{-1}$$

are skew. Differentiating the identity

$$\tilde{\varphi}(t)\circ\varphi(t)=\iota$$

we obtain

$$\dot{\tilde{\varphi}}(t)\circ\varphi(t)+\tilde{\varphi}(t)\circ\dot{\varphi}(t)=0.$$

Inserting

$$\dot{\varphi}(t)=\psi(t)\circ\varphi(t)$$

and

$$\dot{\tilde{\varphi}}(t)=\tilde{\dot{\varphi}}(t)=\tilde{\varphi}(t)\circ\tilde{\psi}(t)$$

into this equation we find

$$\tilde{\varphi}(t)\circ(\tilde{\psi}(t)+\psi(t))\circ\varphi(t)=0,$$

whence

$$\tilde{\psi}(t)+\psi(t)=0.$$

Conversely, let the family of linear automorphisms $\varphi(t)$ be defined by the differential equation

$$\dot{\varphi}(t)=\psi(t)\circ\varphi(t),\quad \varphi(t_0)=\iota$$

where $\psi(t)$ is a continuous family of *skew* mappings. Then every automorphism $\varphi(t)$ is a proper rotation. To prove this, define the family $\chi(t)$ by

$$\chi(t)=\tilde{\varphi}(t)\circ\varphi(t).$$

Then

$$\begin{aligned}\dot{\chi}(t)&=\dot{\tilde{\varphi}}(t)\circ\varphi(t)+\tilde{\varphi}(t)\circ\dot{\varphi}(t)\\&=-\tilde{\varphi}(t)\circ\psi(t)\circ\varphi(t)+\tilde{\varphi}(t)\circ\psi(t)\circ\varphi(t)=0\end{aligned}$$

and

$$\chi(t_0)=\iota.$$

Now the uniqueness theorem implies that $\chi(t)\equiv\iota$, whence

$$\tilde{\varphi}(t)\circ\varphi(t)=\iota.$$

This equation shows that the mappings $\varphi(t)$ are rotations.

8.32. Angular velocity. As an example, let $\varphi(t)$ be a differentiable family of rotations of the 3-space such that $\varphi(0)=\iota$. If t is interpreted as the time, the family $\varphi(t)$ can be considered as a rigid motion of the space E. Given a vector x, the curve

$$x(t) = \varphi(t)x$$

describes its orbit. The corresponding velocity-vector is determined by

$$\dot{x}(t) = \dot{\varphi}(t)x = \psi(t)\varphi(t)x = \psi(t)x(t). \tag{8.76}$$

Now assume that an orientation is defined in E. Then every mapping $\psi(t)$ can be written as

$$\psi(t)y = u(t) \times y. \tag{8.77}$$

The vector $u(t)$ is uniquely determined by $\psi(t)$ and hence by t. Equations (8.76) and (8.77) yield

$$\dot{x}(t) = u(t) \times x(t). \tag{8.78}$$

The vector $u(t)$ is called the *angular velocity* at the time t. To obtain a physical interpretation of the angular velocity, fix a certain instant t and assume that $u(t)\neq 0$. Then equation (8.78) shows that $\dot{x}(t)=0$ if and only if $x(t)$ is a multiple of $u(t)$. In other words, the straight line generated by $u(t)$ consists of all vectors having the velocity zero at the instant t. This straight line is called the *instantaneous axis*. Equation (8.78) implies that the velocity-vector $\dot{x}(t)$ is orthogonal to the instantaneous axis.

Passing over to the norm in equation (8.78) we find that

$$|\dot{x}(t)| = |u(t)|\,|h(t)|$$

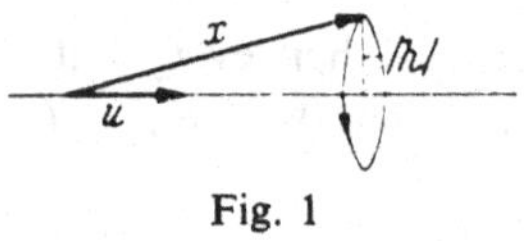

Fig. 1

where $|h(t)|$ is the distance of the vector $x(t)$ from the instantaneous axis (fig. 1). Consequently, the norm of $u(t)$ is equal to the magnitude of the velocity of a vector having the distance 1 from the instantaneous axis.

The uniqueness theorem in sec. 8.29 implies that the rigid motion $\varphi(t)$ is uniquely determined by $\varphi(t_0)$ if the angular velocity is a given function of t.

8.33. The trigonometric functions. In this concluding section we shall apply our general results about families of rotations to the Euclidean plane and show that this leads to the trigonometric functions cos and sin. This definition has the advantage that the addition theorems can be proved in a simple fashion, without making use of geometric intuition.

Let E be an oriented Euclidean plane and Δ be the normed determinant function representing the given orientation. Consider the skew mapping ψ which is defined by the equation

$$(\psi x, y) = \Delta(x, y). \tag{8.79}$$

First of all we notice that ψ is a proper rotation. In fact, the identity 7.24 yields

$$(\psi x, y)^2 = \Delta(x, y)^2 = (x, x)(y, y) - (x, y)^2.$$

Inserting $y = \psi x$ we find

$$(\psi x, \psi x)^2 = (x, x)(\psi x, \psi x).$$

Now ψ is regular as follows from (8.79). Hence the above equation implies that

$$(\psi x, \psi x) = (x, x).$$

Replacing x and y by ψx and ψy respectively in (8.79) we obtain the relation

$$\Delta(\psi x, \psi y) = (\psi^2 x, \psi y) = (\psi x, y) = \Delta(x, y)$$

showing that

$$\det \psi = +1.$$

Let $\varphi(t)(-\infty < t < \infty)$ be the family of rotations defined by the differential equation

$$\dot{\varphi}(t) = \psi \circ \varphi(t) \tag{8.80}$$

and the initial condition

$$\varphi(0) = \iota.$$

Then it follows from the result of sec. 8.30 that

$$\varphi(t + \tau) = \varphi(t) \circ \varphi(\tau). \tag{8.81}$$

We now define functions c and s by

$$\left.\begin{aligned} c(t) &= \tfrac{1}{2} \operatorname{tr} \varphi(t) \\ \text{and} \qquad & \\ s(t) &= -\tfrac{1}{2} \operatorname{tr}(\psi \circ \varphi(t)) \end{aligned}\right\} \quad -\infty < t < \infty. \tag{8.82}$$

These functions are the well-known functions cos and sin. In fact, all the properties of the trigonometric functions can easily be derived from (8.82). Select an arbitrary unit-vector e. Then the vectors e and ψe form an orthonormal basis of E. Consequently,

$$\operatorname{tr}\varphi(t) = (\varphi(t)e, e) + (\varphi(t)\psi e, \psi e). \tag{8.83}$$

Since ψ is itself a proper rotation, the mappings $\varphi(t)$ and ψ commute. Hence, the second term in (8.83) can be written as

$$(\varphi(t)\psi e, \psi e) = (\psi\varphi(t)e, \psi e) = (\varphi(t)e, e).$$

We thus obtain

$$c(t) = (\varphi(t)e, e). \tag{8.84}$$

In the same way it is shown that

$$s(t) = (\varphi(t)e, \psi e). \tag{8.85}$$

Equations (8.84) and (8.85) imply that

$$\varphi(t)e = c(t)e + s(t)\psi e. \tag{8.86}$$

Replacing t by $t+\tau$ in (8.84) and using the formulae (8.81) and (8.86) we obtain

$$\begin{aligned} c(t+\tau) &= (\varphi(t+\tau)e, e) = (\varphi(t)\varphi(\tau)e, e) \\ &= c(t)(\varphi(\tau)e, e) - s(t)(\varphi(\tau)e, \psi e). \end{aligned} \tag{8.87}$$

Equations (8.87), (8.84) and (8.85) yield the addition theorem of the function c:

$$c(t+\tau) = c(t)c(\tau) - s(t)s(\tau).$$

In the same way it is shown that

$$s(t+\tau) = s(t)c(\tau) + c(t)s(\tau).$$

Problems

1. Let ψ be a linear transformation of the inner product space E. Define the linear automorphism $\exp\psi$ by

$$\exp\psi = \varphi(1)$$

where $\varphi(t)$ is the family of linear automorphisms defined by

$$\dot\varphi(t) = \psi\circ\varphi(t), \varphi(0) = \iota.$$

Prove that

$$\varphi(t) = \exp(t\psi) \qquad (-\infty < t < \infty).$$

2. Show that the mapping $\psi \to \exp\psi$ defined in problem 1 has the following properties:

1. $\exp(\psi_1+\psi_2)=\exp\psi_1\circ\exp\psi_2$ if $\psi_2\circ\psi_1=\psi_1\circ\psi_2$.
2. $\exp(-\psi)=(\exp\psi)^{-1}$.
3. $\exp 0=1$.
4. $\widetilde{\exp\psi}=\exp\tilde{\psi}$.
5. $\det\exp\psi=e^{\operatorname{tr}\psi}$.

From these formulas derive that $\exp\psi$ is selfadjoint if ψ is selfadjoint and that $\exp\psi$ is a proper rotation if ψ is skew.

3. Consider the family of rotations $\varphi(t)$ defined by (8.80).

a) Assuming that there is a real number $p\neq 0$ such that $\varphi(p)=\iota$, prove that $\varphi(t+p)=\varphi(t)(-\infty<t<\infty)$.

b) Prove that $\varphi(t_0)=\iota$ if and only if

$$t_0 = 4k\int_0^1 \frac{d\tau}{\sqrt{1-\tau^2}} \qquad (k=0,\pm 1,\pm 2,\ldots).$$

c) Show that the family $\varphi(t)$ has derivatives of every order and that

$$\varphi^{(\nu+2)}(t) = -\varphi^{(\nu)}(t) \qquad (\nu=0,1\ldots).$$

d) Define the curve $x(t)$ by

$$x(t)=\varphi(t)e$$

where e is a fixed unit-vector. Show that

$$\int_0^t |\dot{x}(s)|\,ds = t.$$

4. Derive from formulae (8.82) that the function c is even and that the function s is odd.

5. Let ψ be the skew mapping defined by (8.79). Prove *De Moivre's formula*

$$\exp(t\psi)=c(t)\iota+s(t)\psi.$$

6. Let ψ a skew mapping of an n-dimensional inner product space and $\varphi(t)$ the corresponding family of rotations. Consider the normal form (8.35) of the matrix of ψ. Prove that the function $\varphi(t)(-\infty<t<\infty)$ is periodic if and only if all the ratios $\kappa_\nu:\kappa_\mu$ are rational.

7. Let A be a finite dimensional associative real algebra.

a) Consider a differentiable family of endomorphisms $\varphi_t: A \to A$ such that $\varphi_0 = \iota$. Prove that $\dot{\varphi}_0$ is a derivation in A.

b) Let θ be a derivation in A and define the family φ_t of linear transformations by

$$\dot{\varphi}_t = \theta \varphi_t, \quad \varphi_0 = \iota.$$

Prove that every φ_t is an automorphism of A. Show that every φ_t commutes with θ.

8. The quaternionic exponential function. Fix a quaternion a and consider the differential equation

$$\dot{x}(t) = a\,x(t) \qquad t \in \mathbb{R}$$

with the initial condition $x(0) = e$. Define $\exp a$ by

$$\exp a = x(1).$$

Decompose a in the form $a = \lambda e + b$ where $\lambda \in \mathbb{R}$ and $(b, e) = 0$.

(i) Show that

a) $\exp(\lambda e) = \exp \lambda \cdot e$

where $\exp \lambda$ is the exponential function for the reals.

b) $\exp b = e \cos |b| + \frac{b}{|b|} \sin |b|$.

c) $\exp a = \exp \lambda \left(e \cos |b| + \frac{b}{|b|} \sin |b| \right)$.

(ii) Show that

$$|\exp a| = \exp \lambda.$$

(iii) Prove that $\exp a = e$ if and only if

$$\lambda = 0 \quad \text{and} \quad |b| = 2k\pi, \qquad k \in \mathbb{Z}.$$

(iv) Let $a_1 = \lambda_1 e + b_1$ and $a_2 = \lambda_2 e + b_2$ be two quaternions with $b_1 \neq 0$ and $b_2 \neq 0$. Show that $\exp a_2 = \exp a_1$ if and only if

$$\lambda_2 = \lambda_1 \quad \text{and} \quad b_2 = b_1 + 2k\pi \frac{b_1}{|b_1|}, \qquad k \in \mathbb{Z}.$$

(v) Let B^3 be the closed 3-ball given by $(x, e) = 0$, $|x| \leqq \pi$. Define a map $\varphi: \mathbb{R} \times B^3 \to E$ by

$$\varphi(\lambda, y) = \exp(\lambda e + y) \qquad \lambda \in \mathbb{R},\ y \in B^3.$$

Show that a) φ maps B^3 onto $E - 0$.

b) φ is injective in the interior of B^3.

c) If S^3 denotes the boundary of B^3 then

$$\varphi(\lambda, y) = -\exp \lambda \cdot e \qquad y \in S^3.$$

Chapter IX

Symmetric bilinear functions

All the properties of an inner product space discussed in Chapter VII are based upon the bilinearity, the symmetry and the definiteness of the inner product. The question arises which of these properties do not depend on the definiteness and hence can be carried over to a real linear space with an indefinite inner product. Linear spaces of this type will be discussed in § 4. First of all, the general properties of a symmetric bilinear function will be investigated. It will be assumed throughout the chapter that all linear spaces are real.

§ 1. Bilinear and quadratic functions

9.1. Definition. Let E be a real vector space and Φ be a bilinear function in $E \times E$. The bilinear function Φ is called *symmetric* if

$$\Phi(x, y) = \Phi(y, x) \qquad x, y \in E.$$

Given a symmetric bilinear function Φ consider the (non-linear) function Ψ defined by

$$\Psi(x) = \Phi(x, x). \tag{9.1}$$

Then Φ is uniquely determined by Ψ.

In fact, replacing x by $x+y$ in (9.1) we obtain

$$\Psi(x + y) = \Phi(x + y, x + y) = \Psi(x) + 2\Phi(x, y) + \Psi(y), \tag{9.2}$$

whence

$$\Phi(x, y) = \tfrac{1}{2}\{\Psi(x + y) - \Psi(x) - \Psi(y)\}. \tag{9.3}$$

Equation (9.3) shows that different symmetric bilinear functions Φ lead to different functions Ψ.

Replacing y by $-y$ in (9.2) we find

$$\Psi(x - y) = \Psi(x) - 2\Phi(x, y) + \Psi(y). \tag{9.4}$$

Adding the equations (9.2) and (9.4) we obtain the so-called *parallelogram-identity*

$$\Psi(x+y)+\Psi(x-y)=2(\Psi(x)+\Psi(y)). \tag{9.5}$$

9.2. Quadratic functions. A continuous function Ψ of one vector which satisfies the parallelogram-identity will be called a *quadratic function*. Every symmetric bilinear function yields a quadratic function by setting $x=y$. We shall now prove that, conversely, every quadratic function can be obtained in this way.

Substituting $x=y=0$ in the parallelogram-identity we find that

$$\Psi(0)=0. \tag{9.6}$$

Now the same identity yields for $x=0$

$$\Psi(-y)=\Psi(y)$$

showing that a quadratic function is an even function.

If there exists at all a symmetric bilinear function Φ such that

$$\Phi(x,x)=\Psi(x)$$

this function is given by the equation

$$\Phi(x,y)=\tfrac{1}{2}\{\Psi(x+y)-\Psi(x)-\Psi(y)\}. \tag{9.7}$$

Therefore it remains to be shown that the function Φ defined by (9.7) is indeed bilinear and symmetric. The symmetry is an immediate consequence of (9.7). Next, we prove the relation

$$\Phi(x_1+x_2,y)=\Phi(x_1,y)+\Phi(x_2,y). \tag{9.8}$$

Equation (9.7) yields

$$\begin{aligned}2\Phi(x_1+x_2,y)&=\Psi(x_1+x_2+y)-\Psi(x_1+x_2)-\Psi(y)\\ 2\Phi(x_1,y)&=\Psi(x_1+y)-\Psi(x_1)-\Psi(y)\\ 2\Phi(x_2,y)&=\Psi(x_2+y)-\Psi(x_2)-\Psi(y),\end{aligned}$$

whence

$$\begin{aligned}2\{\Phi(x_1+x_2,y)-\Phi(x_1,y)-\Phi(x_2,y)\}&=\{\Psi(x_1+x_2+y)+\Psi(y)\}-\\ -\{\Psi(x_1+y)+\Psi(x_2+y)\}&-\{\Psi(x_1+x_2)-\Psi(x_1)-\Psi(x_2)\}.\end{aligned} \tag{9.9}$$

It follows from (9.5) that

$$\Psi(x_1+x_2+y)+\Psi(y)=\tfrac{1}{2}\{\Psi(x_1+x_2+2y)+\Psi(x_1+x_2)\} \tag{9.10}$$

and

$$\Psi(x_1 + y) + \Psi(x_2 + y) = \tfrac{1}{2}\{\Psi(x_1 + x_2 + 2y) + \Psi(x_1 - x_2)\}. \qquad (9.11)$$

Subtracting (9.11) from (9.10) and using the parallelogram-identity again we find that

$$\begin{aligned} &\{\Psi(x_1 + x_2 + y) + \Psi(y)\} - \{\Psi(x_1 + y) + \Psi(x_2 + y)\} \qquad (9.12)\\ &= \tfrac{1}{2}\{\Psi(x_1 + x_2) - \Psi(x_1 - x_2)\} = -\Psi(x_1) - \Psi(x_2) + \Psi(x_1 + x_2). \end{aligned}$$

Now equations (9.9) and (9.12) imply (9.8). Inserting $x_1 = x$ and $x_2 = -x$ into (9.8) we obtain

$$\Phi(-x, y) = -\Phi(x, y). \qquad (9.13)$$

It remains to be shown that

$$\Phi(\lambda x, y) = \lambda \Phi(x, y) \qquad (9.14)$$

for every real number λ. First of all it follows from (9.8) that

$$\Phi(k x, y) = k \Phi(x, y)$$

for a positive integer k. Equation (9.13) shows that (9.14) is also correct for negative integers. Now consider a rational number

$$\lambda = \frac{p}{q} \qquad (p, q \text{ integers}).$$

Then

$$q \Phi\left(\frac{p}{q} x, y\right) = \Phi(p x, y) = p \Phi(x, y),$$

whence

$$\Phi\left(\frac{p}{q} x, y\right) = \frac{p}{q} \Phi(x, y).$$

To prove (9.14) for an irrational factor λ we note first that Φ is a continuous function of x and y, as follows from the continuity of Ψ. Now select a sequence of rational numbers λ_n such that

$$\lim_{n \to \infty} \lambda_n = \lambda.$$

Then we have that

$$\Phi(\lambda_n x, y) = \lambda_n \Phi(x, y). \qquad (9.15)$$

For $n \to \infty$ we obtain from (9.15) the relation (9.14).

Our result shows that the relations (9.1) and (9.7) define a one-to-one correspondence between all symmetric bilinear functions and all quadratic functions. If no ambiguity is possible we shall designate a symmetric bilinear function and the corresponding quadratic function by the same symbol, i.e., we shall simply write

$$\Phi(x,x) = \Phi(x).$$

9.3. Bilinear and quadratic forms. Now assume that E has dimension n and let $x_\nu (\nu = 1 \dots n)$ be a basis of E. Then a symmetric bilinear function Φ can be expressed as a bilinear form

$$\Phi(x,y) = \sum_{\nu,\mu} \alpha_{\nu\mu} \xi^\nu \eta^\mu \tag{9.16}$$

where

$$x = \sum_\nu \xi^\nu x_\nu, \quad y = \sum_\nu \eta^\nu x_\nu$$

and the matrix $\alpha_{\nu\mu}$ is defined by

$$\alpha_{\nu\mu} = \Phi(x_\nu, x_\mu) \;{}^{*)}.$$

It follows from the symmetry of Φ that the matrix $\alpha_{\nu\mu}$ is symmetric:

$$\alpha_{\nu\mu} = \alpha_{\mu\nu}.$$

Replacing y by x in (9.16) we obtain the corresponding *quadratic form*

$$\Phi(x) = \sum_{\nu,\mu} \alpha_{\nu\mu} \xi^\nu \xi^\mu.$$

Problems

1. Let f and g be two linearly independent linear functions in E and let $\theta: \mathbb{R} \to \mathbb{R}$ be an additive map such that $\theta(\lambda\mu) = \theta(\lambda)\mu + \lambda\,\theta(\mu)$; $\lambda, \mu \in \mathbb{R}$. Show that the function

$$\Psi(x) = f(x)\,\theta[g(x)] - g(x)\,\theta[f(x)]$$

satisfies the paralelogram identity and the relation $\Psi(\lambda x) = \lambda^2 \Psi(x)$. Prove that the function Φ obtained from Ψ by (9.7) is bilinear if and only if $\theta = 0$.

2. Prove that a symmetric bilinear function in E defines a quadratic function in the direct sum $E \oplus E$.

3. Denote by A and by $\bar{A}$ the matrices of the bilinear function Φ with respect to two bases x_ν and $\bar{x}_\nu (\nu = 1 \dots n)$. Show that

$$\bar{A} = T A T^*$$

where T is the matrix of the basis transformation $x_\nu \to \bar{x}_\nu$.

---

*) The first index counts the row.

§ 2. The decomposition of E

9.4. Rank. Let E be a vector space of dimension n and Φ a symmetric bilinear function in $E \times E$. Recall that the nullspace E_0 of Φ is defined to be the set of all vectors $x_0 \in E$ such that

$$\Phi(x_0, y) = 0 \qquad \text{for every } y \in E. \tag{9.17}$$

The difference of the dimensions of E and E_0 is called the *rank* of Φ. Hence Φ is non-degenerate if and only it has rank n.

Now let E^* be a dual space and consider the linear mapping $\varphi: E \to E^*$ defined by

$$\Phi(x, y) = \langle \varphi x, y \rangle \qquad x, y \in E. \tag{9.18}$$

Then the null-space of Φ obviously coincides with the kernel of φ,

$$E_0 = \ker \varphi .$$

Consequently, the rank of Φ is equal to the rank of the mapping φ. Let $(\alpha_{\nu\mu})$ be the matrix of Φ relative to a basis x_ν $(\nu = 1 \dots n)$ of E. Then relation (9.18) yields

$$\langle \varphi x_\nu, x_\mu \rangle = \Phi(x_\nu, x_\mu) = \alpha_{\nu\mu}$$

showing that $\alpha_{\nu\mu}$ is the matrix of the mapping φ. This implies that the rank of the matrix $(\alpha_{\nu\mu})$ is equal to the rank of φ and hence equal to the rank of Φ. In particular, a symmetric bilinear function is non-degenerate if and only if the determinant of $(\alpha_{\nu\mu})$ is different from zero.

9.5. Definiteness. A symmetric bilinear function Φ is called *positive definite* if

$$\Phi(x) > 0$$

for all vectors $x \neq 0$. As has been shown in sec. 7.4, a positive definite bilinear function satisfies the Schwarz-inequality

$$\Phi(x, y)^2 \leqq \Phi(x)\Phi(y) \qquad x, y \in E.$$

Equality holds if and only if the vectors x and y are linearly dependent. A positive definite function Φ is non-degenerate.

If $\Phi(x) \geqq 0$ for all vectors $x \in E$, but $\Phi(x) = 0$ for some vectors $x \neq 0$, the function Φ is called *positive semidefinite*. The Schwarz inequality is still valid for a semidefinite function. But now equality may hold without the vectors x and y being linearly dependent. A semidefinite function is always degenerate. In fact, consider a vector $x_0 \neq 0$ such that $\Phi(x_0)=0$. Then the Schwarz inequality implies that

$$\Phi(x_0, y)^2 \leqq \Phi(x_0)\Phi(y) = 0$$

whence $\Phi(x_0, y)=0$ for all vectors y.

In the same way negative definite and negative semidefinite bilinear functions are defined.

The bilinear function Φ is called *indefinite* if the function $\Phi(x)$ assumes positive and negative values. An indefinite function may be degenerate or non-degenerate.

9.6. The decomposition of E. Let a non-degenerate indefinite bilinear function Φ be given in the n-dimensional space E. It will be shown that the space E can be decomposed into two subspaces E^+ and E^- such that Φ is positive definite in E^+ and is negative definite in E^-.

Since Φ is indefinite, there is a non-trivial subspace of E in which Φ is positive definite. For instance, every vector a for which $\Phi(a)>0$ generates such a subspace.

Let E^+ be a subspace of maximal dimension such that Φ is positive definite in E^+. Consider the orthogonal complement E^- of E^+ with respect to the scalar product defined by Φ. Since Φ is positive definite in E^+, the intersection $E^+ \cap E^-$ consists only of the zero-vector. At the same time we have the relation (cf. Proposition II, sec. 2.33)

$$\dim E^+ + \dim E^- = \dim E.$$

This yields the direct decomposition

$$E = E^+ \oplus E^-.$$

Now it will be shown that Φ is negative definite in E^-. Given a vector $z \neq 0$ of E^-, consider the subspace E_1 generated by E^+ and z. Every vector of this subspace can be written as

$$x = y + \lambda z \qquad y \in E^+.$$

This implies that

$$\Phi(x) = \Phi(y) + \lambda^2 \Phi(z). \tag{9.19}$$

Now assume that $\Phi(z)>0$. Then equation (9.19) shows that Φ is positive definite in the subspace E_1 which is a contradiction to the maximum-property of E^+. Consequently,

$$\Phi(z) \leqq 0 \quad \text{for all vectors} \quad z \in E^-$$

i.e., Φ is negative semidefinite in E^-. Using the Schwarz inequality

$$\Phi(z_1, z)^2 \leqq \Phi(z_1)\Phi(z) \qquad z_1 \in E^-, z \in E^- \tag{9.20}$$

we can prove that Φ is even negative definite in E^-. Assume that $\Phi(z_1)=0$ for a vector $z_1 \in E^-$. Then the inequality (9.20) yields

$$\Phi(z_1, z) = 0$$

for all vectors $z \in E^-$. At the same time we know that

$$\Phi(z_1, y) = 0$$

for all vectors $y \in E^+$. These two equations imply that

$$\Phi(z_1, x) = 0$$

for all vectors $x \in E$, whence $z_1 = 0$.

9.7. The decomposition in the degenerate case. If the bilinear function Φ is degenerate, select a subspace E_1 complementary to the nullspace E_0,

$$E = E_0 \oplus E_1 .$$

Then Φ is non-degenerate in E_1. In fact, assume that

$$\Phi(x_1, y_1) = 0$$

for a fixed vector $x_1 \in E_1$ and all vectors $y_1 \in E_1$. Consider an arbitrary vector $y \in E$. This vector can be written as

$$y = y_0 + y_1 \qquad y_0 \in E_0, y_1 \in E_1$$

whence

$$\Phi(x_1, y) = \Phi(x_1, y_0) + \Phi(x_1, y_1) = 0 . \tag{9.21}$$

This equation shows that x_1 is contained in E_0 and hence it is contained in the intersection $E_0 \cap E_1$. This implies that $x_1 = 0$.

Now the construction of sec. 9.6 can be applied to the subspace E_1. We thus obtain altogether a direct decomposition

$$E = E_0 \oplus E^+ \oplus E^- \tag{9.22}$$

of E such that Φ is positive definite in E^+ and negative definite in E^-.

9.8. Diagonalization of the matrix. Let $(x_1 \dots x_s)$ be a basis of E^+, which is orthonormal with respect to Φ, $(x_{s+1} \dots x_r)$ be a basis of E^- which is orthonormal with respect to $-\Phi$, and $(x_{r+1} \dots x_n)$ be an arbitrary basis of E_0. Then

$$\Phi(x_\nu, x_\mu) = \varepsilon_\nu \delta_{\nu\mu} \quad \text{where } \varepsilon_\nu = \begin{cases} +1 \, (\nu = 1 \dots s) \\ -1 \, (\nu = s+1 \dots r) \\ 0 \, (\nu = r+1 \dots n) \end{cases}$$

The vectors $(x_1 \dots x_n)$ then form a basis of E in which the matrix of Φ has the following diagonal-form:

$$\begin{pmatrix} \underbrace{\begin{matrix} 1 & & \\ & \ddots & \\ & & 1 \end{matrix}}_{s} & & & 0 \\ & \underbrace{\begin{matrix} -1 & & \\ & \ddots & \\ & & -1 \end{matrix}}_{r-s} & & \\ & & \underbrace{\begin{matrix} 0 & & \\ & \ddots & \\ & & 0 \end{matrix}}_{n-r} & \\ 0 & & & \end{pmatrix}.$$

9.9. The index. It is clear from the above construction that there are infinitely many different decompositions of the form (9.22). However, the dimensions of E^+ and E^- are uniquely determined by the bilinear function Φ. To prove this, consider two decompositions

$$E = E_1^+ \oplus E_1^- \oplus E_0 \tag{9.23}$$

and

$$E = E_2^+ \oplus E_2^- \oplus E_0 \tag{9.24}$$

such that Φ is positive definite in E_1^+ and E_2^+ and negative definite in E_1^- and E_2^-. This implies that

$$E_2^+ \cap (E_1^- \oplus E_0) = 0$$

whence

$$\dim E_2^+ + \dim E_1^- + \dim E_0 \leqq n. \tag{9.25}$$

Comparing the dimensions in (9.23) we find

$$\dim E_1^+ + \dim E_1^- + \dim E_0 = n. \tag{9.26}$$

Equations (9.25) and (9.26) yield

$$\dim E_2^+ \leqq \dim E_1^+ .$$

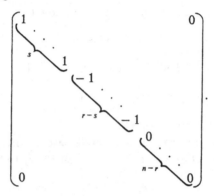

Interchanging E_1^+ and E_2^+ we obtain

$$\dim E_1^+ \leqq \dim E_2^+ ,$$

whence

$$\dim E_1^+ = \dim E_2^+ .$$

Consequently, the dimension of E^+ is uniquely determined by Φ. This number is called the *index* of the bilinear function Φ and the number $\dim E^+ - \dim E^- = 2s - r$ is called the *signature* of Φ.

Now suppose that $x_\nu (\nu = 1 \dots n)$ is a basis of E in which the quadratic function Φ has diagonal form

$$\Phi(x) = \sum_\nu \lambda_\nu \xi^\nu \xi^\nu$$

and assume that

$$\lambda_\nu > 0 (\nu = 1 \dots p) \quad \text{and} \quad \lambda_\nu \leqq 0 (\nu = p+1 \dots n).$$

Then p is the index of Φ. In fact, the vectors $x_\nu (\nu = 1 \dots p)$ generate a subspace of maximal dimension in which Φ is positive definite.

From the above result we obtain *Sylvester's law of inertia* which asserts that the number of positive coefficients is the same for every diagonal form.

9.10. The rank and the index of a symmetric bilinear function can be determined explicitly from the corresponding quadratic form

$$\Phi(x) = \sum_{\nu, \mu} \alpha_{\nu\mu} \xi^\nu \xi^\mu .$$

We can exclude the case $\Phi = 0$. Then at least one coefficient α_{ij} is different from zero. If $i \neq j$, apply the substitution

$$\xi^i = \bar{\xi}^i + \bar{\xi}^j, \xi^j = \bar{\xi}^i - \bar{\xi}^j .$$

Then

$$\Phi(x) = \sum_{\nu, \mu} \bar{\alpha}_{\nu\mu} \bar{\xi}^\nu \bar{\xi}^\mu$$

where $\bar{\alpha}_{ii} \neq 0$ and $\bar{\alpha}_{jj} \neq 0$. Thus, we may assume that at least one coefficient α_{ii}, say α_{11}, is different from zero. Then $\Phi(x)$ can be written as

$$\Phi(x) = \alpha_{11} \left\{ (\xi^1)^2 + \frac{2}{\alpha_{11}} \sum_{\mu=2}^n \alpha_{1\mu} \xi^1 \xi^\mu \right\} + \sum_{\nu, \mu = 2}^n \alpha_{\nu\mu} \xi^\nu \xi^\mu .$$

The substitution

$$\eta^1 = \xi^1 + \frac{1}{\alpha_{11}} \sum_{\mu=2}^n \alpha_{1\mu} \xi^\mu$$

$$\eta^\nu = \xi^\nu \qquad (\nu = 2 \dots n)$$

yields

$$\Phi(x) = \alpha_{11}(\eta^1)^2 + \sum_{\nu,\mu=2}^{n} \beta_{\nu\mu}\eta^\nu\eta^\mu . \tag{9.27}$$

The sum in (9.27) is a symmetric bilinear form in $(n-1)$ variables and hence the same reduction can be applied to this sum. Continuing this way we finally obtain an expression of the form

$$\Phi(x) = \sum_\nu \lambda^\nu \zeta^\nu \zeta^\nu .$$

Rearranging the variables we can achieve that

$$\begin{aligned} \lambda^\nu &> 0 \quad (\nu = 1 \dots s) \\ \lambda^\nu &< 0 \quad (\nu = s+1 \dots r) \\ \lambda^\nu &= 0 \quad (\nu = r+1 \dots n). \end{aligned}$$

Then r is the rank and s is the index of Φ.

Problems

1. Let $\Phi \neq 0$ be a given quadratic function. Prove that Φ can be written in the form

$$\Phi(x) = \varepsilon f(x)^2, \varepsilon = \pm 1$$

where f is a linear function, if and only if the corresponding bilinear function has rank 1.

2. Given a non-degenerate symmetric bilinear form Φ in E, let J be a subspace of maximal dimension such that $\Phi(x, x) = 0$ for every $x \in J$. Prove that

$$\dim J = \min(s, n-s).$$

Hint: Introduce two dual spaces E^* and F^* and linear mappings

$$\varphi_1 : E \to E^* \quad \text{and} \quad \varphi_2 : E \to F^*$$

defined by

$$\Phi(x, y) = \langle \varphi_1 x, y \rangle \quad \text{and} \quad \Phi(x, y) = \langle x, \varphi_2 y \rangle .$$

3. Define the bilinear function Φ in the space $L(E; E)$ by

$$\Phi(\varphi, \psi) = \operatorname{tr}(\psi \circ \varphi).$$

Let $S(E; E)$ be the space of all selfadjoint mappings and $A(E; E)$ be the space of all skew mappings with respect to a positive definite inner product. Prove:

a) $\Phi(\varphi, \varphi) > 0$ for every $\varphi \neq 0$ in $S(E; E)$,
b) $\Phi(\varphi, \varphi) < 0$ for every $\varphi \neq 0$ in $A(E; E)$,
c) $\Phi(\varphi, \psi) = 0$ if $\varphi \in S(E; E)$ and $\psi \in A(E; E)$,
d) The index of Φ is $\frac{n(n+1)}{2}$, where $n = \dim E$.

4. Find the index of the bilinear function

$$\Phi(\varphi, \psi) = \operatorname{tr}(\psi \circ \varphi) - \operatorname{tr} \varphi \operatorname{tr} \psi$$

in the space $L(E; E)$.

5. Find the index of the quadratic form

$$\Phi(x) = \sum_{i<j} \xi^i \xi^j .$$

6. Let Φ be a bilinear function in E. Assume that E_1 is a subspace of E such that Φ is non-degenerate in E_1. Define the subspace E_2 as follows: A vector $x_2 \in E$ is contained in E_2 if

$$\Phi(x_1, x_2) = 0 \quad \text{for all vectors } x_1 \in E_1 .$$

Prove that

$$E = E_1 \oplus E_2 .$$

7. Consider a (not necessarily symmetric) bilinear function Φ such that $\Phi(x, x) > 0$ for all vectors $x \neq 0$. Construct a basis of E in which the matrix of Φ has the form

$$\begin{pmatrix} 1 & \kappa_1 & & & & & & \\ -\kappa_1 & 1 & & & & & & \\ & & \ddots & & & & & \\ & & & 1 & \kappa_p & & & \\ & & & -\kappa_p & 1 & & & \\ & & & & & 1 & & \\ & & & & & & \ddots & \\ & & & & & & & 1 \end{pmatrix}.$$

Hint: Decompose Φ in the form

$$\Phi = \Phi_1 + \Phi_2 ,$$

where

$$\Phi_1(x, y) = \tfrac{1}{2}(\Phi(x, y) + \Phi(y, x))$$

and

$$\Phi_2(x, y) = \tfrac{1}{2}(\Phi(x, y) - \Phi(y, x)) .$$

8. Let E be a 2-dimensional vector space, and consider the 4-dimensional space $L(E; E)$. Prove that there exists a 3-dimensional subspace $F \subset L(E; E)$ and a symmetric bilinear function Φ in F such that the nilpotent transformations (cf. problem 7, Chap. IV, § 6) are precisely the transformations τ satisfying $\Phi(\tau)=0$ (In other words, the nilpotent transformations form a cone in F).

§ 3. Pairs of symmetric bilinear functions

9.11. In this paragraph we shall investigate the question under which conditions two symmetric bilinear functions Φ and Ψ can be simultaneously reduced to diagonal form.

To obtain a first criterion we consider the case that one of the bilinear functions, say Ψ, is non-degenerate. Then the vector space E is self-dual with respect to Ψ and hence there exists a linear transformation $\varphi: E \to E$ satisfying

$$\Phi(x, y) = \Psi(\varphi x, y) \qquad x, y \in E$$

(cf. Prop. III, sec. 2.33). Suppose now that x_1 and x_2 are eigenvectors of φ such that the corresponding eigenvalues λ_1 and λ_2 are different. Then we have that

$$\Phi(x_1, x_2) = \lambda_1 \Psi(x_1, x_2)$$

and

$$\Phi(x_2, x_1) = \lambda_2 \Psi(x_2, x_1)$$

whence in view of the symmetry of Φ and Ψ

$$(\lambda_1 - \lambda_2)\Psi(x_1, x_2) = 0.$$

Since $\lambda_1 \neq \lambda_2$ it follows that $\Psi(x_1, x_2)=0$ and hence $\Phi(x_1, x_2)=0$.

Proposition: Assume that Ψ is non-degenerate. Then Φ and Ψ are simultaneously diagonalizable if and only if the linear transformation φ has n linearly independent eigenvectors.

Proof: If φ has n linearly independent eigenvectors consider the distinct eigenvalues $\lambda_1 \ldots \lambda_r$ of φ. Then it follows that

$$E = E_1 \oplus \cdots \oplus E_r$$

where E_i is the eigenspace of λ_i. Then we have for $x_i \in E_i$ and $x_j \in E_j$, $i \neq j$

$$\Psi(x_i, x_j) = 0 \quad \text{and} \quad \Phi(x_i, x_j) = 0.$$

Now choose a basis in each space E_i such that Ψ has diagonal form (cf.

sec. 9.8). Since

$$\Phi(x, y) = \lambda_i \Psi(x, y) \qquad x, y \in E_i$$

it follows that Φ has also diagonal form in this basis. Combining all these bases of the E_i we obtain a basis of E such that Φ and Ψ have diagonal form.

Conversely, let $e_i (i=1 \ldots n)$ be a basis of E such that $\Phi(e_i, e_j)=0$ and $\Psi(e_i, e_j)=0$ if $i \neq j$. Then we have that

$$\Psi(\varphi e_i, e_j) = 0 \quad i \neq j .$$

This equation shows that the vector φe_i is contained in the orthogonal complement (with respect to the scalar product defined by Ψ) of the subspace F_i generated by the vectors e_ν, $\nu \neq i$. But $F_i^\perp$ is the 1-dimensional subspace generated by e_i, and so it follows that $\varphi e_i = \lambda e_i$. In other words, the e_i are eigenvectors of φ.

As an example let E be a plane with basis a, b and consider the bilinear functions Φ, Ψ given by

$$\Phi(a, a) = 1 , \quad \Phi(a, b) = 0 , \quad \Phi(b, b) = -1$$

and

$$\Psi(a, a) = 0 , \quad \Psi(a, b) = 1 , \quad \Psi(b, b) = 0 .$$

It is easy to verify that then the linear transformation φ is given by

$$\varphi a = b , \quad \varphi b = -a .$$

Since the characteristic polynomial of φ is $\lambda^2 + 1$ it follows that φ has no eigenvectors. Hence, the bilinear functions Φ and Ψ are not simultaneously diagonalizable.

Theorem: Let E be a vector space of dimension $n \geqq 3$ and let Φ and Ψ be two symmetric bilinear functions such that

$$\Phi(x)^2 + \Psi(x)^2 \neq 0 \quad \text{if} \quad x \neq 0 .$$

Then Φ and Ψ are simultaneously diagonalizable.

Before giving the proof we comment that the theorem is *not* correct for dimension 2 as the example above shows.

9.12. To prove the above theorem we employ a similar method as in sec. 8.6. If one of the functions Φ and Ψ, say Ψ, is positive definite the desired basis-vectors are those for which the function

$$F(x) = \frac{\Phi(x)}{\Psi(x)} \qquad x \neq 0 . \tag{9.28}$$

assumes a relative minimum. However, if Ψ is indefinite, the denominator in (9.28) assumes the value zero for certain vectors $x \neq 0$ and hence the function F is no longer defined in the entire space $x \neq 0$. The method of sec. 8.6 can still be carried over to the present case if the function F is replaced by the function

$$\operatorname{arc\,tan} F(x). \tag{9.29}$$

To avoid difficulties arising from the fact that the function arc tan is not single-valued, we shall write the function as a line-integral. At this point the hypothesis $n \geqq 3$ will be essential*).

Let $\dot{E}$ be the deleted space $x \neq 0$ and $x = x(t)\,(0 \leqq t \leqq 1)$ be a differentiable curve in $\dot{E}$. Consider the line-integral

$$J = \int_0^1 \frac{\Phi(x)\,\Psi(x,\dot{x}) - \Phi(x,\dot{x})\,\Psi(x)}{\Phi(x)^2 + \Psi(x)^2}\,dt \tag{9.30}$$

taken along the curve $x(t)$. First of all it will be shown that the integral J depends only on the initial point $x_0 = x(0)$ and the endpoint $x = x(1)$ of the curve $x(t)$. For this purpose define the following mapping of E into the complex w-plane:

$$\omega(x) = \Phi(x) + i\,\Psi(x).$$

The image of the curve $x(t)$ under this mapping is the curve

$$\omega(t) = \Phi(x(t)) + i\,\Psi(x(t)) \qquad (0 \leqq t \leqq 1) \tag{9.31}$$

in the w-plane. The hypothesis $\Phi(x)^2 + \Psi(x)^2 \neq 0$ implies that the curve $\omega(t)\,(0 \leqq t \leqq 1)$ does not go through the point $\omega = 0$. The integral (9.30) can now be written as

$$J = \frac{1}{2}\int_0^1 \frac{u\,\dot{v} - \dot{u}\,v}{u^2 + v^2}\,dt$$

where the integration is taken along the curve (9.31).

Now let $\theta(t)$ be an angle-function for the curve $\omega(t)$ i.e. a continuous function of t such that

$$\cos\theta(t) = \frac{u(t)}{|\omega(t)|} \quad \text{and} \quad \sin\theta(t) = \frac{v(t)}{|\omega(t)|} \tag{9.32}$$

*) The proof given is due to John Milnor.

(cf. fig. 2)*). It follows from the differentiability of the curve $\omega(t)$ that the angle-function θ is differentiable and we thus obtain from (9.32)

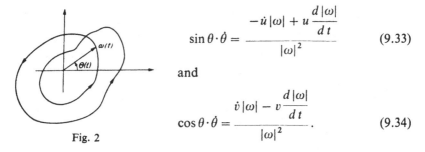

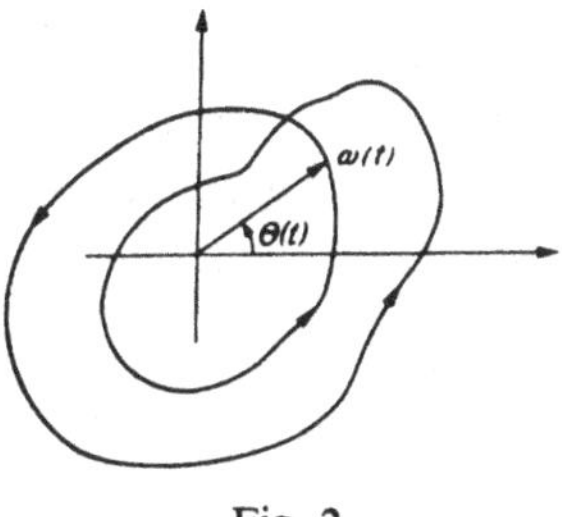

Fig. 2

$$\sin\theta\cdot\dot\theta = \frac{-\dot u\,|\omega| + u\,\dfrac{d\,|\omega|}{d\,t}}{|\omega|^2} \tag{9.33}$$

and

$$\cos\theta\cdot\dot\theta = \frac{\dot v\,|\omega| - v\,\dfrac{d\,|\omega|}{d\,t}}{|\omega|^2}. \tag{9.34}$$

Multiplying (9.33) by $\sin\theta$ and (9.34) by $\cos\theta$ and adding these equations we find that

$$\dot\theta = \frac{u\,\dot v - \dot u\,v}{u^2 + v^2}.$$

Integration from $t=0$ to $t=1$ gives

$$\int_0^1 \frac{u\,\dot v - \dot u\,v}{u^2 + v^2}\,d\,t = \theta(1) - \theta(0)$$

showing that the integral J is equal to the change of the angle-function θ along the curve $\omega(t)$,

$$2J = \theta(1) - \theta(0). \tag{9.35}$$

Now consider another differentiable curve $x=\bar x(t)\,(0\leqq t\leqq 1)$ in $\dot E$ with the initial point x_0 and the endpoint x and denote by $\bar J$ the integral (9.30) taken along the curve $\bar x(t)$. Then formula (9.35) yields

$$2\bar J = \bar\theta(1) - \bar\theta(0) \tag{9.36}$$

where $\bar\theta$ is an angle-function for the curve

$$\bar\omega(t) = \Phi(\bar x(t)) + i\,\Psi(\bar x(t)) \qquad (0\leqq t\leqq 1).$$

Since the curves $\omega(t)$ and $\bar\omega(t)\,(0\leqq t\leqq 1)$ have the same initial point and the same endpoint it follows that

$$\bar\theta(0) - \theta(0) = 2k_0\pi \quad\text{and}\quad \bar\theta(1) - \theta(1) = 2k_1\pi \tag{9.37}$$

*) For more details cf. P. S. ALEXANDROV. Combinatorial Topology, Vol. I, chapter II, § 2.

where k_0 and k_1 are integers. Equations (9.35), (9.36) and (9.37) show that the difference $\bar{J}-J$ is a multiple of π,

$$\bar{J} - J = k\pi .$$

It remains to be shown that $k=0$. The hypothesis $n \geqq 3$ implies that the space $\dot{E}$ is simply connected. In other words, there exists a continuous mapping $x=x(t,\tau)$ of the square $0 \leqq t \leqq 1$, $0 \leqq \tau \leqq 1$ into $\dot{E}$ such that

$$x(t,0) = x(t), \quad x(t,1) = \bar{x}(t) \qquad 0 \leqq t \leqq 1$$

and

$$x(0,\tau) = x_0, \quad x(1,\tau) = x \qquad 0 \leqq \tau \leqq 1 .$$

The mapping $x(t,\tau)$ can even be assumed to be differentiable. Then, for every fixed τ, we can form the integral (9.30) along the curve

$$x(t,\tau) \qquad (0 \leqq t \leqq 1).$$

This integral is a continuous function $J(\tau)$ of τ. At the same time we know that the difference $J(\tau)-J$ is a multiple of π,

$$J(\tau) - J = \pi k(\tau). \tag{9.38}$$

Hence, $k(\tau)$ is a continuous integer-valued function in the interval $0 \leqq \tau \leqq 1$ and thus $k(\tau)$ must be a constant. Since $k(0)=0$ it follows that $k(\tau)=0\,(0 \leqq \tau \leqq 1)$. Now equation (9.38) yields

$$J(\tau) = J \qquad (0 \leqq \tau \leqq 1).$$

Inserting $\tau = 1$ we obtain the relation

$$\bar{J} = J$$

showing that the integral (9.30) is indeed independent of the curve $x(t)$.

9.13. The function *F*. We now can define a single-valued function F in the space $\dot{E}$ by

$$F(x) = \int_{x_0}^{x} \frac{\Phi(x)\,\Psi(x,\dot{x}) - \Phi(x,\dot{x})\,\Psi(x)}{\Phi(x)^2 + \Psi(x)^2}\,dt \tag{9.39}$$

where the integration is taken along an arbitrary differentiable curve $x(t)$ in $\dot{E}$ leading from x_0 to x. The function F is homogeneous of degree zero,

$$F(\lambda x) = F(x), \qquad \lambda > 0 . \tag{9.40}$$

To prove this, observe that

$$F(\lambda x) - F(x) = \int_{x}^{\lambda x} \frac{\Phi(x)\Psi(x,\dot{x}) - \Phi(x,\dot{x})\Psi(x)}{\Phi(x)^2 + \Psi(x)^2}\, dt.$$

Choosing the straight segment

$$x(t) = (1-t)x + t\lambda x \qquad (0 \leqq t \leqq 1)$$

as path of integration we find that

$$\dot{x} = (\lambda - 1)x$$

whence

$$\Phi(x)\Psi(x,\dot{x}) - \Phi(x,\dot{x})\Psi(x) = 0.$$

This implies the equation (9.40).

9.14. The construction of eigenvectors. From now on our proof will follow the same lines as in sec. 8.6. We consider first the case that Ψ is non-degenerate. Introduce a positive definite inner product in E. Then the continuous function F assumes a minimum on the sphere $|x| = 1$. Let e_1 be a vector on $|x| = 1$ such that

$$F(e_1) \leqq F(x)$$

for all vectors $|x| = 1$. Then the homogeneity of F implies that

$$F(e_1) \leqq F(x)$$

for all vectors $x \neq 0$.

Consequently, the function

$$f(t) = F(e_1 + t y),$$

where y is an arbitrary vector, assumes a minimum at $t = 0$, whence

$$f'(0) = 0. \tag{9.41}$$

Carrying out the differentiation we find that

$$f'(0) = \frac{\Phi(e_1)\Psi(e_1, y) - \Phi(e_1, y)\Psi(e_1)}{\Phi(e_1)^2 + \Psi(e_1)^2}. \tag{9.42}$$

Equations (9.41) and (9.42) imply that

$$\Phi(e_1, y)\Psi(e_1) - \Phi(e_1)\Psi(e_1, y) = 0 \tag{9.43}$$

for all vectors $y \in E$. In this equation $\Psi(e_1) \neq 0$. In fact, assume that

$\Psi(e_1)=0$. Then $\Phi(e_1)\neq 0$ and hence equation (9.43) yields $\Psi(e_1,y)=0$ for all vectors $y\in E$. This is a contradiction to our assumption that Ψ is non-degenerate.

Define the number λ_1 by

$$\lambda_1=\frac{\Phi(e_1)}{\Psi(e_1)};$$

then equation (9.43) can be written as

$$\Phi(e_1,y)=\lambda_1\Psi(e_1,y)\qquad y\in E.$$

9.15. Now consider the subspace E_1 defined by the equation

$$\Psi(e_1,z)=0.$$

Since Ψ is non-degenerate, E_1 has the dimension $n-1$. Moreover, the restriction of Ψ to E_1 is again non-degenerate: Assume that z_1 is a vector of E_1 such that

$$\Psi(z_1,z)=0 \tag{9.44}$$

for all vectors $z\in E_1$. Equation (9.44) implies that

$$\Psi(z_1,x)=0 \tag{9.45}$$

for every vector $x\in E$ because x can be decomposed in the form

$$x=\xi e_1+z\qquad z\in E_1.$$

Now it follows from (9.45) that $z_1=0$, and so Ψ is non-degenerate in E_1. Therefore, the construction of sec. 9.14 can be applied to E_1. We thus obtain a vector $e_2\in E_1$ with the property that

$$\Phi(e_2,z)=\lambda_2\Psi(e_2,z)\quad\text{for every vector } z\in E_1 \tag{9.46}$$

where

$$\lambda_2=\frac{\Phi(e_2)}{\Psi(e_2)}.$$

Equation (9.46) implies that

$$\Phi(e_2,y)=\lambda_2\Psi(e_2,y) \tag{9.47}$$

for every vector $y\in E$; in fact, y can be decomposed in the form

$$y=\xi e_1+z\qquad z\in E_1$$

and we thus obtain

$$\Phi(e_2, y) = \xi\,\Phi(e_2, e_1) + \Phi(e_2, z) = \xi\,\Phi(e_1, e_2) + \Phi(e_2, z) \qquad (9.48)$$
$$= \xi\,\lambda_1\,\Psi(e_1, e_2) + \Phi(e_2, z) = \Phi(e_2, z)$$

and

$$\Psi(e_2, y) = \xi\,\Psi(e_2, e_1) + \Psi(e_2, z) = \Psi(e_2, z). \qquad (9.49)$$

Equations (9.46), (9.48) and (9.49) yield (9.47).

Continuing this construction we obtain after n steps a system of n vectors e_ν subject to the following conditions:

$$\begin{aligned} &\Phi(e_\nu, y) = \lambda_\nu\,\Psi(e_\nu, y) \qquad y \in E \\ &\Psi(e_\nu, e_\nu) \neq 0 \\ &\Psi(e_\nu, e_\mu) = 0 \qquad (\nu \neq \mu). \end{aligned} \qquad (9.50)$$

Rearranging the vectors e_ν and multiplying them with appropriate scalars we can achieve that

$$\Psi(e_\nu, e_\mu) = \varepsilon_\nu\,\delta_{\nu\mu} \quad \varepsilon_\nu = \begin{cases} +1\,(\nu = 1 \ldots s) \\ -1\,(\nu = s+1 \ldots n) \end{cases} \qquad (9.51)$$

where s denotes the index of Ψ. It follows from the above relations that the vectors e_ν form a basis of E.

Inserting $y = e_\mu$ in the first equation (9.50) we find

$$\Phi(e_\nu, e_\mu) = \lambda_\nu\,\varepsilon_\nu\,\delta_{\nu\mu}. \qquad (9.52)$$

Equations (9.51) and (9.52) show that the bilinear functions Φ and Ψ have diagonal form in the basis $e_\nu (\nu = 1 \ldots n)$.

9.16. The degenerate case. The degenerate case now remains to be considered. We may assume that $\Psi \neq 0$. Then it will be shown that there exists a scalar λ_0 such that the bilinear function $\Phi + \lambda_0 \Psi$ is non-degenerate

Let E^* be a dual space of E and consider the linear mappings

$$\varphi : E \to E^* \quad \text{and} \quad \psi : E \to E^*$$

defined by the equations

$$\Phi(x, y) = \langle \varphi x, y \rangle \quad \text{and} \quad \Psi(x, y) = \langle \psi x, y \rangle.$$

Then

$$\operatorname{Im}\psi \cap \varphi(\ker\psi) = 0. \qquad (9.53)$$

To prove this relation, let $y^* \in \operatorname{Im}\psi \cap \varphi(\ker\psi)$ be any vector. Then $y^* = \varphi x$

for some $x \in \ker \psi$. Hence

$$\Psi(x) = \langle x, \psi x \rangle = 0 \tag{9.54}$$

and

$$\Phi(x) = \langle \varphi x, x \rangle = \langle y^*, x \rangle = 0 \tag{9.55}$$

because $y^* \in \operatorname{Im} \psi$ and $x \in \ker \psi$.

Equations (9.54) and (9.55) imply that $x = 0$ and hence that $y^* = \varphi x = 0$.

Now let $x_\nu (\nu = 1 \dots n)$ be a basis of E such that the vectors $(x_{r+1} \dots x_n)$ form a basis of $\ker \psi$. Employing a determinant-function $\Delta \neq 0$ in E we obtain

$$\begin{aligned} &\Delta(\varphi x_1 + \lambda \psi x_1 \dots \varphi x_n + \lambda \psi x_n) \\ &= \Delta(\varphi x_1 + \lambda \psi x_1 \dots \varphi x_r + \lambda \psi x_r, \varphi x_{r+1} \dots \varphi x_n). \end{aligned}$$

The expansion of this expression yields a polynomial $f(\lambda)$ starting with the term

$$\lambda^r \Delta(\psi x_1 \dots \psi x_r, \varphi x_{r+1} \dots \varphi x_n).$$

The coefficient of λ^r is not identically zero. This follows from the relation (9.53) and the fact that the r vectors $\psi x_\varrho \in \operatorname{Im} \psi (\varrho = 1 \dots r)$ and the $(n-r)$ vectors $\varphi x_\sigma \in \varphi(\ker \psi)$ $(\sigma = r+1 \dots n)$ are linearly independent.

Hence, f is a polynomial of degree r. Our assumption $\Psi \neq 0$ implies that $r \geqq 1$. Consequently, a number λ_0 can be chosen such that $f(\lambda_0) \neq 0$. Then $\Phi + \lambda_0 \Psi$ is non-degenerate (cf. sec. 9.4).

By the previous theorem there exists a basis $e_\nu (\nu = 1 \dots n)$ of E in which the bilinear functions Φ and $\Phi + \lambda_0 \Psi$ both have diagonal form. Then the functions Φ and Ψ have also diagonal form in this basis.

Problems

1. Let Φ and Ψ be two symmetric bilinear functions in E. Prove that the condition

$$\Phi(x)^2 + \Psi(x)^2 > 0, \qquad x \neq 0$$

is equivalent to the following one: There exist real numbers λ and μ such that

$$\lambda \Phi(x) + \mu \Psi(x) > 0$$

for every $x \neq 0$.

2. Let $A = (\alpha_{\nu\mu})$ and $B = (\beta_{\nu\mu})$ be two symmetric $n \times n$-matrices and assume that the equations

$$\sum_{\nu,\mu} \alpha_{\nu\mu} \xi^\nu \xi^\mu = 0 \quad \text{and} \quad \sum_{\nu,\mu} \beta_{\nu\mu} \xi^\nu \xi^\mu = 0$$

together imply that $\xi^\nu = 0$ $(\nu = 1 \ldots n)$. Prove that the polynomial

$$f(\lambda) = \det(A + \lambda B)$$

is of degree r and has r real roots where r is the rank of B.

§ 4. Pseudo-Euclidean spaces

9.17. Definition. A *pseudo-Euclidean* space is a real linear space in which a non-degenerate indefinite bilinear function is given. As in the positive definite case, this bilinear function is called the *inner product* and is denoted by $(,)$. The index of the inner product is called the *index of the pseudo-Euclidean space.*

Since the inner product is indefinite, the number (x, x) can be positive, negative or zero, depending on the vector x. A vector $x \neq 0$ is called

space-like, if $(x, x) > 0$

time-like, if $(x, x) < 0$

a *light-vector*, if $(x, x) = 0$

The *light-cone* is the set of all light-vectors.

As in the definite case two vectors x and y are called *orthogonal* if $(x, y) = 0$. The light-cone consists of all vectors which are orthogonal to themselves.

A basis e_ν $(\nu = 1 \ldots n)$ is called *orthonormal* if

$$(e_\nu, e_\mu) = \varepsilon_\nu \delta_{\nu\mu}$$

where

$$\varepsilon_\nu = \begin{cases} +1 \, (\nu = 1 \ldots s) \\ -1 \, (\nu = s + 1 \ldots n). \end{cases}$$

In sec. 9.8 we have shown that an orthonormal basis can always be constructed.

If an orthonormal basis e_ν $(\nu = 1 \ldots n)$ is chosen, the inner product of two vectors

$$x = \sum_\nu \xi^\nu e_\nu \quad \text{and} \quad y = \sum_\nu \eta^\nu e_\nu$$

is given by the bilinear form

$$(x, y) = \sum_{\nu=1}^{n} \varepsilon_\nu \xi^\nu \eta^\nu = \sum_{\nu=1}^{s} \xi^\nu \eta^\nu - \sum_{\nu=s+1}^{n} \xi^\nu \eta^\nu \tag{9.56}$$

and the equation of the light-cone reads

$$\sum_{\nu=1}^{s} \xi^\nu \xi^\nu - \sum_{\nu=s+1}^{n} \xi^\nu \xi^\nu = 0.$$

9.18. Orthogonal complements. Since the inner product in E is non-degenerate the space E is dual to itself. Hence, every subspace $E_1 \subset E$ determines an orthogonal complement $E_1^\perp$ which is again a subspace of E and has complementary dimension.

However, the intersection $E_1 \cap E_1^\perp$ does not necessarily consist of the zero-vector alone, as in the positive definite case. Assume, for instance, that E_1 is the 1-dimensional subspace generated by a light-vector l. Then E_1 is contained in $E_1^\perp$.

It will be shown that $E_1 \cap E_1^\perp = 0$ if and only if the inner product is non-degenerate in E_1. Assume first that this condition is fulfilled. Let x_1 be a vector of $E_1 \cap E_1^\perp$. Then

$$(x_1, y_1) = 0 \quad \text{for all vectors } y_1 \in E_1, \tag{9.57}$$

and thus $x_1 = 0$. Conversely, assume that $E_1 \cap E_1^\perp = 0$. Then it follows that

$$E = E_1 \oplus E_1^\perp \tag{9.58}$$

since E_1 and $E_1^\perp$ have complementary dimension.

Now let x_1 be a vector of E_1 such that

$$(x_1, y_1) = 0 \quad \text{for all vectors } y_1 \in E_1.$$

It follows from (9.58) that every vector y of E can be written as

$$y = y_1 + y_1^\perp \qquad y_1 \in E_1, \; y_1^\perp \in E_1^\perp,$$

whence

$$(x_1, y) = (x_1, y_1) + (x_1, y_1^\perp) = 0 \quad \text{for all vectors} \quad y \in E.$$

This equation implies that $x_1 = 0$. Consequently, the inner product is non-degenerate in E_1.

9.19. Normed determinant functions. Let $\Delta_0 \neq 0$ be a determinant function in E. Since E is dual to itself, the identity (4.21) applies to E yielding

$$\Delta_0(x_1, \dots, x_n)\,\Delta_0(y_1, \dots, y_n) = \alpha \det(x_i, y_j) \qquad \begin{matrix} \alpha \in \mathbb{R} \\ \alpha \neq 0. \end{matrix} \tag{9.59}$$

Substituting $x_\nu = y_\nu = e_\nu$ in (9.59), where $e_\nu\,(\nu = 1 \dots n)$ is an orthonormal basis, we obtain

$$\Delta_0(e_1 \dots e_n)^2 = (-1)^{n-s}\alpha.$$

This equation shows that

$$\alpha(-1)^{n-s} > 0.$$

Consequently, another determinant function, Δ, can be defined by

$$\Delta = \pm \frac{\Delta_0}{\sqrt{(-1)^{n-s}\alpha}}. \tag{9.60}$$

Then the identity (9.59) assumes the form

$$\Delta(x_1 \ldots x_n)\,\Delta(y_1 \ldots y_n) = (-1)^{n-s} \det(x_i, y_j). \tag{9.61}$$

A determinant function satisfying the relation (9.61) is called a *normed determinant function*. Equation (9.60) shows that there exist exactly two normed determinant functions Δ and $-\Delta$ in E.

9.20. The pseudo-Euclidean plane. The simplest example of a pseudo-Euclidean space is a 2-dimensional linear space with an inner product of index 1. Then the light-cone consists of two straight lines. Selecting two vectors l_1 and l_2 which generate these lines we have the equations

$$(l_1, l_1) = 0 \quad \text{and} \quad (l_2, l_2) = 0. \tag{9.62}$$

But

$$(l_1, l_2) \neq 0$$

because otherwise the inner product would be identically zero. We can therefore assume that

$$(l_1, l_2) = -1. \tag{9.63}$$

Given a vector

$$x = \xi^1 l_1 + \xi^2 l_2$$

of E the equations (9.62) and (9.63) yield

$$(x, x) = -2\xi^1 \xi^2$$

showing that x is space-like if $\xi^1 \xi^2 < 0$ and x is time-like if $\xi^1 \xi^2 > 0$. In other words, the space-like vectors are contained in the two sectors S_1 and S_2 of fig. 3 and the time-like vectors are contained in T^+ and T^-.

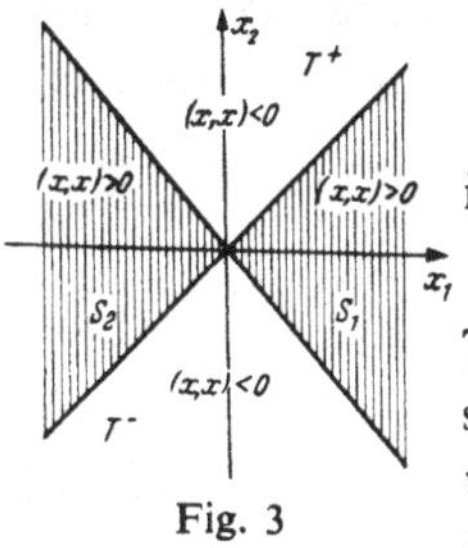

Fig. 3

The inner product of two vectors

$$x = \xi^1 l_1 + \xi^2 l_2 \quad \text{and} \quad y = \eta^1 l_1 + \eta^2 l_2$$

is given by

$$(x, y) = -(\xi^1 \eta^2 + \xi^2 \eta^1).$$

This formula shows that the inner product of two space-like vectors is positive if and only if these vectors are contained in the same one of the sectors S_1 and S_2.

Let an orientation be defined in E by the normed determinant function Δ.

Then the identity (9.61) yields ($n=2$, $s=1$)

$$(x,y)^2 - \Delta(x,y)^2 = (x,x)(y,y). \tag{9.64}$$

If x and y are not light-vectors equation (9.64) may be written in the form

$$\frac{(x,y)^2}{(x,x)(y,y)} - \frac{\Delta(x,y)^2}{(x,x)(y,y)} = 1. \tag{9.65}$$

Now assume in addition, that the vectors x and y are space-like and are contained in the same one of the sectors S_1 and S_2. Then

$$(x,y) > 0. \tag{9.66}$$

Relations (9.65) and (9.66) imply that there exists exactly one real number $\theta(-\infty<\theta<\infty)$ such that

$$\cosh\theta = \frac{(x,y)}{|x|\,|y|} \quad \text{and} \quad \sinh\theta = \frac{\Delta(x,y)}{|x|\,|y|}. \tag{9.67}$$

This number is called the pseudo-Euclidean angle between the space-like vectors x and y.

We finally note that the vectors

$$e_1 = \frac{1}{\sqrt{2}}(l_1 - l_2) \quad \text{and} \quad e_2 = \frac{1}{\sqrt{2}}(l_1 + l_2)$$

form an orthonormal basis of E. Relative to this basis the equation of the light-cone assumes the form

$$(\xi^1)^2 - (\xi^2)^2 = 0.$$

9.21. Pseudo-Euclidean spaces of index $n-1$. More generally let us consider an n-dimensional pseudo-Euclidean space with index $n-1$. Then every fixed time-like unit vector z determines an orthogonal decomposition of E into an $(n-1)$-dimensional subspace consisting of space-like vectors and the 1-dimensional subspace generated by z. In fact, every vector $x \in E$ can be uniquely decomposed in the form

$$x = \lambda z + y \qquad (z,y) = 0$$

where the scalar λ is given by

$$\lambda = -(x,z).$$

Passing over to the norm we obtain the equation

$$(x,x) = -\lambda^2 + (y,y)$$

showing that

$$\begin{aligned}&\lambda^2 < (y, y) \text{ if } x \text{ is space-like}\\&\lambda^2 > (y, y) \text{ if } x \text{ is time-like}\\&\lambda^2 = (y, y) \text{ if } x \text{ is a light-vector}.\end{aligned} \tag{9.68}$$

From this decomposition we shall now derive the following properties:

(1) Two time-like vectors are never orthogonal.

(2) A time-like vector is never orthogonal to a light-vector.

(3) Two light-vectors are orthogonal if and only if they are linearly dependent.

(4) The orthogonal complement of a light-vector is an $(n-1)$-dimensional subspace of E in which the inner product is positive semidefinite and has rank $n-2$.

To prove (1), consider another time-like vector z_1. This vector z_1 can be written as

$$z_1 = \lambda z + y_1 \qquad (z, y_1) = 0. \tag{9.69}$$

Then

$$\lambda^2 > (y_1, y_1),$$

whence $\lambda \neq 0$. Inner multiplication of (9.69) by z yields

$$(z, z_1) = \lambda(z, z) \neq 0.$$

Next, consider a light-vector l. Then

$$l = \lambda z + y \qquad (z, y) = 0$$

and

$$\lambda^2 = (y, y) > 0.$$

These two relations imply that

$$(l, z) = \lambda(z, z) \neq 0$$

which proves (2).

Now let l_1 and l_2 be two orthogonal light-vectors. Then we have the decompositions

$$l_1 = \lambda_1 z + y_1 \quad \text{and} \quad l_2 = \lambda_2 z + y_2,$$

whence

$$-\lambda_1 \lambda_2 + (y_1, y_2) = 0. \tag{9.70}$$

Observing that

$$\lambda_1^2 = (y_1, y_1) \quad \text{and} \quad \lambda_2^2 = (y_2, y_2)$$

we obtain from (9.71) the equation

$$(y_1, y_1)(y_2, y_2) = (y_1, y_2)^2. \tag{9.71}$$

The vectors y_1 and y_2 are contained in the orthogonal complement of z. In this space the inner product is positive definite and hence equation. (9.71) implies that y_1 and y_2 are linearly dependent, $y_2=\lambda y_1$. Inserting this into (9.70) we find $\lambda_2=\lambda\lambda_1$, whence $l_2=\lambda l_1$.

Finally, let l be a light-vector and E_1 be the orthogonal complement of l. It follows from property (2) that E_1 does not contain time-like vectors. In other words, the inner product is positive semidefinite in E_1. To find the null-space of the inner product, assume that y_1 is a vector of E_1 such that

$$(y_1, y) = 0 \quad \text{for all vectors } y \in E_1 .$$

This implies that $(y_1, y_1)=0$ showing that y_1 is a light-vector. Now it follows from property (3) that y_1 is a multiple of l. Consequently, the null-space of the inner product in E_1 is generated by l.

9.22. Fore-cone and past-cone. As another consequence of the properties established in the last section it will now be shown that the set of all time-like vectors consists of two disjoint sectors T^+ and T^- (cf. fig. 4.) To this purpose we define an equivalence relation in the set T of all time-like vectors in the following way:

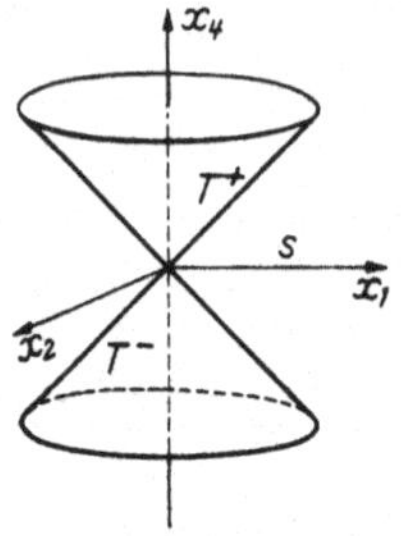

Fig. 4

$$z_1 \sim z_2 \quad \text{if} \quad (z_1, z_2) < 0 . \tag{9.72}$$

Relation (9.72) is obviously symmetric and reflexive. To prove the transitivity, consider three time-like vectors $z_i (i=1, 2, 3)$ and assume that

$$(z_1, z_3) < 0 \quad \text{and} \quad (z_2, z_3) < 0 .$$

We have to show that

$$(z_1, z_2) < 0 .$$

We may assume that z_3 is a time-like *unit-vector*. Then the vectors z_1 and z_2 can be decomposed in the form

$$z_i = \lambda_i z_3 + y_i , \quad \lambda_i = -(z_i, z_3) \qquad (i = 1,2) \tag{9.73}$$

where the vectors y_1 and y_2 are contained in the orthogonal complement F of z_3. Equations (9.73) yield

$$(z_i, z_i) = -\lambda_i^2 + (y_i, y_i) \qquad (i = 1,2) \tag{9.74}$$

and

$$(z_1, z_2) = -\lambda_1 \lambda_2 + (y_1, y_2). \tag{9.75}$$

It follows from (9.68) that

$$(y_i, y_i) < \lambda_i^2 \qquad (i = 1, 2).$$

Now observe that the inner product is positive definite in the subspace F. Consequently, the Schwarz inequality applies to the vectors y_1 and y_2, yielding

$$(y_1, y_2)^2 \leqq (y_1, y_1)(y_2, y_2) \leqq \lambda_1^2 \lambda_2^2 .$$

This inequality shows that the first term determines the sign on the right-hand side of (9.75). But this term is negative because $\lambda_i = -(z_i, z_3) > 0$ $(i = 1, 2)$ and we thus obtain

$$(z_1, z_2) < 0 .$$

The equivalence relation (9.72) induces a decomposition of the set T into two classes T^+ and T^- which are obtained from each other by the reflection $x \to -x$.

9.23. The two subsets T^+ and T^- are *convex*, i.e., they contain with any two vectors z_1 and z_2 the straight segment

$$z(t) = (1 - t) z_1 + t z_2 \qquad (0 \leqq t \leqq 1).$$

In fact, assume that $z_1 \in T^+$ and $z_2 \in T^+$. Then

$$(z_1, z_1) < 0, (z_2, z_2) < 0 \quad \text{and} \quad (z_1, z_2) < 0 ,$$

whence

$$(z(t), z(t)) = (1 - t)^2 (z_1, z_1) + 2t(1 - t)(z_1, z_2) + t^2 (z_2, z_2) < 0 ,$$
$$(0 \leqq t \leqq 1).$$

In the special theory of relativity the sectors T^+ and T^- are called the *fore-cone* and the *past-cone*.

The set S of the space-like vectors is not convex as fig. 4 shows.

Problems

1. Let E be a pseudo-Euclidean plane and g_1, g_2 be the two straight lines generated by the light-vectors. Introduce a Euclidean metric in E such that g_1 and g_2 are orthogonal. Prove that two vectors $x \neq 0$ and $y \neq 0$ are orthogonal with respect to the pseudo-Euclidean metric if and only if they generate the Euclidean bisectors of g_1 and g_2.

2. Consider a pseudo-Euclidean space of dimension 3 and index 2. Assume that an orientation is defined in E by the normed determinant

function Δ. As in a Euclidean space define the cross product of two vectors x_1 and x_2 by the relation

$$(x_1 \times x_2, x_3) = \Delta(x_1, x_2, x_3).$$

Prove: a) $x_1 \times x_2 = 0$ if and only if the vectors x_1 and x_2 are linearly dependent.
b) $(x_1 \times x_2, x_1 \times x_2) = (x_1, x_2)^2 - (x_1, x_1)(x_2, x_2)$
c) If e_1, e_2, e_3 is a positive orthonormal basis of E, then

$$e_1 \times e_2 = -e_3, \quad e_1 \times e_3 = -e_2, \quad e_2 \times e_3 = e_1.$$

3. Let E be an n-dimensional pseudo-Euclidean space of index $n-1$. Given two time-like unit vectors z_1 and z_2 prove: a) The vector $z_1 + z_2$ is time-like or space-like depending on whether z_1 and z_2 are contained in the same cone or in different cones. b) The Schwarz inequality holds in the reversed form

$$(z_1, z_2)^2 \geqq (z_1, z_1)(z_2, z_2).$$

Equality holds if and only if z_1 and z_2 are linearly dependent.

4. Denote by S the set of all space-like vectors. Prove that the set S is connected if $n \geqq 3$. More precisely: Given two vectors $x_0 \in S$ and $x_1 \in S$ there exists a continuous curve $x = x(t)(0 \leqq t \leqq 1)$ in S such that $x(0) = x_0$ $x(1) = x_1$.

§ 5. Linear mappings of pseudo-Euclidean spaces

9.24. The adjoint mapping. Let φ a linear transformation of the n-dimensional pseudo-Euclidean space E. Since E is dual to itself with respect to the inner product the adjoint mapping $\tilde{\varphi}$ can be defined as in sec. (8.1). The mappings φ and $\tilde{\varphi}$ are connected by the relation

$$(\varphi x, y) = (x, \tilde{\varphi} y) \qquad x, y \in E. \tag{9.76}$$

The duality of the mappings φ and $\tilde{\varphi}$ implies that

$$\det \tilde{\varphi} = \det \varphi \quad \text{and} \quad \operatorname{tr} \tilde{\varphi} = \operatorname{tr} \varphi.$$

Let (α_ν^μ) and $(\tilde{\alpha}_\nu^\mu)$ $(\nu, \mu = 1 \dots n)$ be the matrices of φ and $\tilde{\varphi}$ relative to an orthonormal basis e_ν. Inserting $x = e_\nu$ and $y = e_\mu$ into (9.76) we find that

$$\varepsilon_\mu \alpha_\nu^\mu = \varepsilon_\nu \tilde{\alpha}_\mu^\nu \qquad (\nu, \mu = 1 \dots n)$$

where

$$\varepsilon_\nu = \begin{cases} +1 \, (\nu = 1 \ldots s) \\ -1 \, (\nu = s+1 \ldots n). \end{cases}$$

Now assume that the mapping φ is selfadjoint, $\tilde{\varphi} = \varphi$. In the positive definite case we have shown that there exists a system of n orthonormal eigenvectors. This result can be carried over to pseudo-Euclidean spaces of dimension $n \geqq 3$ if we add the hypothesis that $(x, \varphi x) \neq 0$ for all light-vectors. To prove this, consider the symmetric bilinear functions

$$\Phi(x, y) = (\varphi x, y) \quad \text{and} \quad \Psi(x, y) = (x, y).$$

It follows from the above assumption that

$$\Phi(x)^2 + \Psi(x)^2 > 0 \quad \text{for all vectors} \quad x \neq 0.$$

Hence the theorem of sec. 9.11 applies to Φ and Ψ, showing that there exists an orthonormal basis e_ν $(\nu = 1 \ldots n)$ such that

$$(\varphi e_\nu, e_\mu) = \lambda_\nu \varepsilon_\nu \delta_{\nu\mu} \qquad (\nu, \mu = 1 \ldots n). \tag{9.77}$$

Equations (9.77) imply that

$$\varphi e_\nu = \lambda_\nu e_\nu \quad (\nu = 1 \ldots n)$$

showing that the e_ν are eigenvectors of φ.

9.25. Pseudo-Euclidean rotations. A linear transformation φ of the pseudo-Euclidean space E which preserves the inner product,

$$(\varphi x, \varphi y) = (x, y) \tag{9.78}$$

is called a *pseudo-Euclidean rotation*. Replacing y by x in (9.78) we obtain the equation

$$(\varphi x, \varphi x) = (x, x) \qquad x \in E$$

showing that a pseudo-Euclidean rotation sends space-like vectors into space-like vectors, time-like vectors into time-like vectors and light-vectors into light-vectors. A rotation is always regular. In fact, assume that $\varphi x = 0$ for a vector $x \in E$. Then it follows from (9.78) that

$$(x, y) = (\varphi x, \varphi y) = 0$$

for all vectors $y \in E$, whence $x = 0$.

Comparing the relations (9.76) and (9.78) we see that the adjoint and the inverse of a pseudo-Euclidean rotation coincide,

$$\tilde{\varphi} = \varphi^{-1}. \tag{9.79}$$

Equation (9.79) shows that the determinant of φ must be ± 1, as in the Euclidean case.

Now let e be an eigenvector of φ and λ be the corresponding eigenvalue,

$$\varphi e = \lambda e .$$

Passing over to the norms we obtain

$$(\varphi e, \varphi e) = \lambda^2 (e, e).$$

This equation shows that $\lambda = \pm 1$ provided that e is not a light-vector. An eigenvector which is contained in the light-cone may have an eigenvalue $\lambda \neq \pm 1$ as can be seen from examples.

If an orthonormal basis is chosen in E the matrix of φ satisfies the relations

$$\sum_\lambda \varepsilon_\lambda \alpha_\nu^\lambda \alpha_\mu^\lambda = \varepsilon_\nu \delta_{\nu\mu} .$$

A matrix of this kind is called *pseudo-orthogonal*.

9.26. Pseudo-Euclidean rotations of the plane. In particular, consider a pseudo-Euclidean rotation φ of a 2-dimensional space with index 1. Then the light-cone consists of two straight lines. Since the light-cone is preserved under the rotation φ, it follows that these straight lines are either transformed into themselves or they are interchanged. Now assume that φ is a *proper* rotation i.e. det $\varphi = +1$. Then the second case is impossible because the inner product is preserved under φ. Thus we can write

$$\varphi l_1 = \lambda l_1 \quad \text{and} \quad \varphi l_2 = \frac{1}{\lambda} l_2 , \tag{9.80}$$

where l_1, l_2 is the basis of E defined in sec. 9.20. The number λ is positive or negative depending on whether the sectors T^+ and T^- are mapped onto themselves or interchanged.

Now consider an arbitrary vector

$$x = \xi^1 l_1 + \xi^2 l_2 . \tag{9.81}$$

Then equations (9.80) and (9.81) yield

$$\varphi x = \lambda \xi^1 l_1 + \frac{1}{\lambda} \xi^2 l_2 ,$$

whence

$$(x, \varphi x) = -\left(\lambda + \frac{1}{\lambda}\right) \xi^1 \xi^2 = \frac{1}{2}\left(\lambda + \frac{1}{\lambda}\right)(x, x) . \tag{9.82}$$

This equation shows, that the inner product of x and φx depends only on the norm of x as in the case of a Euclidean rotation (cf. sec. 8.21).

To find the "rotation-angle" of φ, introduce an orientation in E such that the basis l_1, l_2 is positive. Let Δ be a normed determinant function which represents this orientation. Then identity (9.64) yields

$$\Delta(l_1, l_2)^2 = (l_1, l_2)^2 = 1,$$

whence

$$\Delta(l_1, l_2) = 1.$$

Inserting the vectors x and φx into Δ we find that

$$\Delta(x, \varphi x) = \frac{1}{2}\left(\lambda - \frac{1}{\lambda}\right)(x, x)'\Delta(l_1, l_2) = \frac{1}{2}\left(\lambda - \frac{1}{\lambda}\right)(x, x).$$

Now assume in addition that φ transforms the sectors T^+ and T^- into themselves (i. e., that φ does not interchange T^+ and T^-). Then $\lambda > 0$ and equation (9.82) shows that $(x, \varphi x) > 0$ for every space-like vector x. Using formulae (9.67) we obtain the equations

$$\cosh\theta = \frac{1}{2}\left(\lambda + \frac{1}{\lambda}\right) \quad \text{and} \quad \sinh\theta = \frac{1}{2}\left(\lambda - \frac{1}{\lambda}\right), \tag{9.83}$$

where θ denotes the pseudo-Euclidean angle between the vectors x and φx.

Now consider the orthonormal basis of E which is determined by the vectors

$$e_1 = \frac{1}{\sqrt{2}}(l_1 - l_2) \quad \text{and} \quad e_2 = \frac{1}{\sqrt{2}}(l_1 + l_2).$$

Then equations (9.80) yield

$$\varphi e_1 = \frac{1}{2}\left(\lambda + \frac{1}{\lambda}\right)e_1 + \frac{1}{2}\left(\lambda - \frac{1}{\lambda}\right)e_2$$

$$\varphi e_2 = \frac{1}{2}\left(\lambda - \frac{1}{\lambda}\right)e_1 + \frac{1}{2}\left(\lambda + \frac{1}{\lambda}\right)e_2.$$

We thus obtain the following representation of φ, which corresponds to the representation of a Euclidean rotation given in sec. 8.21:

$$\varphi e_1 = e_1\cosh\theta + e_2\sinh\theta$$
$$\varphi e_2 = e_1\sinh\theta + e_2\cosh\theta.$$

9.27. Lorentz-transformations. A 4-dimensional pseudo-Euclidean space with index 3 is called *Minkowski-space*. A *Lorentz-transformation* is a rotation of the Minkowski-space. The purpose of this section is to show that *a proper Lorentz-transformation* φ *possesses always at least one eigenvector on the light-cone**). We may restrict ourselves to Lorentz-transformations which do not interchange fore-cone and past-cone because this can be achieved by multiplication with -1. These transformations are called *orthochronous*. First of all we observe that a light-vector l is an eigenvector of φ if and only if $(l, \varphi l)=0$. In fact, the equation $\varphi l=\lambda l$ yields

$$(l,\varphi l)=\lambda(l,l)=0 .$$

Conversely, assume that l is a light-vector with the property that $(l, \varphi l)=0$. Then it follows from sec. (9.21) property (3) that the vectors l and φl are linearly dependent. In other words, l is an eigenvector of φ.

Now consider the selfadjoint mapping

$$\psi=\tfrac{1}{2}(\varphi+\tilde{\varphi})=\tfrac{1}{2}(\varphi+\varphi^{-1}). \tag{9.84}$$

Then

$$(x,\psi x)=\tfrac{1}{2}(x,\varphi x)+\tfrac{1}{2}(x,\tilde{\varphi} x)=(x,\varphi x) \quad x\in E . \tag{9.85}$$

It follows from the above remark and from (9.85) that a light-vector l is an eigenvector of φ if and only if $(l, \psi l)=0$. We now preceed indirectly and assume that φ does not have an eigenvector on the lightcone. Then $(x, \psi x)=(x, \varphi x)\neq 0$ for all light-vectors and hence we can apply the result of sec. 9.24 to the mapping ψ: There exist four eigenvectors e_ν $(\nu=1\ldots4)$ such that

$$(e_\nu,e_\mu)=\varepsilon_\nu\delta_{\nu\mu} \quad \varepsilon_\nu=\begin{cases}+1\,(\nu=1,2,3)\\ -1\,(\nu=4).\end{cases}$$

Let us denote the time-like eigenvector e_4 by e and the corresponding eigenvalue by λ. Then $\psi e=\lambda e$ and hence it follows from (9.84) that

$$\varphi^2 e=2\lambda\varphi\, e-e .$$

Next, we wish to construct a time-like eigenvector of the mapping φ. If φe is a multiple of e, e is such a vector. We thus may assume that the vectors e and φe are linearly independent. Then these two vectors generate

---

*) Observe that a proper *Euclidean* rotation of a 4-dimensional space need not have eigenvectors.

a plane F which is invariant under φ. This plane intersects the light-cone in two straight lines. Since the plane F and the light-cone are both invariant under φ, these two lines are either interchanged or transformed into themselves. In the second case we have two eigenvectors of φ on the light-cone, in contradiction to our assumption. In the first case select two generating vectors l_1 and l_2 on these lines such that $(l_1, l_2)=1$. Then

$$\varphi l_1 = \alpha l_2 \quad \text{and} \quad \varphi l_2 = \beta l_1 .$$

The condition

$$(\varphi l_1, \varphi l_2) = (l_1, l_2)$$

implies that $\alpha\beta = 1$. Now consider the vector z

$$z = l_1 + \alpha l_2 .$$

Then

$$\varphi z = \alpha l_2 + \alpha\beta l_1 = \alpha l_2 + l_1 = z$$

To show that z is timelike observe that

$$(z, z) = 2\alpha(l_1, l_2) = 2\alpha$$

and that the vector $t = l_1 - l_2$ is time-like. Moreover,

$$(t, \varphi t) = \alpha + \frac{1}{\alpha} . \tag{9.86}$$

Now $(t, \varphi t) < 0$ because φ leaves the fore-cone and the past-cone invariant (cf. sec. 9.22). Hence equation (9.86) implies that $\alpha < 0$, showing that z is time-like.

Using the time-like vector z we shall now construct an eigenvector on the light-cone which will give us a contradiction. Let E_1 be the orthogonal complement of z. E_1 is a 3-dimensional Euclidean subspace of E which is invariant under φ. Since φ is a proper Lorentz-transformation it induces a proper Euclidean rotation in E_1. Consequently, there exists an invariant axis in E_1 (cf. sec. 8.22). Let y be a vector of this axis such that $(y, y) = -2\alpha$. Then $l = y + z$ is a light-vector and

$$\varphi l = \varphi y + \varphi z = y + z = l$$

i. e. l is an eigenvector of φ.

Hence, the assumption that there are no eigenvectors on the light-cone leads to a contradiction and the assertion in the beginning of this section is proved.

We finally note that every eigenvalue λ of φ whose eigenvector l lies on the light-cone is positive. In fact, select a space-like unit-vector y such that $(l, y)=1$ and consider the vector $z=l+\tau y$ where τ is a real parameter. Then we have the relation

$$(z, z) = 2\tau + \tau^2$$

showing that z is time-like if $-2<\tau<0$. Since φ preserves fore-cone and past-cone it follows that

$$(z, \varphi z) < 0 \qquad (-2 < \tau < 0).$$

But

$$(z, \varphi z) = (l + \tau y, \lambda l + \tau \varphi y) = \tau\left(\lambda + \frac{1}{\lambda}\right) + \tau^2 (y, \varphi y)$$

and we thus obtain

$$\tau\left(\lambda + \frac{1}{\lambda}\right) + \tau^2 (y, \varphi y) < 0 \qquad (-2 < \tau < 0).$$

Letting $\tau \to 0$ we see that λ must be positive.

Problems

1. Let φ be a linear automorphism of the plane E. Prove that an inner product of index 1 can be introduced in E such that φ becomes a proper pseudo-Euclidean rotation if and only if the following conditions are satisfied:

   1. There are two linearly independent eigenvectors.
   2. $\det\varphi = 1$.
   3. $|\operatorname{tr}\varphi| \geqq 2$.

2. Find the eigenvectors of the Lorentz-transformation defined by the matrix

$$\begin{pmatrix} \frac{1}{2} & 0 & -1 & -\frac{1}{2} \\ 0 & 1 & 0 & 0 \\ 1 & 0 & 1 & 1 \\ \frac{1}{2} & 0 & 1 & \frac{3}{2} \end{pmatrix}.$$

Verify that there exists an eigenvector on the light-cone.

3. Let a and b be two linearly independent light-vectors in the pseudo-Euclidean plane E. Then a linear transformation ψ of E is defined by

$$\psi a = a, \quad \psi b = -b.$$

Consider the family of linear automorphisms $\varphi(t)$ which is defined by the differential equation

$$\dot{\varphi}(t) = \psi \circ \varphi(t)$$

and the initial condition

$$\varphi(0) = \iota .$$

a) Prove that $\varphi(t)$ is a family of proper rotations carrying fore-cone and past-cone into themselves.

b) Define the functions $C(t)$ and $S(t)$ by

$$C(t) = \tfrac{1}{2}\operatorname{tr}\varphi(t) \quad \text{and} \quad S(t) = \tfrac{1}{2}\operatorname{tr}(\psi \circ \varphi(t)).$$

Prove the functional-equations

$$C(t_1 + t_2) = C(t_1)C(t_2) + S(t_1)S(t_2)$$

and

$$S(t_1 + t_2) = S(t_1)C(t_2) + S(t_2)C(t_1).$$

c) Prove that

$$\varphi(t)a = e^{-t}a \quad \text{and} \quad \varphi(t)b = e^{-t}b.$$

4. Let E be a pseudo-Euclidean space and consider an orthogonal decomposition

$$E = E^+ \oplus E^-$$

such that the restriction of the inner product to E^+ (E^-) is positive (negative) definite. Let ω be a selfadjoint involution of E such that E^+ (and hence E^-) is stable under ω. Define a symmetric bilinear function Ψ by

$$\Psi(x, y) = (\omega x, y) \qquad x, y \in E.$$

Prove that the signature of Ψ is given by

$$\operatorname{sig}\Psi = \operatorname{tr}\omega^+ - \operatorname{tr}\omega^-$$

where ω^+ and ω^- denote the restrictions of ω to E^+ and E^- respectively.

Chapter X

Quadrics *

In the present Chapter the theory of the bilinear functions developed in Chapter IX will be applied to the discussion of quadrics. In this context we shall have to deal with *affine spaces.*

§ 1. Affine spaces

10.1. Points and vectors. Let E be a real n-dimensional linear space and let A be a set of elements $P, Q\ldots$ which will be called *points.* Assume that a relation between points and vectors is defined in the following way:

1. To every ordered pair P, Q of A there is assigned a vector of E, called the *difference vector* and denoted by $\overrightarrow{PQ}$.

2. To every point $P\in A$ and every vector $x\in E$ there exists exactly one point $Q\in A$ such that $\overrightarrow{PQ}=x$.

3. If P, Q, R are three arbitrary points, then

$$\overrightarrow{PQ}+\overrightarrow{QR}=\overrightarrow{PR}. \tag{10.1}$$

A is called an *n-dimensional affine space* with the *difference space E.*

Insertion of $Q=P$ in (10.1) yields $\overrightarrow{PP}+\overrightarrow{PR}=\overrightarrow{PR}$, whence $\overrightarrow{PP}=0$ for every point $P\in A$. Using this relation we obtain from (10.1)

$$\overrightarrow{QP}=-\overrightarrow{PQ}.$$

The equation $\overrightarrow{P_1Q_1}=\overrightarrow{P_2Q_2}$ implies that $\overrightarrow{P_1P_2}=\overrightarrow{Q_1Q_2}$ (parallelogram-law). In fact,

$$\overrightarrow{P_1P_2}=\overrightarrow{P_1Q_2}-\overrightarrow{P_2Q_2}$$

$$\overrightarrow{Q_1Q_2}=\overrightarrow{P_1Q_2}-\overrightarrow{P_1Q_1}.$$

Subtraction of these equations yields $\overrightarrow{P_1P_2}=\overrightarrow{Q_1Q_2}$.

For any given linear space E, an affine space can be constructed which possesses E as difference space:

Define the points as the vectors of E and the difference-vector of two points x and y as the vector $y-x$. Then the above conditions are obviously satisfied.

Let A be a given affine space. If a fixed point O is distinguished as origin, every point P is uniquely determined by the vector $x=\overrightarrow{OP}$. x is called the *position-vector* of P and every point P can be identified with the corresponding position-vector x. The difference-vector of two points x and y is simply the vector $y-x$.

10.2. Affine coordinate systems. An *affine coordinate-system* $(O; x_1 \dots x_n)$ consists of a fixed point $O \in A$, the *origin*, and a basis $x_\nu (\nu=1 \dots n)$ of the difference-space E. Then every point $P \in A$ determines a system of n numbers ξ^ν $(\nu=1 \dots n)$ by

$$\overrightarrow{OP} = \sum_\nu \xi^\nu x_\nu .$$

The numbers $\xi^\nu (\nu=1 \dots n)$ are called the *affine coordinates* of P relative to the given system. The origin O has the coordinates $\xi^\nu=0$.

Now consider two affine coordinate-systems

$$(O; x_1 \dots x_n) \quad \text{and} \quad (O'; y_1 \dots y_n).$$

Denote by α_ν^μ the matrix of the basis-transformation $x_\nu \to y_\nu$ and by β^ν the affine coordinates of O' relative to the system $(O; x_1 \dots x_n)$,

$$y_\nu = \sum_\mu \alpha_\nu^\mu x_\mu , \quad \overrightarrow{OO'} = \sum_\nu \beta^\nu x_\nu .$$

The affine coordinates ξ^ν and η^ν of a point P, corresponding to the systems $(O; x_1 \dots x_n)$ and $(O'; y_1 \dots y_n)$ respectively, are given by

$$\overrightarrow{OP} = \sum_\nu \xi^\nu x_\nu \quad \text{and} \quad \overrightarrow{O'P} = \sum_\nu \eta^\nu y_\nu . \tag{10.2}$$

Inserting $\overrightarrow{O'P} = \overrightarrow{OP} - \overrightarrow{OO'}$ in the second equation (10.2) we obtain

$$\sum_\nu \eta^\nu y_\nu = \sum_\nu (\xi^\nu - \beta^\nu) x_\nu ,$$

whence

$$\sum_\nu \alpha_\nu^\mu \eta^\nu = \xi^\mu - \beta^\mu \qquad (\mu = 1 \dots n).$$

Multiplication by the inverse matrix yields the transformation-formula for the affine coordinates:

$$\eta^\nu = \sum_\mu \check{\alpha}_\mu^\nu (\xi^\mu - \beta^\mu) \qquad (\nu = 1 \dots n).$$

10.3. Affine subspaces. An *affine subspace* of A is a subset A_1 of A such that the vectors $\overrightarrow{PQ}(P\in A_1, Q\in A_1)$ form a subspace of E. If O is the origin of A and $(O_1; x_1 \dots x_p)$ is an affine coordinate-system of A_1, the points of A_1 can be represented as

$$\overrightarrow{OP} = \overrightarrow{OO_1} + \sum_{\nu=1}^{p} \xi^\nu x_\nu . \tag{10.3}$$

For $p=1$ we obtain a straight line through O_1 with the "direction vector" x,

$$\overrightarrow{OP} = \overrightarrow{OO_1} + \xi x .$$

In the case $p=2$ equation (10.3) reads

$$\overrightarrow{OP} = \overrightarrow{OO_1} + \xi^1 x_1 + \xi^2 x_2 .$$

It then represents the plane through O_1 generated by the vectors x_1 and x_2. An affine subspace of dimension $n-1$ is called a *hyperplane*.

Two affine subspaces A_1 and A_2 of A are called *parallel* if the difference-space E_1 of A_1 is contained in the difference-space E_2 of A_2, or conversely. Parallel subspaces are either disjoint or contained in each other. Assume, for instance, that E_2 is contained in E_1. Let Q be a point of the intersection $A_1 \cap A_2$ and P_2 be an arbitrary point of A_2. Then $\overrightarrow{QP_2}$ is contained in E_2 and hence is contained in E_1. This implies that $P_2 \in A_1$, whence $A_2 \subset A_1$.

10.4. Affine mappings. Let $P \to P'$ be a mapping of A into itself subject to the following conditions:

1. $\overrightarrow{P_1Q_1} = \overrightarrow{P_2Q_2}$ implies that $\overrightarrow{P_1'Q_1'} = \overrightarrow{P_2'Q_2'}$.

2. The mapping $\varphi: E \to E$ defined by $\varphi(\overrightarrow{PQ}) = \overrightarrow{P'Q'}$ is linear.

Then $P \to P'$ is called an *affine mapping*. Given two points O and O' and a linear mapping $\varphi: E \to E$, there exists exactly one affine mapping which sends O into O' and induces φ. This mapping is defined by

$$\overrightarrow{OP'} = \overrightarrow{OO'} + \varphi(\overrightarrow{OP}).$$

If a fixed origin is used in A, every affine mapping $x \to x'$ can be written in the form

$$x' = \varphi x + b ,$$

where φ is the induced linear mapping and $b = \overrightarrow{OO'}$.

A *translation* is an affine mapping which induces the identity in E,

$$\overrightarrow{P'Q'} = \overrightarrow{PQ}.$$

For two arbitrary points P and P' there obviously exists exactly one translation which sends P into P'.

10.5. Euclidean space. Let A be an n-dimensional affine space and assume that a positive definite inner product is defined in the difference-space E. Then A is called a *Euclidean space*. The distance of two points P and Q is defined by

$$\varrho(P,Q) = |\overrightarrow{PQ}|.$$

It follows from this definition that the distance has the following properties:

1. $\varrho(P,Q) \geqq 0$ and $\varrho(P,Q)=0$ if and only if $P=Q$.
2. $\varrho(P,Q)=\varrho(Q,P)$.
3. $\varrho(PQ) \leqq \varrho(P,R)+\varrho(R,Q)$.

All the metric concepts defined for an inner product space (cf. Chap. VII) can be applied to a Euclidean space. Given a point x_1 of A and a vector $p \neq 0$ there exists exactly one hyperplane through x_1 whose difference-space is orthogonal to p. This hyperplane is represented by the equation

$$(x - x_1, p) = 0.$$

A *rigid motion* of a Euclidean space is an affine mapping $P \to P'$ which preserves the distance,

$$\varrho(P',Q') = \varrho(P,Q). \tag{10.4}$$

Condition (10.4) implies that the linear mapping, which is induced in the difference-space by a rigid motion, is a rotation. Conversely, given a rotation φ and two points $O \in A$ and $O' \in A$, there exists exactly one rigid motion which induces φ and maps O into O'.

Problems

1. $(p+1)$ points $P_\nu\,(\nu=0\ldots p)$ in an affine space are said to be *in general position*, if the points P_ν are not contained in a $(p-1)$-dimensional subspace. Prove that the points $P_\nu\,(\nu=0\ldots p)$ are in general position if and only if the vectors $\overrightarrow{P_0P_\nu}$ are linearly independent.

2. Given $(p+1)$ points $P_\nu(\nu=0\ldots p)$ in general position, the set of all points P satisfying

$$\overrightarrow{P_0P}=\sum_{\nu=1}^{p}\xi^\nu\overrightarrow{P_0P_\nu}\quad \xi^\nu\geqq 0,\quad \sum_{\nu=1}^{p}\xi^\nu\leqq 1$$

is called the *p-simplex* spanned by the points $P_\nu(\nu=0\ldots p)$. If O is the origin of A, prove that a point P of the above simplex can be uniquely represented as

$$\overrightarrow{OP}=\sum_{\nu=0}^{p}\xi^\nu\overrightarrow{OP_\nu},\quad \xi^\nu\geqq 0,\quad \sum_{\nu=0}^{p}\xi^\nu=1.$$

The numbers $\xi^\nu\ (\nu=0\ldots p)$ are called the *barycentric coordinates* of P. The point B with the barycentric coordinates $\xi^\nu=\dfrac{1}{p+1}$ $(\nu=0\ldots p)$ is called the *center* of S.

3. Given a p-simplex $(P_0\ldots P_p)$ $(p\geqq 2)$ consider the $(p-1)$ simplex S_i defined by the points $(P_0\ldots\hat{P}_i\ldots P_p)$ $(0\leqq i\leqq p)$ and denote by B_i the center of $S_i(0\leqq i\leqq p)$. Show that the straight lines (P_i,B_i) and (P_j,B_j) $(i\neq j)$ intersect each other at the center B of S and that

$$\overrightarrow{B_iB}=\frac{1}{p+1}\overrightarrow{B_iP_i}.$$

4. An *equilateral simplex* of length a in a Euclidean space is a simplex $(P_0\ldots P_p)$ with the property that $|\overrightarrow{P_\nu P_\mu}|=a(\nu\neq\mu)$. Find the angle between the vectors $\overrightarrow{P_\nu P_\mu}$ and $\overrightarrow{P_\nu P_\lambda}(\mu\neq\nu,\ \lambda\neq\nu)$ and between the vectors $\overrightarrow{BP_\nu}$ and $\overrightarrow{BP_\mu}$ where B is the center of $(P_0\ldots P_p)$.

5. Assume that an orientation is defined in the difference-space E by the determinant function Δ. An ordered system of $(n+1)$ points $(P_0\ldots P_n)$ in general position is called *positive* with respect to the given orientation, if

$$\Delta(\overrightarrow{P_0P_1}\ldots\overrightarrow{P_0P_n})>0.$$

a) If the system $(P_0\ldots P_n)$ is positive and σ is a permutation of the numbers $(0,1\ldots n)$, show that the system $(P_{\sigma(0)}\ldots P_{\sigma(n)})$ is again positive if and only if the permutation σ is even.

b) Let A_i be the $(n-1)$-dimensional subspace spanned by the points $P_0\ldots\hat{P}_i\ldots P_n$. Introduce an orientation in the difference-space of A_i with the help of the determinant function

$$\Delta_i(x_1\ldots x_{n-1})=\Delta(\overrightarrow{P_iQ},x_1\ldots x_{n-1}),$$

where Q is an arbitrary point of A_i. Prove that the ordered n-tuple $(P_0 \dots \hat{P}_i \dots P_n)$ is positive with respect to the determinant function $(-1)^i \Delta_i$.

6. Let $(P_0 \dots P_n)$ be an n-simplex and S be its center. Denote by $S_{i_1 \dots i_k}$ the center of the $(n-k)$-simplex obtained by deleting the vertices $P_{i_1}, \dots, P_{i_k}$. Now select an ordered system of n integers $i_1, \dots, i_n$ $(0 \leqq i_\nu \leqq n)$ and define the affine mapping α by

$$\alpha: S \to P_0, S_{i_1} \to P_1, S_{i_2 i_2} \to P_2, \dots, S_{i_1 \dots in} \to P_n .$$

Prove that $\det \varphi = \dfrac{1}{(n+1)!} \varepsilon_\sigma$ where φ is the corresponding linear map. In this equation σ denotes the permutation $\sigma(\nu) = i_\nu$ $(\nu = 1 \dots n)$, $\sigma(0) = k$ where k is the integer not appearing among the numbers $(i_1 \dots i_n)$.

7. Let g_1 and g_2 be two straight lines in a Euclidean space which are not parallel and do not intersect. Prove that there exists exactly one point P_i on g_i $(i = 1, 2)$ such that $\overrightarrow{P_1 P_2}$ is orthogonal to g_1 and to g_2.

8. Let A_1 and A_2 be two subspaces of the affine space such that the difference-spaces E_1 and E_2 form a direct decomposition of E. Prove that the intersection $A_1 \cap A_2$ consists of exactly one point.

9. Prove that a rigid motion $x' = \tau x + a$ has a fixed point if and only if the vector a is orthogonal to all the vectors invariant under τ.

A rigid motion is called *proper* if $\det \tau = +1$. Prove that every proper rigid motion of the Euclidean plane without fixed points is a translation.

10. Consider a proper rigid motion $x' = \tau x + a$ $(\tau \neq \iota)$ of the Euclidean plane. Prove that there is exactly one fixed point x_0 and that

$$x_0 = \tfrac{1}{2} |a| \left(a + b \cot \frac{\theta}{2} \right).$$

In this equation, b is a vector of the same length as a and orthogonal to a. θ designates the rotation-angle relative to the orientation defined by the basis (a, b).

11. Prove that two proper rigid motions $\alpha \neq \iota$ and $\beta \neq \iota$ of the plane commute, if and only if one of the two conditions holds:

1. α and β are both translations
2. α and β have the same fixed point.

§ 2. Quadrics in the affine space

10.6. Definition. From elementary analytic geometry it is well known that every conic section can be represented by an equation of the form

$$\sum_{\nu,\mu=1}^{2} \alpha_{\nu\mu}\xi^{\nu}\xi^{\mu} + 2\sum_{\nu=1}^{2} \beta_{\nu}\xi^{\nu} = \alpha,$$

where $\alpha_{\nu\mu}$, β_ν and α are constants. Generalizing this to higher dimensions we define a *quadric* Q in an n-dimensional affine space A as the set of all points satisfying an equation of the form

$$\Phi(x) + 2f(x) = \alpha, \tag{10.5}$$

where $\Phi \neq 0$ is a quadratic function, f a linear function and α a constant.

For the following discussion it will be convenient to introduce a dual space E^* of the difference-space E. Then the bilinear function Φ can be written in the form

$$\Phi(x, y) = \langle \varphi x, y\rangle \qquad x, y \in E,$$

where φ is a linear mapping of E into E^* which is dual to itself: $\varphi^* = \varphi$. Moreover, the linear function f determines a vector $a^* \in E^*$ such that

$$f(x) = \langle a^*, x\rangle \qquad x \in E.$$

Hence, equation (10.5) can be written in the form

$$\langle \varphi x, x\rangle + 2\langle a^*, x\rangle = \alpha. \tag{10.6}$$

We recall that the null-space of the bilinear function Φ coincides with the kernel of the linear mapping φ.

10.7. Cones. Let us assume that there exists a point $x_0 \in Q$ such that $\varphi x_0 + a^* = 0$. Then (10.6) can be written as

$$\langle \varphi x, x\rangle - 2\langle \varphi x_0, x\rangle = \alpha \tag{10.7}$$

and the substitution $x = x_0$ gives

$$\alpha = -\langle \varphi x_0, x_0\rangle.$$

Inserting this into (10.7) we obtain

$$\langle \varphi x, x\rangle - 2\langle \varphi x_0, x\rangle + \langle \varphi x_0, x_0\rangle = 0.$$

Hence, the equation of Q assumes the form

$$\Phi(x - x_0) = 0.$$

A quadric of this kind is called a *cone with the vertex* x_0. For the sake of simplicity, cones will be excluded in the following discussion. In other words, it will be assumed that

$$\varphi x + a^* \neq 0 \quad \text{for all points } x \in Q. \tag{10.8}$$

10.8. Tangent-space. Consider a fixed point $x_0 \in Q$. It follows from condition (10.8) that the orthogonal complement of the vector $\varphi x_0 + a^*$ is an $(n-1)$-dimensional subspace T_{x_0} of E. This subspace is called the *tangent-space* of Q at the point x_0. A vector $y \in E$ is contained in T_{x_0} if and only if

$$\langle a^* + \varphi x_0, y \rangle = 0. \tag{10.9}$$

In terms of the functions Φ and f equation (10.9) can be written as

$$\Phi(x_0, y) + f(y) = 0. \tag{10.10}$$

The $(n-1)$-dimensional affine subspace which is determined by the point x_0 and the tangent-space T_{x_0} is called the *tangent-hyperplane* of Q at x_0. It consists of all points

$$x = x_0 + y \qquad y \in T_{x_0}.$$

Inserting $y = x - x_0$ into equation (10.10) we obtain

$$\Phi(x_0, x - x_0) + f(x - x_0) = 0. \tag{10.11}$$

Observing that

$$\Phi(x_0) + 2f(x_0) = \alpha$$

we can write equation (10.11) of the tangent-hyperplane in the form

$$\Phi(x_0, x) + f(x_0 + x) = \alpha. \tag{10.12}$$

To obtain a geometric picture of the tangent-space, consider a 2-dimensional plane

$$F: x = x_0 + \xi a + \eta b \tag{10.13}$$

through x_0 where a and b are two linearly independent vectors. Inserting (10.13) into equation (10.5) we obtain the relation

$$\begin{aligned} &\xi^2 \Phi(a) + 2\xi\eta\,\Phi(a,b) + \eta^2 \Phi(b) + \\ &\quad + 2\xi(\Phi(x_0, a) + f(a)) + 2\eta(\Phi(x_0, b) + f(b)) = 0 \end{aligned} \tag{10.14}$$

showing that the plane F intersects Q in a conic γ. Define a linear function g by setting

$$g(x) = 2(\Phi(x_0, x) + f(x)). \tag{10.15}$$

Then, in view of (10.8), $g \neq 0$. Now the equation of the conic γ can be written in the form

$$\xi^2 \Phi(a) + 2\xi\eta\,\Phi(a,b) + \eta^2 \Phi(b) + \xi g(a) + \eta g(b) = 0. \qquad (10.16)$$

Now assume that the vectors a and b are chosen such that $g(a)$ and $g(b)$ are not both equal to zero. Then the conic has a unique tangent at the point $\xi = \eta = 0$ and this tangent is generated by the vector

$$t = -g(b)a + g(a)b. \qquad (10.17)$$

The vector t is contained in the tangent-space T_{x_0}; this follows from the equation

$$g(t) = -g(b)g(a) + g(a)g(b) = 0.$$

Every vector $y \neq 0$ of the tangent-space T_{x_0} can be obtained in this way. In fact, let a be a vector such that $g(a) = 1$ and consider the plane through x_0 spanned by a and y. Then equation (10.17) yields

$$t = -g(y)a + g(a)y = y \qquad (10.18)$$

showing that y is the tangent-vector of the intersection $Q \cap F$ at the point $\xi = \eta = 0$.

Note: If $g(a) = 0$ and $g(b) = 0$ equation (10.16) reduces to

$$\xi^2 \Phi(a) + 2\xi\eta\,\Phi(a,b) + \eta^2 \Phi(b) = 0.$$

Then the intersection of Q and F consists of

a) two straight lines intersecting at x_0, if

$$\Phi(a,b)^2 - \Phi(a)\Phi(b) > 0,$$

b) the point x_0 only, if

$$\Phi(a,b)^2 - \Phi(a)\Phi(b) < 0,$$

c) one straight line through x_0, if

$$\Phi(a,b)^2 - \Phi(a)\Phi(b) = 0,$$

but not all three coefficients $\Phi(a)$, $\Phi(b)$ and $\Phi(a,b)$ are zero

d) The entire plane F, if

$$\Phi(a) = \Phi(b) = \Phi(a,b) = 0.$$

10.9. Uniqueness of the representation. Assume that a quadric Q is represented in two ways

$$\Phi_1(x) + 2f_1(x) = \alpha_1 \qquad (10.19)$$

and

$$\Phi_2(x) + 2f_2(x) = \alpha_2 . \tag{10.20}$$

It will be shown that

$$\Phi_2 = \lambda \Phi_1, f_2 = \lambda f_1, \alpha_2 = \lambda \alpha_1$$

where $\lambda \neq 0$ is a real number. Let x_0 be a fixed point of Q. It follows from hypothesis (10.8) that the linear functions g_1 and g_2, defined by

$$g_1(x) = \Phi_1(x, x_0) + f_1(x) \quad \text{and} \quad g_2(x) = \Phi_2(x, x_0) + f_2(x) \tag{10.21}$$

are not identically zero.

Choose a vector a such that $g_1(a) \neq 0$ and $g_2(a) \neq 0$, and a vector $b \neq 0$ such that $g_1(b) = 0$. Obviously a and b are linearly independent. The plane

$$x = x_0 + \xi a + \eta b$$

then intersects the quadric Q in a conic γ whose equation is given by each one of the equations

$$\xi^2 \Phi_1(a) + 2\xi\eta \Phi_1(a, b) + \eta^2 \Phi_1(b) + \xi g_1(a) = 0 \tag{10.22}$$

and

$$\xi^2 \Phi_2(a) + 2\xi\eta \Phi_2(a, b) + \eta^2 \Phi_2(b) + \xi g_2(a) + \eta g_2(b) = 0 . \tag{10.23}$$

The tangent of this curve at the point $\xi = \eta = 0$ is generated by the vector

$$t_1 = g_1(a) b$$

and also by

$$t_2 = -g_2(b) a + g_2(a) b .$$

This implies that

$$g_2(b) = 0 . \tag{10.24}$$

But $b \neq 0$ was an arbitrary vector of the kernel of g_1. Hence equation (10.24) shows that $g_2(b) = 0$ whenever $g_1(b) = 0$. In other words, the linear functions g_1 and g_2 have the same kernel. Consequently g_2 is a constant multiple of g_1,

$$g_2 = \lambda g_1 \quad \lambda \neq 0 . \tag{10.25}$$

Multiplying equation (10.22) by λ and subtracting it from (10.23) we obtain in view of (10.24) that

$$\xi^2 (\Phi_2 - \lambda \Phi_1)(a) + 2\xi\eta (\Phi_2 - \lambda \Phi_1)(a, b) + \eta^2 (\Phi_2 - \lambda \Phi_1)(b) = 0 . \tag{10.26}$$

In this equation all three coefficients must be zero. In fact, if at least one coefficient is different from zero, equation (10.26) implies that the conic γ consists of two straight lines, one straight line or the point x_0 only. But this is impossible because $g_1(a) \neq 0$. We thus obtain from (10.26) that

$$\Phi_2(a) = \lambda\Phi_1(a), \Phi_2(a,b) = \lambda\Phi_1(a,b), \Phi_2(b) = \lambda\Phi_1(b). \tag{10.27}$$

These equations show that

$$\Phi_2(x) = \lambda\Phi_1(x) \tag{10.28}$$

for all vectors $x \in E$: If $g_1(x)=0$, (10.28) follows from the third equation (10.27); if $g_1(x) \neq 0$, then $g_2(x) \neq 0$ [in view of (10.25)] and (10.28) follows from the first equation (10.27).

Altogether we thus obtain the identities

$$\Phi_2 = \lambda\Phi_1 \quad \text{and} \quad g_2 = \lambda g_1 \qquad \lambda \neq 0.$$

Now relations (10.21) imply that

$$f_2 = \lambda f_1$$

and equation (10.20) can be written as

$$\lambda(\Phi_1(x) + 2f_1(x)) = \alpha_2. \tag{10.29}$$

Comparing equations (10.19) and (10.29) we finally obtain $\alpha_2 = \lambda a_1$. This completes the proof of the uniqueness theorem.

10.10. Centers. Let

$$Q: \Phi(x) + 2f(x) = \alpha$$

be a given quadric and c be an arbitrary point of the space A. If we introduce c as a new origin,

$$x = c + x',$$

the equation of Q is transformed into

$$\Phi(x') + 2(\Phi(c,x') + f(x')) = \alpha - \Phi(c) - 2f(c). \tag{10.30}$$

Here the question arises whether the point c can be chosen such that the linear terms in (10.30) disappear, i. e. that

$$\Phi(c,x') + f(x') = 0 \tag{10.31}$$

for all vectors $x' \in E$. If this is possible, c is called a *center* of Q. Writing equation (10.31) in the form

$$\langle \varphi c + a^*, x' \rangle = 0 \qquad x' \in E$$

we see that c is a center of Q if and only if (cf. sec. 10.7)

$$\varphi c = -a^* . \tag{10.32}$$

This implies that the quadric Q has a center if and only if the vector a^* is contained in the image space Im φ. Observing that Im φ is the orthogonal complement of the kernel of φ we obtain the following criterion: *A quadric Q has a center if and only if the vector a^* is orthogonal to the kernel of φ.*

If this condition is satisfied, the center is determined up to a vector of ker φ. In other words, the set of all centers is an affine subspace of A with ker φ as difference-space.

Now assume that the bilinear function Φ is non-degenerate. Then φ is a regular mapping and hence equation (10.32) has exactly one solution. Thus it follows from the above criterion that *a non-degenerate quadric has exactly one center.*

10.11. Normal-form of a quadric with center. Suppose that Q is a quadric with centers. If a center is used as origin the equation of Q assumes the form

$$\Phi(x) = \beta \qquad \beta \neq 0 . \tag{10.33}$$

Then the tangent-vectors y at a point $x_0 \in Q$ are characterized by the equation $\langle \varphi x_0, y \rangle = 0$. Observing that $\langle \varphi x_0, y \rangle = \langle x_0, \varphi y \rangle$ we see that every tangent-space T_{x_0} contains the null-space of Φ.

The equation of the tangent-hyperplane of Q at x_0 is given by

$$\Phi(x_0, y) = \beta . \tag{10.34}$$

It follows from (10.34) that a center of Q is never contained in a tangent-hyperplane.

Dividing (10.33) by β and replacing the quadratic function Φ by $\frac{1}{\beta}\Phi$ we can write the equation of Q in the *normal-form*

$$\Phi(x) = 1 . \tag{10.35}$$

Now select a basis $x_\nu (\nu = 1 \ldots n)$ of E such that

$$\Phi(x_\nu, x_\beta) = \varepsilon_\nu \delta_{\nu\beta} \qquad \varepsilon_\nu = \begin{cases} +1 (\nu = 1 \ldots s) \\ -1 (\nu = s+1 \ldots r) \\ 0 (\nu = r+1 \ldots n) \end{cases} \tag{10.36}$$

where r denotes the rank and s denotes the index of Φ. Then the normal-form (10.36) can be written as

$$\sum_{\nu=1}^{r} \varepsilon_\nu \xi^\nu \xi^\nu = 1. \tag{10.37}$$

10.12. Normal-form of a quadric without center. Now consider a quadric Q without a center. If a point of Q is chosen as origin the constant α in (10.5) becomes zero and the equation of Q reads

$$\Phi(x) + 2\langle a^*, x\rangle = 0. \tag{10.38}$$

By multiplying equation (10.38) with -1 if necessary we can achieve that $2s \geqq r$. In other words, we can assume that the signature of Φ is not negative

To reduce equation (10.38) to a normal form consider the tangent-space T_0 at the origin. Equation (10.9) shows that T_0 is the orthogonal complement of a^*. Hence, a^* is contained in the orthogonal complement $T_0^\perp$. On the other hand, a^* is not contained in the orthogonal complement $K^\perp$ ($K = \ker \varphi$) because otherwise Q would have a center (cf. sec. 10.10). The relations $a^* \in T_0^\perp$ and $a^* \notin K^\perp$ show that $T_0^\perp \not\subset K^\perp$. Taking the orthogonal complement we obtain the relation $T_0 \not\supset K$ showing that there exists a vector $a \in K$ which is not contained in T_0 (cf. fig. 5). Then $\langle a^*, a\rangle \neq 0$ and hence we may assume that $\langle a^*, a\rangle = 1$.

Now T_0 has dimension $n-1$ and hence every vector $x \in E$ can be written in the form

$$x = y + \xi a \qquad y \in T_0. \tag{10.39}$$

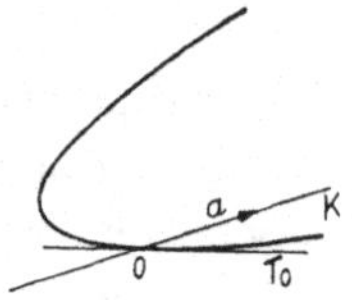

Fig. 5

Inserting (10.39) into equation (10.38) we obtain

$$\Phi(y) + 2\xi\,\Phi(y,a) + \xi^2\,\Phi(a) + 2\langle a^*, y + \xi a\rangle = 0. \tag{10.40}$$

Now

$$\Phi(y,a) = 0 \quad \text{and} \quad \Phi(a) = 0,$$

because $a \in K$, and

$$\langle a^*, y\rangle = 0,$$

because $y \in T_0$. Hence, equation (10.40) reduces to the following *normal-form*:

$$\Phi(y) + 2\xi = 0. \tag{10.41}$$

Since a is contained in the null space of Φ it follows from the decomposition (10.39) that the restriction of Φ to the tangent-space T_0 has again rank r and index s. Therefore we can select a basis $x_\nu(\nu=1\dots n-1)$ of T_0 such that

$$\Phi(x_\nu, x_\mu) = \varepsilon_\nu \delta_{\nu\mu} \qquad (\nu, \mu = 1 \dots n-1).$$

Then the vectors $x_\nu(\nu=1\dots n-1)$ and a form a basis of E in which the normal form (10.41) can be written as

$$\sum_{\nu=1}^{r} \varepsilon_\nu \xi^\nu \xi^\nu + 2\xi = 0. \tag{10.42}$$

Problems

1. Let E be a 3-dimensional pseudo-Euclidean space with index 2. Given an orientation in E define the cross product $x \times y$ by

$$(x \times y, z) = \Delta(x, y, z) \qquad x, y, z \in E$$

where Δ is a normed determinant function (cf. sec. 9.19) which represents the orientation. Consider a point $x_0 \neq 0$ of the light-cone $(x,x)=0$ and a plane

$$F: x = x_0 + \xi a + \eta b$$

which does not contain the point O. Prove that the intersection of the plane F and the light-cone is

an *ellipse* if $a \times b$ is time-like

a *hyperbola* if $a \times b$ is space-like

a *parabola* if $a \times b$ is a light-vector.

2. Consider the quadric $Q: \Phi(x)=1$ where Φ is a non-degenerate quadratic function. Then every point $x_1 \neq 0$ defines an $(n-1)$-dimensional subspace $P(x_1)$ by the equation

$$\Phi(x, x_1) = 1.$$

This subspace is called the *polar* of x_1. It follows from the above equation that the polar $P(x_1)$ does not contain the center O.

a) Prove that $x_2 \in P(x_1)$ if and only if $x_1 \in P(x_2)$.

b) Given an $(n-1)$-dimensional affine subspace A_1 of A which does not contain O, show that there exists exactly one point x_1 such that $A_1 \subset P(x_1)$.

c) Show that $P(x_1)$ is a tangent-plane of Q if and only if $x_1 \in Q$.

3. Let x_1 be a point of the quadric $\Phi(x)=1$. Prove that the restriction of the bilinear function Φ to the tangent-space T_{x_1} has the rank $r-1$ and index $s-1$.

4. Show that a center of a quadric Q lies on Q only if Q is a cone.

5. Let x_0 be a point of the quadric

$$\Phi(x) + 2f(x) + \alpha = 0$$

and consider the skew bilinear mapping $\omega: E \times E \to E$ defined by

$$\omega(x, y) = g(x)y - g(y)x \qquad x, y \in E$$

where the linear function g is defined by (10.15). Show that the linear closure of the set $\omega(x,y)$ under this mapping is the tangent-space T_{x_0}.

§ 3. Affine equivalence of quadrics

10.13. Definition. Let an affine mapping $x' = \tau x + b$ of A onto itself be given. Then the image of a quadric

$$Q: \Phi(x) + 2f(x) = \alpha$$

is the quadric Q' defined by the equation

$$Q': \Psi(x) + 2g(x) = \beta,$$

where

$$\Psi(x) = \Phi(\tau^{-1}x), \tag{10.43}$$

$$g(x) = -\Phi(\tau^{-1}x, \tau^{-1}b) + f(\tau^{-1}x) \tag{10.44}$$

and

$$\beta = -\Phi(\tau^{-1}b) + 2f(\tau^{-1}b) + \alpha. \tag{10.45}$$

In fact, relations (10.43), (10.44) and (10.45) yield

$$\Psi(\tau x + b) + 2g(\tau x + b) - \beta = \Phi(x) + 2f(x) - \alpha$$

showing that a point $x \in A$ is contained in Q if and only if the point $x' = \tau x + b$ is contained in Q'.

Two quadrics Q_1 and Q_2 are called *affine equivalent* if there exists a one-to-one affine mapping of A onto itself which carries Q_1 into Q_2. The affine equivalence induces a decomposition of all possible quadrics into affine equivalence classes. It is the purpose of this paragraph to construct a complete system of representatives of these equivalence classes.

10.14. The affine classification of quadrics is based upon the following theorem: Let E and F be two n-dimensional linear spaces and Φ and Ψ two symmetric bilinear functions in E and in F. Then there exists a linear isomorphism $\tau\colon E \to F$ with the property that

$$\Phi(x, y) = \Psi(\tau x, \tau y) \qquad x, y \in E \tag{10.46}$$

if and only if Φ and Ψ have the same rank and the same index.

To prove this assume first that the relation (10.46) holds. Select a basis $a_\nu\,(\nu=1\ldots n)$ of E such that

$$\Phi(a_\nu, a_\mu) = \varepsilon_\nu \delta_{\nu\mu} \qquad \varepsilon_\nu = \begin{cases} +1\,(\nu = 1 \cdots s) \\ -1\,(\nu = s+1 \ldots r) \\ 0\,(\nu = r+1 \ldots n). \end{cases} \tag{10.47}$$

Then equations (10.46) and (10.47) yield

$$\Psi(\tau a_\nu, \tau a_\mu) = \Phi(a_\nu, a_\mu) = \varepsilon_\nu \delta_{\nu\mu}.$$

showing that Ψ has rank r and index s.

Conversely, assume that this condition is satisfied. Then there exist bases a_ν and $b_\nu\,(\nu=1\ldots n)$ of E and of F such that

$$\Phi(a_\nu, a_\mu) = \varepsilon_\nu \delta_{\nu\mu} \quad \text{and} \quad \Psi(b_\nu, b_\mu) = \varepsilon_\nu \delta_{\nu\mu}.$$

Define the isomorphism $\tau\colon E \to F$ by the equations

$$\tau a_\nu = b_\nu \qquad (\nu = 1 \ldots n).$$

Then

$$\Phi(a_\nu, a_\mu) = \Psi(\tau a_\nu, \tau a_\mu) \qquad (\nu, \mu = 1 \ldots n)$$

and consequently

$$\Phi(x, y) = \Psi(\tau x, \tau y) \qquad x, y \in E.$$

10.15. Affine classification. First of all it will be shown that the centers are invariant under an affine mapping. In fact, let

$$Q\colon \Phi(x - c) = \beta$$

be a quadric with c as center and $x' = \tau x + b$ an affine mapping of A onto itself. Then the image Q' of Q is given by the equation

$$Q'\colon \Psi(x - c') = \beta,$$

where

$$\Psi(x) = \Phi(\tau^{-1} x)$$

and

$$c' = b + \tau c.$$

This equation shows that c' is a center of Q'.

Now consider two quadrics with center

$$Q_1 : \Phi_1(x - c_1) = 1 \tag{10.48}$$

and

$$Q_2 : \Phi_2(x - c_2) = 1 \tag{10.49}$$

and assume that $x \to x'$ is an affine mapping carrying Q_1 into Q_2. Since centers are transformed into centers we may assume the mapping $x \to x'$ sends c_1 into c_2 and hence it has the form

$$x' = \tau(x - c_1) + c_2 .$$

By hypothesis, Q_1 is mapped onto Q_2 and hence the equation

$$\Phi_2(\tau(x - c_1)) = 1 \tag{10.50}$$

must represent the quadric Q_1. Comparing (10.48) and (10.50) and applying the uniqueness theorem of sec. 10.9 we find that

$$\Phi_1(x) = \Phi_2(\tau x).$$

This relation implies that

$$r_1 = r_2 \quad \text{and} \quad s_1 = s_2 . \tag{10.51}$$

Conversely, the relations (10.51) imply that there exists a linear automorphism τ of E such that

$$\Phi_1(x) = \Phi_2(\tau x).$$

Then the affine mapping $x \to x'$ defined by

$$x' = \tau(x - c_1) + c_2$$

transforms Q_1 into Q_2. We thus obtain the following criterion: *The two normal forms* (10.48) *and* (10.49) *represent affine equivalent quadrics if and only if the bilinear functions Φ_1 and Φ_2 have the same rank and the same index.*

10.16. Next, let

$$Q_1 : \Phi_1(x - q_1) + 2\langle a_1^*, x - q_1\rangle = 0 \qquad q_1 \in Q_1 \tag{10.52}$$

and

$$Q_2 : \Phi_2(x - q_2) + 2\langle a_2^*, x - q_2\rangle = 0 \qquad q_2 \in Q_2 \tag{10.53}$$

be two quadrics without a center. It is assumed that the equations (10.52) and (10.53) are written in such a way that $2s_1 \geqq r_1$ and $2s_2 \geqq r_2$. If $x' = \tau(x - q_1) + q_2$ is an affine mapping transforming Q_1 into Q_2, the equation of Q_1 can be written in the form

$$\Phi_2(\tau(x-q_1)) + 2\langle a_2^*, \tau(x-q_1)\rangle = 0.$$

Now the uniqueness theorem yields

$$\Phi_1(x) = \lambda\,\Phi_2(\tau x)$$

where $\lambda \neq 0$ is a constant. This relation implies that the bilinear functions Φ_1 and Φ_2 have the same rank r and that $s_2 = s_1$ or $s_2 = r - s_1$ depending on whether $\lambda > 0$ or $\lambda < 0$. But the equation $s_2 = r - s_1$ is only compatible with the inequalities $2s_1 \geqq r_1$ and $2s_2 \geqq r_2$ if $s_1 = s_2 = \frac{r}{2}$ and hence we see that $s_1 = s_2$ in either case.

Conversely, assume that $r_1 = r_2 = r$ and $s_1 = s_2 = s$. To find an affine mapping which transforms Q_1 into Q_2 consider the tangent-spaces $T_{q_1}(Q_1)$ and $T_{q_2}(Q_2)$. As has been mentioned in sec. 10.12 the restriction of Φ_i to the subspace $T_{q_i}(Q)_i$ $(i=1, 2)$ has the same rank and the same index as Φ_i. Consequently, there exists an isomorphism $\varrho: T_{q_1}(Q_1) \to T_{q_2}(Q_2)$ such that

$$\Phi_1(y) = \Phi_2(\varrho y) \qquad y \in T_{q_1}(Q_1).$$

Now select a vector a_i in the nullspace of Φ_i $(i = 1, 2)$ such that

$$\langle a_i^*, a_i\rangle = 1 \quad (i = 1, 2)$$

and define the linear automorphism τ of E by the equations

$$\tau y = \varrho y \qquad y \in T_{q_1}(Q_1)$$

and

$$\tau a_1 = a_2. \tag{10.54}$$

Then

$$\Phi_2(\tau x) + \langle a_2^*, \tau x\rangle = \Phi_1(x) + \langle a_1^*, x\rangle \qquad x \in E. \tag{10.55}$$

In fact, every vector $x \in E$ can be decomposed in the form

$$x = y + \xi a_1 \qquad y \in T_{q_1}(Q_1). \tag{10.56}$$

Equations (10.54), (10.55) and (10.56) imply that

$$\Phi_2(\tau x) = \Phi_2(\tau y + \xi a_2) = \Phi_2(\tau y) = \Phi_1(y) = \Phi_1(y + \xi a_1) = \Phi_1(x) \tag{10.57}$$

and

$$\langle a_2^*, \tau x\rangle = \langle a_2^*, \varrho y + \xi a_2\rangle = \xi\langle a_2^*, a_2\rangle = \xi = \langle a_1^*, x\rangle. \tag{10.58}$$

Adding (10.57) and (10.58) we obtain (10.55). Relation (10.55) shows that the affine mapping $x'=\tau(x-q_1)+q_2$ sends Q_1 into Q_2 and we have the following result: *The normal-forms* (10.52) *and* (10.53) *represent affine equivalent quadrics if and only if the bilinear functions* Φ_1 *and* Φ_2 *have the same rank and the same index.*

10.17. The affine classes. It follows from the two criteria in sec. 10.15 and 10.16 that the normal forms

$$\xi^1 \xi^1 + \cdots + \xi^s \xi^s - \xi^{s+1} \xi^{s+1} - \cdots - \xi^r \xi^r = 1 \qquad (1 \leqq s \leqq r)$$

and

$$\xi^1 \xi^1 + \cdots + \xi^s \xi^s - \xi^{s+1} \xi^{s+1} - \cdots - \xi^r \xi^r + 2\xi = 0 \qquad (r \leqq 2s)$$

form a complete system of representatives of the affine classes. Denote by $N_1(r)$ and by $N_2(r)$ the total number of affine classes with center and without center respectively of a given rank r. Then the above equations show that

$$N_1(r) = r \quad \text{and} \quad N_2(r) = \begin{cases} \dfrac{r+1}{2} \text{ if } r \text{ is odd} & \\ & 1 \leqq r \leqq n-1 \\ \dfrac{r+2}{2} \text{ if } r \text{ is even} & \\ 0 & r = n. \end{cases}$$

The following list contains a system of representatives of the affine classes in the plane and in 3-space*):

Plane:

I. Quadrics with center:

1. $r=2$: a) $s=2$: $\xi^2+\eta^2=1$ ellipse,
   b) $s=1$: $\xi^2-\eta^2=1$ hyperbola.
2. $r=1$: $s=1$: $\xi=\pm 1$ two parallel lines.

II. Quadrics without center:

$r=1$, $s=1$: $\xi^2-2\eta=0$ parabola.

3-space:

I. Quadrics with center:

1. $r=3$: a) $s=3$: $\xi^2+\eta^2+\zeta^2=1$ ellipsoid,
   b) $s=2$: $\xi^2+\eta^2-\zeta^2=1$ hyperboloid with one shell,
   c) $s=1$: $\xi^2-\eta^2-\zeta^2=1$ hyperboloid with two shells.

*) In the following equations the coordinates are denoted by ξ, η, ζ and the superscripts indicate exponents.

2. $r=2$: a) $s=2$: $\xi^2+\eta^2=1$ elliptic cylinder,
b) $s=1$: $\xi^2-\eta^2=1$ hyperbolic cylinder.
3. $r=1$: $s=1$: $\xi=\pm 1$ two parallel planes.

II. Quadrics without center:

1. $r=2$: a) $s=2$: $\xi^2+\eta^2-2\zeta=0$ elliptic paraboloid,
b) $s=1$: $\xi^2-\eta^2-2\zeta=0$ hyperbolic paraboloid.
2. $r=1$, $s=1$: $\xi^2-2\zeta=0$ parabolic cylinder.

Problems

1. Let Q be a given quadric and C be a given point. Show that C is a center of Q if and only if the affine mapping $P \to P'$ defined by $\overrightarrow{CP'} = -\overrightarrow{CP}$ transforms Q into itself.

2. If Φ is an indefinite quadratic function, show that the quadrics

$$\Phi(x)=1 \quad \text{and} \quad \Phi(x)=-1$$

are equivalent if and only if the signature of Φ is zero.

3. Denote by N_1 and by N_2 the total number of affine classes with center and without center respectively. Prove that

$$N_1 = \frac{n(n+1)}{2}$$

$$N_2 = \begin{cases} k^2+k-1 & \text{if} \quad n=2k \\ k^2+2k & \text{if} \quad n=2k+1 . \end{cases}$$

4. Let x_1 and x_2 be two points of the quadric

$$Q: \Phi(x)=1 .$$

Assume that an isomorphism $\tau: T_{x_1} \to T_{x_2}$ is given such that

$$\Phi(\tau y, \tau z) = \Phi(y, z) \qquad y, z \in T_{x_1} .$$

Construct an affine mapping $A \to A$ which transforms Q into itself and which induces the isomorphism τ in the tangent-space T_{x_1}.

5. Prove the assertion of problem 4 for the quadric

$$\Phi(x) + 2\langle a^*, x\rangle = 0 .$$

§ 4. Quadrics in the Euclidean space

10.18. Normal-vector. Let A be an n-dimensional Euclidean space and

$$Q: \Phi(x) + 2f(x) = \alpha$$

be a quadric in A. The bilinear function Φ determines a selfadjoint linear transformation φ of E by the equation

$$\Phi(x, y) = (\varphi x, y).$$

The linear function f can be written as

$$f(x) = (a, x)$$

where a is a fixed vector of E. Cones will again be excluded; i. e. we shall assume that

$$\varphi x \neq -a$$

for all points $x \in Q$. Let x_0 be a fixed point of Q. Then equation (10.10) shows that the tangent-space T_{x_0} consists of all vectors y satisfying the relation

$$(\varphi x_0 + a, y) = 0.$$

In other words, the tangent-space T_{x_0} is the orthogonal complement of the *normal-vector*

$$p(x_0) = \varphi x_0 + a.$$

The straight line determined by the point x_0 and the vector $p(x_0)$ is called the *normal* of Q at x_0.

10.19. Quadrics with center. Now consider a quadric with center

$$Q: \Phi(x) = 1. \tag{10.59}$$

Then the normal-vector $p(x_0)$ is simply given by

$$p(x_0) = \varphi x_0.$$

This equation shows that the linear mapping φ associates with every point $x_0 \in Q$ the corresponding normal-vector. In particular, let x_0 be a point of Q whose position-vector is an eigenvector of φ. Then we have the relation

$$\varphi x_0 = \lambda x_0$$

showing that the normal-vector is a multiple of the position-vector x_0.

Inserting this into equation (10.59) we see that the corresponding eigenvalue is equal to

$$\lambda = \frac{1}{|x_0|^2}.$$

As has been shown in sec. 8.7 there exists an orthonormal system of n eigenvectors $e_\nu (\nu = 1 \ldots n)$. Then

$$\varphi e_\nu = \lambda_\nu e_\nu \qquad (\nu = 1 \ldots n) \tag{10.60}$$

whence

$$\Phi(e_\nu, e_\mu) = \lambda_\nu \delta_{\nu\mu}.$$

Let us enumerate the eigenvectors e_ν such that

$$\begin{aligned} 0 < \lambda_1 &\leqq \lambda_2 \leqq \cdots \leqq \lambda_s \\ 0 > \lambda_{s+1} &\geqq \lambda_{s+2} \geqq \cdots \geqq \lambda_r \\ \lambda_{r+1} &= \cdots = \lambda_n = 0 \end{aligned} \tag{10.61}$$

where r is the rank and s is the index of Φ. Then equation (10.59) can be written as

$$\sum_{\nu=1}^{r} \lambda_\nu \xi^\nu \xi^\nu = 1. \tag{10.62}$$

The vectors

$$a_\nu = \frac{e_\nu}{\sqrt{\lambda_\nu}} \quad (\nu = 1 \ldots s) \quad \text{and} \quad a_\nu = \frac{e_\nu}{\sqrt{-\lambda_\nu}} \quad (\nu = s + 1 \ldots r)$$

are called the *principal axes* and the *conjugate principal axes* of Q. Inserting

$$\lambda_\nu = \frac{1}{|a_\nu|^2} \quad (\nu = 1 \ldots s) \quad \text{and} \quad \lambda_\nu = -\frac{1}{|a_\nu|^2} \quad (\nu = s + 1 \ldots r)$$

into (10.62) we obtain the *metric normal-form* of Q:

$$\sum_{\nu=1}^{s} \frac{\xi^\nu \xi^\nu}{|a_\nu|^2} - \sum_{\nu=s+1}^{r} \frac{\xi^\nu \xi^\nu}{|a_\nu|^2} = 1. \tag{10.63}$$

Every principal axis a_ν generates a straight line which intersects the quadric Q in the points a_ν and $-a_\nu$. The straight lines generated by the conjugate axes have no points in common with Q but they intersect the conjugate quadric

$$Q': \Phi(x) = -1$$

at the points a_ν and $-a_\nu (\nu = s+1 \ldots r)$.

10.20. Quadrics without center. Now consider a quadric Q without center. Using an arbitrary point of Q as origin, we can write the equation of Q in the form

$$\Phi(x) + 2(a, x) = 0 \tag{10.64}$$

where a is a normal vector of Q at the point $x=0$.
For every point $x \in Q$ the vector

$$p(x) = \varphi x + a \tag{10.65}$$

is contained in the normal of Q. A point $x \in Q$ is called a *vertex* if the corresponding normal is contained in the null-space K of Φ.

It will be shown that every quadric without center has at least one vertex.

Applying φ to the equation (10.65) we obtain

$$\varphi\, p(x) = \varphi^2 x + \varphi\, a$$

showing that a point $x \in Q$ is a vertex if and only if

$$\varphi^2 x = -\varphi\, a\,. \tag{10.66}$$

To find all vertices of Q we thus have to determine all the solutions of equations (10.64) and (10.66). The self-adjointness of φ implies that the mappings φ and φ^2 have the same image-space and the same kernel (cf. sec. 8.7). Consequently, equation (10.66) has at least one solution x. The general solution of (10.66) can be written in the form

$$x = x_0 + z$$

where z is an arbitrary vector of the kernel K. Inserting this into equation (10.64) we obtain

$$\tfrac{1}{2}\Phi(x_0) + (a, x_0) + (a, z) = 0\,. \tag{10.67}$$

Now $a \notin K^{\perp}$ (otherwise Q would have a center) and consequently (10.67) has a solution $z \in K$. This solution is determined up to an arbitrary vector of the intersection $K \cap T_0$. In other words, the set of all vertices of Q forms an affine subspace with the difference-space $K \cap T_0$. This subspace has dimension $(n-r-1)$.

Now we are ready to construct the normal form of the quadric (10.64). First of all we select a vertex of Q as origin. Then the vector a in (10.64) is contained in the kernel K. Multiplying equation (10.64) by an appropriate scalar we can achieve that $|a|=1$ and that $2s \geqq r$. Now let $e_\nu\,(\nu=1$

$\ldots n-1$) be a basis of T_0 consisting of eigenvectors of Φ. Then the vectors $e_\nu(\nu=1\ldots n-1)$ and a form an orthonormal basis of E such that

$$\Phi(e_\nu, e_\mu) = \lambda_\nu \delta_{\nu\mu} \qquad (\nu, \mu = 1 \ldots n-1)$$

and

$$\Phi(e_\nu, a) = (e_\nu, \varphi a) = 0 \qquad (\nu = 1 \ldots n-1).$$

In this basis the equation of Q assumes the *metric normal-form*

$$\sum_{\nu=1}^{r} \lambda_\nu \xi^\nu \xi^\nu + 2\xi = 0. \tag{10.68}$$

Upon introduction of the principal axes and the principal conjugate axes

$$a_\nu = \frac{e_\nu}{\sqrt{\lambda_\nu}} \quad (\nu = 1 \ldots s) \quad \text{and} \quad a_\nu = \frac{e_\nu}{\sqrt{-\lambda_\nu}} \quad (\nu = s+1 \ldots r)$$

the normal-form (10.68) can be also written as

$$\sum_{\nu=1}^{s} \frac{\xi^\nu \xi^\nu}{|a_\nu|^2} - \sum_{\nu=s+1}^{r} \frac{\xi^\nu \xi^\nu}{|a_\nu|^2} + 2\xi = 0. \tag{10.69}$$

10.21. Metric classification of bilinear forms. Two quadrics Q and Q' in the Euclidean space A are called *metrically equivalent*, if there exists a rigid motion $x \to x'$ which transforms Q into Q'. Two metrically equivalent quadrics are a fortiori affine equivalent. Hence, the metric classification of quadrics consists in the construction of the metric subclasses within every affine equivalence class.

It will be shown that the lengths of the principal axes form a complete system of metric invariants. In other words, *two affine equivalent quadrics Q and Q' are metrically equivalent if and only if the principal axes of Q and Q' respectively have the same length.*

We prove first the following criterion: Let E and F be two n-dimensional Euclidean spaces and consider two symmetric bilinear functions Φ and Ψ having the same rank and the same index. Then there exists an isometric mapping $\tau: E \to F$ such that

$$\Phi(x, y) = \Psi(\tau x, \tau y) \qquad x, y \in E \tag{10.70}$$

if and only if Φ and Ψ have the same eigenvalues.

Define linear transformations $\varphi: E \to E$ and $\psi: F \to F$ by

$$\Phi(x, y) = (\varphi x, y) \quad x, y \in E \qquad \text{and} \qquad \Psi(x, y) = (\psi x, y) \quad x, y \in F.$$

Then the eigenvalues of Φ and Ψ are equal to the eigenvalues of φ and ψ respectively (cf. sec. 8.10).

Now assume that τ is an isometric mapping of E onto F such that relation (10.70) holds. Then

$$(\varphi x, y) = (\psi \tau x, \tau y) \tag{10.71}$$

whence

$$\varphi = \tau^{-1} \circ \psi \circ \tau .$$

This relation implies that φ and ψ have the same eigenvalues.

Conversely, assume that φ and ψ have the same eigenvalues. Then there is an orthonormal basis a_ν in E and an orthonormal basis $b_\nu (\nu = 1 \ldots n)$ in F such that

$$\varphi a_\nu = \lambda_\nu a_\nu \quad \text{and} \quad \psi b_\nu = \lambda_\nu b_\nu \qquad (\nu = 1 \ldots n). \tag{10.72}$$

Hence, an isometric mapping $\tau: E \to F$ is defined by

$$\tau a_\nu = b_\nu \qquad (\nu = 1 \ldots n). \tag{10.73}$$

Equations (10.72) and (10.73) imply that

$$(\varphi a_\nu, a_\mu) = (\psi \tau a_\nu, \tau a_\mu) \qquad (\nu, \mu = 1 \ldots n)$$

whence (10.71).

10.22. Metric classification of quadrics. Consider first two quadrics Q and Q' with center. Since a translation does not change the principal axes we may assume that Q and Q' have the common center O. Then the equations of Q and Q' read

$$Q : \Phi(x) = 1$$

and

$$Q' : \Phi'(x) = 1 .$$

Now assume that there exists a rotation of E carrying Q into Q'. Then

$$\Phi(x) = \Phi'(\tau x) \qquad x \in E . \tag{10.74}$$

It follows from the criterion in sec. 10.21 that the bilinear functions Φ and Φ' have the same eigenvalues. This implies that the principal axes of Q and Q' have the same length.

$$|a_\nu| = |a'_\nu| \qquad (\nu = 1 \ldots r). \tag{10.75}$$

Conversely, assume the relations (10.75). Then

$$|\lambda_\nu| = |\lambda'_\nu| \qquad (\nu = 1 \ldots n).$$

Observing the conditions (10.61) we see that $\lambda_\nu = \lambda'_\nu (\nu = 1 \dots n)$. According to the criterion in sec. 10.21 there exists a rotation τ of E such that

$$\Phi(x) = \Phi(\tau x').$$

This rotation obviously transforms Q into Q'.

Now let Q and Q' be two quadrics without center. Without loss of generality we may assume that Q and Q' have the common vertex O. Then the equations of Q and Q' read

$$Q: \Phi(x) + 2(a, x) = 0 \qquad a \in K \quad |a| = 1, \tag{10.76}$$

and

$$Q': \Phi'(x) + 2(a', x) = 0 \qquad a' \in K \quad |a'| = 1. \tag{10.77}$$

If Q and Q' are metrically equivalent there exists a rotation τ such that

$$\Phi(x) = \Phi'(\tau x).$$

Then

$$|a_\nu| = |a'_\nu| \qquad (\nu = 1 \dots r). \tag{10.78}$$

Conversely, equations (10.78) imply that the bilinear functions Φ and Φ' have the same eigenvalues,

$$\lambda_\nu = \lambda'_\nu \qquad (\nu = 1 \dots n). \tag{10.79}$$

Now consider the restriction Ψ of Φ to the subspace $T_0(Q)$. Then every eigenvalue of Ψ is also an eigenvalue of Φ. In fact, assume that

$$\Psi(e, y) = \lambda(e, y)$$

for a fixed vector $e \in T_0(Q)$ and all vectors $y \in T_0(Q)$. Then

$$\Phi(e, x) = \Phi(e, \xi a + y) = \xi \Phi(e, a) + \Psi(e, y) = \xi \Phi(e, a) + \lambda(e, y) \tag{10.80}$$

for an arbitrary vector $x \in E$. Since the point O is a vertex of Q we have that $\Phi(e, a) = 0$. We thus obtain from (10.80) the relation

$$\Phi(e, x) = \lambda(e, y) = \lambda(e, \xi a + y) = \lambda(e, x)$$

showing that λ is an eigenvalue of Φ. Hence we see that the bilinear function Ψ has the eigenvalues $\lambda_1 \dots \lambda_{n-1}$. In the same way we see that the restriction Ψ' of Φ' to the subspace $T_0(Q')$ has the eigenvalues $\lambda'_1 \dots \lambda'_{n-1}$. Now it follows from (10.79) and the criterion in sec. 10.21 that there exists an isometric mapping

$$\varrho: T_0(Q) \to T_0(Q')$$

with the property that

$$\Phi'(\varrho y) = \Phi(y) \qquad y \in T_0(Q).$$

Define the rotation τ of E by

$$\tau y = \varrho y \qquad y \in T_0(Q)$$
$$\tau a = a'.$$

Then

$$\Phi'(\tau x) + 2(a', \tau x) = \Phi(x) + 2(a, x) \qquad x \in E$$

and consequently, τ transforms Q into Q'.

10.23. The metric normal-forms in the plane and in 3-space. Equations (10.63) and (10.69) yield the following metric normal forms for the dimensions $n=2$ and $n=3$:

Plane:

I. Quadrics with center:

1. $\frac{\xi^2}{a^2} + \frac{\eta^2}{b^2} = 1, \quad a \geqq b$ ellipse with the axes a and b.

2. $\frac{\xi^2}{a^2} - \frac{\eta^2}{b^2} = 1$ hyperbola with the axes a and b.

3. $\xi = \pm a$ two parallel lines with the distance $2a$.

II. Quadrics without center:

$\frac{\xi^2}{a^2} = 2\eta$ parabola with latus rectum of length a.

3-space:

I. Quadrics with center:

1. $\frac{\xi^2}{a^2} + \frac{\eta^2}{b^2} + \frac{\zeta^2}{c^2} = 1, \ a \geqq b \geqq c$ ellipsoid with axes a, b, c.

2. $\frac{\xi^2}{a^2} + \frac{\eta^2}{b^2} - \frac{\zeta^2}{c^2} = 1, \ a \geqq b$ hyperboloid with one shell and axes a, b, c.

3. $\frac{\xi^2}{a^2} - \frac{\eta^2}{b^2} - \frac{\zeta^2}{c^2} = 1, \ b \geqq c$ hyperboloid with two shells and axes a, b, c.

4. $\frac{\xi^2}{a^2} + \frac{\eta^2}{b^2} = 1,\ a \geqq b$ — elliptic cylinder with the axes a and b.

5. $\frac{\xi^2}{a^2} - \frac{\eta^2}{b^2} = 1$ — hyperbolic cylinder with the axes a and b.

6. $\xi = \pm a$ — two parallel planes with the distance $2a$.

II. Quadrics without center:

1. $\frac{\xi^2}{a^2} - \frac{\eta^2}{b^2} = 2\zeta,\ a \geqq b$ — elliptic paraboloid with axes a and b.

2. $\frac{\xi^2}{a^2} - \frac{\eta^2}{b^2} = 2\zeta$ — hyperbolic paraboloid with axes a and b.

3. $\frac{\xi^2}{a^2} = 2\zeta$ — parabolic cylinder with latus rectum of length a.

Problems

1. Give the center or vertex, the type and the axes of the following quadrics in the 3-space:

a) $2\xi^2+2\eta^2-\zeta^2+8\xi\eta-4\xi\zeta-4\eta\zeta=2$.
b) $4\xi^2+3\eta^2-\zeta^2-12\xi\eta+4\xi\zeta-8\eta\zeta=1$.
c) $\xi^2+\eta^2+7\zeta^2-16\xi\eta-8\xi\zeta-8\eta\zeta=9$.
d) $3\xi^2+3\eta^2+\zeta^2-2\xi\eta+6\xi-2\eta-2\zeta+3=0$.

2. Given a non-degenerate quadratic function Φ, consider the family (Q_α) of quadrics defined by

$$\Phi(x) = \alpha \qquad (\alpha \neq 0).$$

Show that every point $x \neq 0$ is contained in exactly one quadric Q_α. Prove that the linear transformation φ of E defined by

$$\Phi(x, y) = (\varphi x, y)$$

associates with every point $x \neq 0$ the normal vector of the quadric passing through x.

3. Consider the quadric

$$Q: \Phi(x) = 1,$$

where Φ is a non-degenerate bilinear function. Denote by Q' the image of Q under the mapping φ which corresponds to Φ. Prove that the principal axes of Q' and Q are connected by the relation

$$a'_\nu = \frac{a_\nu}{|a_\nu|^2} \qquad (\nu = 1 \dots n).$$

4. Given two points p, q and a number $2\alpha\,(\alpha > |p-q|)$, consider the locus Q of all points x such that

$$|x - p| + |x - q| = 2\alpha .$$

Prove that Q is a quadric of index n whose principal axes have the length

$$|a_1| = \alpha, \quad |a_\nu| = \sqrt{\alpha^2 - \tfrac{1}{4}|p - q|^2} \qquad (\nu = 2 \dots n).$$

5. Let $\Phi(x) = 1$ be the equation of a non-degenerate quadric Q with the property that x is a normal vector at every point of Q. Prove that Q is a sphere.

Chapter XI

Unitary spaces

§ 1. Hermitian functions

11.1. Sesquilinear functions in a complex space. Let E be an n-dimensional complex linear space and $\Phi: E \times E \to \mathbb{C}$ be a function such that

$$\begin{aligned} \Phi(\lambda x_1 + \mu x_2, y) &= \lambda \Phi(x_1, y) + \mu \Phi(x_2, y) \\ \Phi(x, \lambda y_1 + \mu y_2) &= \bar{\lambda} \Phi(x, y_1) + \bar{\mu} \Phi(x, y_2) \end{aligned} \tag{11.1}$$

where $\bar{\lambda}$ and $\bar{\mu}$ are the complex conjugate coefficients. Then Φ will be called a *sesquilinear function*. Replacing y by x we obtain from Φ the corresponding quadratic function

$$\Psi(x) = \Phi(x, x). \tag{11.2}$$

It follows from (11.1) that Ψ satisfies the relations

$$\Psi(x + y) + \Psi(x - y) = 2(\Psi(x) + \Psi(y)) \tag{11.3}$$

and

$$\Psi(\lambda x) = |\lambda|^2 \Psi(x).$$

The function Φ can be expressed in terms of Ψ. In fact, equation (11.2) yields

$$\Psi(x + y) = \Psi(x) + \Psi(y) + \Phi(x, y) + \Phi(y, x). \tag{11.4}$$

Replacing y by iy we obtain

$$\Psi(x + iy) = \Psi(x) + \Psi(y) - i\Phi(x, y) + i\Phi(y, x). \tag{11.5}$$

Multiplying (11.5) by i and adding it to (11.4) we find

$$\Psi(x + y) + i\Psi(x + iy) = (1 + i)(\Psi(x) + \Psi(y)) + 2\Phi(x, y),$$

whence

$$2\Phi(x, y) = \{\Psi(x + y) - \Psi(x) - \Psi(y)\} + i\{\Psi(x + iy) - \Psi(x) - \Psi(y)\} \tag{11.6}$$

Note: The fact that Φ is uniquely determined by the function Ψ is due to the sesquilinearity. We recall that a bilinear function has to be *symmetric* in order to be uniquely determined by the corresponding quadratic function.

11.2. Hermitian functions. With every sesquilinear function Φ we can associate another sesquilinear function $\tilde{\Phi}$ given by

$$\tilde{\Phi}(x, y) = \overline{\Phi(y, x)}.$$

A sesquilinear function Φ is called *Hermitian* if $\tilde{\Phi} = \Phi$, i. e.

$$\Phi(x, y) = \overline{\Phi(y, x)}. \tag{11.7}$$

Inserting $y = x$ in (11.7) we find that

$$\Psi(x) = \overline{\Psi(x)}. \tag{11.8}$$

Hence the quadratic function Ψ is real valued. Conversely, a sesquilinear function Φ whose quadratic function is real valued is Hermitian. In fact, if Ψ is real valued, both parentheses in (11.6) are real. Interchange of x and y yields

$$2\Phi(y, x) = \{\Psi(x + y) - \Psi(x) - \Psi(y)\} + i\{\Psi(y + ix) - \Psi(x) - \Psi(y)\}. \tag{11.9}$$

Comparison of (11.6) and (11.9) shows that the real parts coincide. The sum of the imaginary parts is equal to

$$\Psi(x + iy) + \Psi(y + ix) - 2\Psi(x) - 2\Psi(y).$$

Replacing y by iy in (11.3) we see that this is equal to zero, whence

$$\Phi(y, x) = \overline{\Phi(x, y)}.$$

A Hermitian function Φ is called *positive definite*, if $\Psi(x) > 0$ for all vectors $x \neq 0$.

11.3. Hermitian matrices. Let x_ν $(\nu = 1 \dots n)$ be a basis of E. Then every sesquilinear function Φ defines a complex $n \times n$-matrix

$$\alpha_{\nu\mu} = \Phi(x_\nu, x_\mu).$$

The function Φ is uniquely determined by the matrix $(\alpha_{\nu\mu})$. In fact, if

$$x = \sum_\nu \xi^\nu x_\nu \quad \text{and} \quad y = \sum_\nu \eta^\nu x_\nu$$

are two arbitrary vectors, we have that

$$\Phi(x, y) = \sum_{\nu,\mu} \alpha_{\nu\mu} \xi^\nu \bar{\eta}^\mu .$$

The matrices $(\alpha_{\nu\mu})$ and $(\tilde{\alpha}_{\nu\mu})$ of Φ and $\tilde{\Phi}$ are obviously connected by the relation

$$\tilde{\alpha}_{\nu\mu} = \bar{\alpha}_{\mu\nu}.$$

If Φ is a Hermitian function it follows that

$$\alpha_{\nu\mu} = \bar{\alpha}_{\mu\nu} .$$

A complex $n \times n$-matrix satisfying this relation is called a *Hermitian matrix*.

Problems

1. Prove that a skew-symmetric sequilinear function is identically zero.
2. Show that the decomposition constructed in sec. 9.6 can be carried over to Hermitian functions.

§ 2. Unitary spaces

11.4. Definition. *A unitary space* is an n-dimensional complex linear space E in which a positive definite Hermitian function, denoted by $(,)$, is distinguished. The number (x, y) is called the *Hermitian inner product* of the vectors x and y. It has the following properties:

1. $(\lambda x_1 + \mu x_2, y) = \lambda(x_1, y) + \mu(x_2, y)$
   $(x, \lambda y_1 + \mu y_2) = \bar{\lambda}(x, y_1) + \bar{\mu}(x, y_2)$.
2. $(x, y) = \overline{(y, x)}$. In particular, (x, x) is real.
3. $(x, x) > 0$ for all vectors $x \neq 0$.

Example I: A Hermitian inner product in the complex number space $\mathbb{C}^n$ is defined by

$$(x, y) = \sum_{\nu=1}^{n} \xi^\nu \bar{\eta}^\nu$$

where

$$x = (\xi^1, \ldots, \xi^n) \quad \text{and} \quad y = (\eta^1, \ldots, \eta^n).$$

Example II: Let E be a Euclidean space and let $E_{\mathbb{C}}$ be the complexification of the vector space E (cf. sec. 2.16). Then a Hermitian inner product is defined in $E_{\mathbb{C}}$ by

$$(x_1 + i y_1, x_2 + i y_2) = (x_1, x_2) + (y_1, y_2) + i\{(y_1, x_2) - (x_1, y_2)\}.$$

The unitary space so obtained is called the *complexification of* the Euclidean space E.

The *norm* of a vector x of a unitary space is defined as the positive square-root

$$|x| = \sqrt{(x, x)}.$$

The Schwarz-inequality

$$|(x, y)| \leqq |x|\,|y| \tag{11.10}$$

is proved in the same way as for real inner product spaces. Equality holds if and only if the vectors x and y are linearly dependent.

From (11.10) we obtain the triangle-inequality

$$|x + y| \leqq |x| + |y|.$$

Equality holds if and only if $y = \lambda x$ where λ is real and non-negative. In fact, assume that

$$|x + y| = |x| + |y|. \tag{11.11}$$

Squaring this equation we obtain

$$(x, y) + \overline{(x, y)} = 2\,|x|\,|y|. \tag{11.12}$$

This can be written as

$$\operatorname{Re}(x, y) = |x|\,|y|$$

where Re denotes the real part. The above relation yields

$$|(x, y)| = |x|\,|y|$$

and hence it implies that the vectors x and y are linearly dependent, $y = \lambda x$. Inserting this into (11.12) we obtain

$$\lambda + \bar{\lambda} = 2\,|\lambda|$$

whence

$$\operatorname{Re}\lambda = |\lambda|.$$

Hence, λ is real and non-negative. Conversely, it is clear that

$$|(1 + \lambda)\,x| = |x| + \lambda\,|x|$$

for every real, non-negative number λ.

Two vectors $x \in E$ and $y \in E$ are called *orthogonal*, if

$$(x, y) = 0.$$

Every subspace $E_1 \subset E$ determines an *orthogonal complement* $E_1^{\perp}$ consisting of all vectors which are orthogonal to E_1. The spaces E_1 and $E_1^{\perp}$ form a direct decomposition of E:

$$E = E_1 \oplus E_1^{\perp}.$$

A basis $x_\nu (\nu = 1 \ldots n)$ of E is called *orthonormal*, if

$$(x_\nu, x_\mu) = \delta_{\nu\mu}.$$

The inner product of two vectors

$$x = \sum_\nu \xi^\nu x_\nu \quad \text{and} \quad y = \sum_\nu \eta^\nu x_\nu$$

is then given by

$$(x, y) = \sum_\nu \xi^\nu \bar{\eta}^\nu.$$

Replacing y by x we obtain

$$|x|^2 = \sum_\nu \xi^\nu \bar{\xi}^\nu.$$

Orthogonal bases can be constructed in the same way as in a real inner product space by the Schmidt-orthogonalization process.

Consider two orthonormal bases x_ν and $\bar{x}_\nu$ $(\nu = 1 \ldots n)$. Then the matrix (α_ν^μ) of the basis-transformation $x_\nu \to \bar{x}_\nu$ satisfies the relations

$$\sum_\mu \alpha_\nu^\mu \bar{\alpha}_\lambda^\mu = \delta_{\nu\lambda}.$$

A complex matrix of this kind is called a *unitary matrix*. Conversely, if an orthonormal basis x_ν and a unitary matrix (α_ν^μ) is given, the basis

$$\bar{x}_\nu = \sum_\mu \alpha_\nu^\mu x_\mu$$

is again orthonormal.

11.5. The conjugate space. To every complex vector space E we can assign a second complex vector space, $\bar{E}$, in the following way: $\bar{E}$ coincides with E as a real vector space. However, scalar multiplication in $\bar{E}$, denoted by $(\lambda, z) \to \lambda \cdot z$ is defined by

$$\lambda \cdot z = \bar{\lambda} z.$$

$\bar{E}$ is called the *conjugate vector space*. Clearly the identity map $\kappa: E \to \bar{E}$ satisfies

$$\kappa(i z) = -i \cdot \kappa(z) \qquad z \in E.$$

Now assume that $(,)$ is a Hermitian inner product in E. Then a complex bilinear function, $\langle\,\rangle$, is defined in $E\times\bar{E}$ by

$$\langle x, y\rangle = (x, \kappa^{-1} y) \qquad x\in E,\ y\in\bar{E}.$$

Clearly this bilinear function is non-degenerate and so it makes E, $\bar{E}$ into a pair of dual complex spaces.

Now all the properties arising from duality can be carried over to unitary spaces. The *Riesz theorem* asserts that every linear function f in a unitary space can be uniquely represented in the form

$$f(x) = (x, a) \qquad x\in E.$$

In fact, consider the conjugate space $\bar{E}$. In view of the duality between E and $\bar{E}$ there is a unique vector $b\in\bar{E}$ such that

$$f(x) = \langle x, b\rangle.$$

Now set $a=\kappa^{-1}(b)$.

11.6. Normed determinant functions. A *normed determinant function* in a unitary space is a determinant function, Δ, which satisfies

$$\Delta(x_1, \dots, x_n)\,\overline{\Delta(y_1, \dots, y_n)} = \det((x_i, y_j)) \qquad x_i\in E,\ y_j\in E. \quad (11.13)$$

We shall show that normed determinant functions always exist and that two normed determinant functions, Δ_1 and Δ_2 are connected by the relation $\Delta_2=\varepsilon\Delta_1$, where ε is a complex number of magnitude 1.

In fact, let $\Delta_0\neq 0$ be any determinant function in E. Then a determinant function, Δ_0^*, in $\bar{E}$ is given by

$$\Delta_0^*(x^{*1}, \dots, x^{*n}) = \overline{\Delta_0(\kappa^{-1}x^{*1}, \dots, \kappa^{-1}x^{*n})}.$$

Since E and $\bar{E}$ are dual with respect to the scalar product $\langle,\rangle$ formula (4.21) yields

$$\Delta_0(x_1, \dots, x_n)\,\Delta_0^*(x^{*1}, \dots, x^{*n}) = \alpha\det(\langle x_i, x^{*j}\rangle)$$

where $\alpha\neq 0$ is a complex constant. Setting

$$x^{*\nu} = \kappa y_\nu \qquad y_\nu\in E$$

we obtain

$$\Delta_0(x_1, \dots, x_n)\cdot\overline{\Delta_0(y_1, \dots, y_n)}=\alpha\cdot\det(\langle x_i, \kappa y_j\rangle)=\alpha\cdot\det((x_i, y_j)). \quad (11.14)$$

Now set $x_\nu=y_\nu=e_\nu$ where e_ν $(\nu=1\dots n)$ is an orthonormal basis of E. It follows that

$$|\Delta_0(e_1, \dots, e_n)|^2 = \alpha$$

and so α is real and positive. Let λ be any complex number satisfying $|\lambda|^2=\alpha$ and define a new determinant function, Δ, by setting

$$\Delta = \frac{1}{\lambda}\Delta_0 .$$

Then (11.14) yields the relation

$$\Delta(x_1, \ldots, x_n)\cdot\overline{\Delta(y_1, \ldots, y_n)} = \det((x_i, y_j)) \qquad x_\nu \in E,\ y_\nu \in E$$

showing that Δ is a normed determinant function.

Finally, assume that Δ_1 and Δ_2 are normed determinant functions in E. Then formula (11.13) implies that $\Delta_2 = \varepsilon\Delta_1$, $|\varepsilon|=1$.

11.7. Real forms. Let E be an n-dimensional complex vector space and consider the underlying $2n$-dimensional real vector space $E_{\mathbb{R}}$. A *real form* of E is a subspace, $F\subset E_{\mathbb{R}}$, such that

$$E_{\mathbb{R}} = F \oplus iF.$$

Clearly, $\dim_{\mathbb{R}} F = n$. If F is a real form of E, then every vector $z\in E$ can be uniquely written in the form $z=x+iy$, $x, y\in F$.
Every complex space has real forms. In fact, let $z_1 \ldots z_n$ be a basis of E and define F to be the subspace of $E_{\mathbb{R}}$ consisting of the vectors

$$x = \sum_{\nu=1}^{n} \xi^\nu z_\nu \qquad \xi^\nu \in \mathbb{R}.$$

Then clearly, F is a real form of E.

A *conjugation* in a complex vector space is an $\mathbb{R}$-linear involution, $z\to\bar z$, of $E_{\mathbb{R}}$ satisfying

$$\overline{iz} = -i\bar z. \tag{11.15}$$

Every real form F of E determines a conjugation given by

$$x + iy \mapsto x - iy \qquad x, y\in F.$$

Conversely, if $z\mapsto\bar z$ is a conjugation in E, then the vectors z which satisfy $\bar z = z$ determine a real form F of E as is easily checked. The vectors of F are called *real* (with respect to the conjugation).

Problems

1. Prove that the *Gram determinant*

$$G(x_1 \ldots x_p) = \det\begin{pmatrix} (x_1, x_1) \ldots (x_1, x_p) \\ \vdots \qquad\qquad \vdots \\ (x_p, x_1) \ldots (x_p, x_p)\end{pmatrix}$$

of p vectors of a unitary space is real and non negative. Show that $G(x_1 \ldots x_p)=0$ if and only if the vectors x_ν are linearly dependent.

2. Let E be a unitary space.

(i) Show that there are conjugations in E satisfying $(\bar{z}_1, \bar{z}_2)=\overline{(z_1, z_2)}$.

(ii) Given such a conjugation show that the Hermitian inner product of E defines a Euclidean inner product in the corresponding real subspace.

3. Let F be a real form of a complex vector space E and assume that a positive definite inner product is defined in F. Show that a Hermitian inner product is given in E by

$$(z_1, z_2) = ((x_1, x_2) + (y_1, y_2)) + i((x_1, y_2) - (x_2, y_1))$$

and that

$$(\bar{z}_1, \bar{z}_2) = \overline{(z_1, z_2)}.$$

4. *Quaternions.* Fix a unit quaternion j (cf. sec. 7.23) such that $(j, e)=0$. Use j to make the space of quaternions into a 2-dimensional complex vector space. Show that this is not an algebra over $\mathbb{C}$.

5. *The complex cross product.* Let E be a 3-dimensional unitary space and choose a normed determinant function Δ.

(i) Show that a skew symmetric bilinear map $E \times E \to E$ (over the reals) is determined by the equation

$$\Delta(x, y, z) = (x, y \times z) \qquad x, y, z \in E.$$

(ii) Prove the relations

$$(i\,x) \times y = x \times (i\,y) = -i(x \times y)$$

$$(x \times y, x) = (x \times y, y) = 0$$

$$(x_1 \times x_2, y_1 \times y_2) = (y_1, x_1)(y_2, x_2) - (y_2, x_1)(y_1, x_2)$$

$$|x_1 \times x_2|^2 = |x_1|^2\,|x_2|^2 - |(x_1, x_2)|^2$$

$$x \times (y \times z) = y(z, x) - z(y, x).$$

6. *Cayley numbers.* Let E be a 4-dimensional Euclidean space and let $E_{\mathbb{C}}$ be the complexification of E (cf. Example II, sec 11.4). Choose a unit vector $e \in E_{\mathbb{C}}$ and let E_1 denote the orthogonal complement of e. Choose a normed determinant function Δ in E_1 and let $\times$ be the corresponding cross product (cf. problem 5). Consider $E_{\mathbb{C}}$ as a real 8-dimen-

sional vector space F and define a bilinear map $x, y \to x \cdot y$ by setting

$$\begin{aligned} x \cdot y &= -(x, y)\, e + x \times y && x, y \in E_1 \\ (\lambda e) \cdot y &= \lambda y && \lambda \in \mathbb{C},\ y \in E_1 \\ x \cdot (\lambda e) &= \bar{\lambda} x && \lambda \in \mathbb{C},\ x \in E_1 \end{aligned}$$

and

$$(\lambda e) \cdot (\mu e) = \lambda \mu e \qquad \lambda, \mu \in \mathbb{C}.$$

Show that this bilinear map makes F into a (non-associative) division algebra over the reals.

Verify the formulae $x \cdot y^2 = (x y) y$ and $x^2 y = x(x y)$.

7. *Symplectic spaces.* Let E be an m-dimensional vector space over a field Γ. A *symplectic inner product* in E is a non-degenerate skew symmetric bilinear function $\langle , \rangle$ in E.

(i) Show that a symplectic inner product exists in E if and only if m is even, $m = 2n$.

(ii) A basis $a_1, \ldots, a_n, b_1, \ldots, b_n$ of a symplectic vector space is called *symplectic*, if

$$\langle a_i, a_j \rangle = 0 \qquad \langle b_i, b_j \rangle = 0$$
$$\langle a_i, b_j \rangle = \delta_{ij}.$$

Show that every symplectic space has a symplectic basis.

(iii) A linear transformation $\varphi: E \to E$ is called *symplectic*, if it satisfies

$$\langle \varphi x, \varphi y \rangle = \langle x, y \rangle \qquad x, y \in E.$$

If Φ and Ψ are symplectic inner products in E, show that there is a linear automorphism α of E such that $\Psi(x, y) = \Phi(\alpha x, \alpha y)$, $x, y \in E$.

8. ***Symplectic adjoint transformations.*** If $\varphi: E \to E$ is a linear transformation of a symplectic space the *symplectic adjoint* transformation, $\tilde{\varphi}$, is defined by the equation

$$\langle \varphi x, y \rangle = \langle x, \tilde{\varphi} y \rangle \qquad x, y \in E.$$

If $\tilde{\varphi} = \varphi$ (respectively $\tilde{\varphi} = -\varphi$) then φ is called *symplectic selfadjoint* (respectively *skew symplectic*).

(i) Show that $\tilde{\tilde{\varphi}} = \varphi$. Conclude that every linear transformation φ of E can be uniquely written in the form $\varphi = \varphi_1 + \varphi_2$ where φ_1 is symplectic selfadjoint and φ_2 is skew symplectic.

(ii) Show that the matrix of a skew symplectic transformation with respect to a symplectic basis has the form

$$\begin{pmatrix} A & B \\ C & D \end{pmatrix}$$

where A, B, C, D are square $(n\times n)$-matrices satisfying

$$B^* = B, \qquad C^* = C, \qquad D = -A^*.$$

Conclude that the dimensions of the spaces of symplectic selfadjoint (respectively skew symplectic) transformations are given by

$$N_1 = n(2n-1) \qquad N_2 = n(2n+1).$$

(iii) Show that the symplectic adjoint of a linear transformation φ of a 2-dimensional space is given by

$$\tilde{\varphi} = \iota \cdot \operatorname{tr}\varphi - \varphi.$$

(iv) Let E be an n-dimensional unitary space with underlying real vector space $E_{\mathbb{R}}$.

Show that a positive definite symmetric inner product and a symplectic inner product is defined in $E_{\mathbb{R}}$ by

$$(x,y) = \operatorname{Re}(x,y)$$

and

$$\langle x,y\rangle = \operatorname{Im}(x,y) \qquad x,y\in E_{\mathbb{R}}.$$

Establish the relations

$$\langle x,y\rangle = -(ix,y)$$

and

$$\langle ix,y\rangle = (x,y).$$

Conclude that the transformation $x\mapsto ix$ is both symplectic and skew symplectic.

§ 3. Linear mappings of unitary spaces

11.8. The adjoint mapping. Let E and F be two unitary spaces and $\varphi\colon E\to F$ a linear mapping of E into F. As in the real case we can associate with φ an adjoint mapping $\tilde{\varphi}$ of F into E. A *conjugation* in a unitary space is a linear automorphism $x\to\bar{x}$ of the underlying real vector space such that $\overline{ix} = -i\bar{x}$ and $(\bar{x},\bar{y}) = \overline{(x,y)}$. If $x\to\bar{x}$ is a conjugation in E, then a non-degenerate complex bilinear function is defined by $\langle x,y\rangle = (x,\bar{y})$, $x,y\in E$. Let $x\to\bar{x}$ and $y\to\bar{y}$ be conjugations in E and in F, respectively. Then E and F are dual to themselves and hence φ determines a dual mapping $\varphi^*\colon F\to E$ by the relation

$$\langle \varphi x, y\rangle = \langle x, \varphi^* y\rangle. \tag{11.16}$$

Replacing y by $\bar{y}$ in (11.16) we obtain

$$\langle \varphi x, \bar{y} \rangle = \langle x, \varphi^* \bar{y} \rangle . \tag{11.17}$$

Observing the relation between inner product and scalar product we can rewrite (11.17) in the form

$$(\varphi x, y) = (x, \overline{\varphi^* \bar{y}}) . \tag{11.18}$$

Now define a mapping $\tilde{\varphi}: E \leftarrow F$ by

$$\tilde{\varphi} y = \overline{\varphi^* \bar{y}} . \tag{11.19}$$

Then relation (11.18) reads

$$(\varphi x, y) = (x, \tilde{\varphi} y) \qquad x \in E, \quad y \in F . \tag{11.20}$$

The mapping $\tilde{\varphi}$ does not depend on the conjugations in E and F. In fact, assume that $\tilde{\varphi}_1$ and $\tilde{\varphi}_2$ are two linear mappings of F into E satisfying (11.20). Then

$$(x, (\tilde{\varphi}_2 - \tilde{\varphi}_1) y) = 0 .$$

This equation holds for every fixed $y \in F$ and all vectors $x \in E$ and hence it implies that $\tilde{\varphi}_2 = \tilde{\varphi}_1$. The mapping $\tilde{\varphi}$ is called the *adjoint* of the mapping φ.

It follows from equation (11.18) that the relations

$$\widetilde{\varphi + \psi} = \tilde{\varphi} + \tilde{\psi} \qquad \widetilde{\lambda \varphi} = \bar{\lambda} \tilde{\Phi} \qquad \text{and} \qquad \widetilde{\psi \circ \varphi} = \tilde{\varphi} \circ \tilde{\psi}$$

hold for any two linear mappings and for every complex coefficient λ.

Equation (11.20) implies that the matrices of φ and $\tilde{\varphi}$ relative to two orthonormal bases of E and F are connected by the relation

$$\tilde{\alpha}_\mu^\nu = \bar{\alpha}_\nu^\mu \qquad (\nu = 1 \dots n, \ \mu = 1 \dots m) .$$

Now consider the case $E = F$. Then the determinants of φ and $\tilde{\varphi}$ are complex conjugates. To prove this, let $\Delta \neq 0$ be a determinant function in E and $\bar{\Delta}$ be the conjugate determinant function. Then it follows from the definition of $\bar{\Delta}$ that

$$\begin{aligned} \bar{\Delta}(\tilde{\varphi} x_1 \dots \tilde{\varphi} x_n) &= \overline{\Delta(\overline{\tilde{\varphi} x_1} \dots \overline{\tilde{\varphi} x_n})} = \overline{\Delta(\varphi^* \bar{x}_1 \dots \varphi^* \bar{x}_n)} \\ &= \overline{\det \varphi}\, \overline{\Delta(\bar{x}_1 \dots \bar{x}_p)} = \overline{\det \varphi}\, \bar{\Delta}(x_1 \dots x_n) . \end{aligned}$$

This equation implies that

$$\det \tilde{\varphi} = \overline{\det \varphi} . \tag{11.21}$$

If φ is replaced by $\varphi-\lambda\iota$, where λ is a complex parameter, relation (11.21) yields

$$\det(\tilde{\varphi}-\lambda\iota)=\overline{\det(\varphi-\bar{\lambda}\iota)}.$$

Expanding both sides with respect to λ we obtain

$$\sum_{\nu}\tilde{\alpha}_{\nu}\lambda^{n-\nu}=\sum_{\nu}\bar{\alpha}_{\nu}\lambda^{n-\nu}.$$

This equation shows that corresponding coefficients in the characteristic polynomials of φ and $\tilde{\varphi}$ are complex conjugates. In particular,

$$\operatorname{tr}\tilde{\varphi}=\overline{\operatorname{tr}\varphi}.$$

11.9. The inner product in the space $L(E;E)$. Consider the space $L(E;E)$. An inner product can be introduced in this space by

$$(\varphi,\psi)=\frac{1}{n}\operatorname{tr}(\varphi\circ\tilde{\psi}). \tag{11.22}$$

It follows immediately from (11.22) that the function (φ,ψ) is sesquilinear. Interchange of φ and ψ yields

$$(\psi,\varphi)=\frac{1}{n}\operatorname{tr}(\psi\circ\tilde{\varphi})=\frac{1}{n}\overline{\operatorname{tr}(\varphi\circ\tilde{\psi})}=\overline{(\varphi,\psi)}.$$

To prove that the Hermitian function (11.22) is positive definite let e_ν $(\nu=1\ldots n)$ be an orthonormal basis. Then

$$\varphi e_\nu=\sum_{\mu}\alpha_\nu^\mu e_\mu \quad\text{and}\quad \tilde{\varphi}e_\nu=\sum_{\mu}\tilde{\alpha}_\nu^\mu e_\mu \tag{11.23}$$

where $\tilde{\alpha}_\nu^\mu=\bar{\alpha}_\mu^\nu$. Equations (11.23) yield

$$\varphi\tilde{\varphi}e_\nu=\sum_{\mu,\nu}\tilde{\alpha}_\nu^\mu\alpha_\mu^\lambda e_\lambda$$

whence

$$\operatorname{tr}(\varphi\circ\tilde{\varphi})=\sum_{\nu,\mu}\tilde{\alpha}_\nu^\mu\alpha_\mu^\nu=\sum_{\nu,\mu}\bar{\alpha}_\mu^\nu\alpha_\mu^\nu=\sum_{\nu,\mu}|\alpha_\mu^\nu|^2.$$

This formula shows that $(\varphi,\varphi)>0$ for every transformation $\varphi\neq 0$.

11.10. Normal mappings. A linear transformation $\varphi:E\to E$ is called *normal*, if the mappings φ and $\tilde{\varphi}$ commute,

$$\tilde{\varphi}\circ\varphi=\varphi\circ\tilde{\varphi}. \tag{11.24}$$

In the same way as for a real inner product (cf. sec. 8.5) it is shown that the condition (11.24) is equivalent to

$$|\varphi x|^2 = |\tilde{\varphi} x|^2 \qquad x \in E. \tag{11.25}$$

It follows from (11.25) that the kernels of φ and $\tilde{\varphi}$ coincide, ker φ = ker $\tilde{\varphi}$. We thus obtain the direct decomposition

$$E = \ker \varphi \oplus \operatorname{Im} \varphi. \tag{11.26}$$

The relation ker φ = ker $\tilde{\varphi}$ implies that the mappings φ and $\tilde{\varphi}$ have the same eigenvectors and that the corresponding eigenvalues are complex conjugates. In fact, assume that e is an eigenvector of φ and that λ is the corresponding eigenvalue,

$$\varphi e = \lambda e.$$

Then e is contained in the kernel of $\varphi - \lambda\iota$. Since the mapping $\varphi - \lambda\iota$ is again normal, e must also be contained in the kernel of $\tilde{\varphi} - \bar{\lambda}\iota$, i. e.

$$\tilde{\varphi} e = \bar{\lambda} e.$$

In sec. 8.7 we have seen that a selfadjoint linear transformation of a real inner product space of dimension n always possesses n eigenvectors which are mutually orthogonal. Now it will be shown that in a complex space the same assertion holds even for normal mappings. Consider the characteristic polynomial of φ. According to the fundamental theorem of algebra this polynomial must have a zero λ_1. Then λ_1 is an eigenvalue of φ. Let e_1 be a corresponding eigenvector and E_1 the orthogonal complement of e_1. The space E_1 is stable under φ. In fact, let y be an arbitrary vector of E_1. Then

$$(\varphi y, e_1) = (y, \tilde{\varphi} e_1) = (y, \bar{\lambda} e_1) = \lambda (y, e_1) = 0$$

and hence φy is contained in E_1. The induced mapping is obviously again normal and hence there exists an eigenvector e_2 in E_1. Continuing this way we finally obtain n eigenvectors e_ν $(\nu = 1 \dots n)$ which are mutually orthogonal,

$$(e_\nu, e_\mu) = 0 \qquad (\nu \neq \mu).$$

If these vectors are normed to length one, they form an orthonormal basis of E. Relative to this basis the matrix of φ has diagonal form with the eigenvalues in the main-diagonal,

$$\varphi e_\nu = \lambda_\nu e_\nu \qquad (\nu = 1 \dots n). \tag{11.27}$$

11.11. Selfadjoint and skew mappings. Let φ be a selfadjoint linear transformation of E; i.e., a mapping such that $\tilde{\varphi}=\varphi$. Then relation (11.20) yields

$$(\varphi x, y) = (x, \varphi y) \qquad x, y \in E.$$

Replacing y by x we obtain

$$(\varphi x, x) = (x, \varphi x) = \overline{(\varphi x, x)}$$

showing that $(\varphi x, x)$ is real for every vector $x \in E$. This implies that all eigenvalues of a selfadjoint transformation are real. In fact, let e be an eigenvector and λ be the corresponding eigenvalue. Then $\varphi e = \lambda e$, whence

$$(\varphi e, e) = \lambda (e, e).$$

Since $(\varphi e, e)$ and $(e, e) \neq 0$ are real, λ must be real.

Every selfadjoint mapping is obviously normal and hence there exists a system of n orthonormal eigenvectors. Relative to this system the matrix of φ has the form (11.27) where all numbers λ_ν are real.

The matrix of a selfadjoint mapping relative to an orthonormal basis is Hermitian.

A linear transformation φ of E is called *skew* if $\tilde{\varphi} = -\varphi$. In a unitary space there is no essential difference between selfadjoint and skew mappings. In fact, the relation

$$\widetilde{i\varphi} = -i\tilde{\varphi}$$

shows that multiplication by i associates with every selfadjoint mapping a skew mapping and conversely.

11.12. Unitary mappings. A *unitary mapping* is a linear transformation of E which preserves the inner product,

$$(\varphi x, \varphi y) = (x, y) \qquad x, y \in E. \tag{11.28}$$

Relation (11.28) implies that

$$|\varphi x| = |x| \qquad x \in E$$

showing that every unitary mapping is regular and hence it is an automorphism of E. If equation (11.28) is written in the form

$$(\varphi x, y) = (x, \varphi^{-1} y)$$

it shows that the inverse mapping of φ is equal to the adjoint mapping,

$$\tilde{\varphi} = \varphi^{-1}. \tag{11.29}$$

Passing over to the determinants we obtain

$$\det\varphi\cdot\overline{\det\varphi}=1$$

whence

$$|\det\varphi|=1\,.$$

Every eigenvalue of a unitary mapping has norm 1. In fact, the equation $\varphi e=\lambda e$ yields

$$|\varphi\, e|=|\lambda|\,|e|$$

whence $|\lambda|=1$.

Equation (11.29) shows that a unitary map is normal. Hence, there exists an orthonormal basis $e_\nu\,(\nu=1\ldots n)$ such that

$$\varphi\, e_\nu=\lambda_\nu\, e_\nu\qquad(\nu=1\ldots n)$$

where the λ_ν are complex numbers of absolute value one.

Problems

1. Given a linear transformation $\varphi\colon E\to E$ show that the bilinear function Φ defined by

$$\Phi(x,y)=(\varphi\, x,y)$$

is sesquilinear. Conversely, prove that every sesquilinear function Φ can be obtained in this way. Prove that the adjoint transformation determines the Hermitian conjugate function.

2. Show that the set of selfadjoint transformations is a real vector space of dimension n^2.

3. Let φ be a linear transformation of a complex vector space E.

a) Prove that a positive definite inner product can be introduced in E such that φ becomes a normal mapping if and only if φ has n linearly independent eigenvectors.

b) Prove that a positive definite inner product can be introduced such that φ is

i) selfadjoint
ii) skew
iii) unitary

if and only if in addition the following conditions are fulfilled in corresponding order:

i) all eigenvalues of φ are real
ii) all eigenvalues of φ are imaginary or zero
iii) all eigenvalues have absolute value 1.

4. Denote by $S(E)$ the space of all selfadjoint mappings and by $A(E)$ the space of all skew mappings of the unitary space E.

Prove that a multiplication is defined in $S(E)$ and $A(E)$ by

$$[\varphi,\psi] = i(\varphi\circ\psi - \psi\circ\varphi) \qquad \varphi\in S(E), \psi\in S(E)$$

and

$$[\varphi,\psi] = \varphi\circ\psi - \psi\circ\varphi \qquad \varphi\in A(E), \psi\in A(E)$$

respectively and that these spaces become Lie algebras under the above multiplications. Show that $S(E)$ and $A(E)$ are real forms of the complex space $L(E; E)$ (cf. sec. 11.7).

§ 4.* Unitary mappings of the complex plane

11.13. Definition. In this paragraph we will study the unitary mappings of a 2-dimensional unitary space C in further detail. Let τ be a unitary mapping of C. Employing an orthonormal basis e_1, e_2 we can represent the mapping τ in the form

$$\begin{aligned} \tau e_1 &= \alpha e_1 + \beta e_2 \\ \tau e_2 &= \varepsilon(-\bar{\beta} e_1 + \bar{\alpha} e_2) \end{aligned} \tag{11.30}$$

where α, β and ε are complex numbers subject to the conditions

$$|\alpha|^2 + |\beta|^2 = 1$$

and

$$|\varepsilon| = 1.$$

These equations show that

$$\det\tau = \varepsilon.$$

We are particularly interested in the unitary mappings with the determinant + 1. For every such mapping equations (11.30) reduce to

$$\begin{aligned} \tau e_1 &= \alpha e_1 + \beta e_2 \\ \tau e_2 &= -\bar{\beta} e_1 + \bar{\alpha} e_2 \end{aligned} \qquad |\alpha|^2 + |\beta|^2 = 1.$$

This implies that

$$\begin{aligned} \tau^{-1} e_1 &= \bar{\alpha} e_1 - \beta e_2 \\ \tau^{-1} e_2 &= \bar{\beta} e_1 + \alpha e_2. \end{aligned}$$

Adding the above relations in the corresponding order we find that

$$(\tau + \tau^{-1}) e_\nu = (\alpha + \bar{\alpha}) e_\nu = \operatorname{tr}\tau\cdot e_\nu \qquad (\nu = 1, 2)$$

whence

$$\tau + \tau^{-1} = \iota\cdot\operatorname{tr}\tau. \tag{11.31}$$

Formula (11.31) implies that

$$(z, \tau z) + (z, \tau^{-1} z) = |z|^2 \operatorname{tr} \tau$$

for every vector $z \in C$. Observing that

$$(z, \tau^{-1} z) = (\tau z, z) = \overline{(z, \tau z)}$$

we thus obtain the relation

$$2 \operatorname{Re}(z, \tau z) = |z|^2 \operatorname{tr} \tau \qquad z \in C \tag{11.32}$$

showing that if $\det \tau = 1$, then the real part of the inner product $(z, \tau z)$ depends only on the norm of z. (11.32) is the complex analogue of the relation (8.40) for a proper rotation of the real plane.

We finally note that the set of all unitary mappings with the determinant $+1$ forms a subgroup of the group of all unitary mappings.

11.14. The algebra Q. Let Q denote the set of complex linear transformations of the complex plane C which satisfy

$$\varphi + \tilde{\varphi} = \iota \cdot \operatorname{tr} \varphi. \tag{11.33}$$

It is easy to see that these transformations form a real 4-dimensional vector space, Q. Since (11.33) is equivalent to $\tilde{\varphi} = \operatorname{ad} \varphi$ (cf. sec. 4.6 and problem 8, § 2, chapter IV) it follows that Q is an algebra under composition. Moreover, we have the relation

$$\varphi \tilde{\varphi} = \varphi \operatorname{ad} \varphi = \iota \cdot \det \varphi \qquad \varphi \in Q. \tag{11.34}$$

Define a positive definite inner product in Q by setting (cf. sec. 11.9)

$$(\varphi, \psi) = \tfrac{1}{2} \operatorname{tr}(\varphi \circ \tilde{\psi}) \qquad \varphi, \psi \in Q.$$

Then we have

$$(\varphi, \varphi) = \tfrac{1}{2} \operatorname{tr}(\varphi \circ \tilde{\varphi}) = \tfrac{1}{2} \det \varphi \cdot \operatorname{tr} \iota = \det \varphi;$$

i.e.,

$$(\varphi, \varphi) = \det \varphi, \qquad \varphi \in Q \tag{11.35}$$

and

$$(\varphi, \iota) = \tfrac{1}{2} \operatorname{tr} \varphi \qquad \varphi \in Q. \tag{11.36}$$

Now we shall establish an isomorphism between the algebra Q and the algebra $\mathbb{H}$ of quaternions defined in sec. 7.23.

Consider the subspace Q_1 of Q consisting of those elements which satisfy $\operatorname{tr} \varphi = 0$ or equivalently $(\varphi, \iota) = 0$. Then Q_1 is a 3-dimensional Euclidean subspace of Q.

For $\varphi \in Q_1$ we have, in view of (11.33), $\tilde{\varphi} = -\varphi$ and thus (11.34) yields

$$\varphi^2 = -\det \varphi \cdot \iota = -(\varphi, \varphi)\,\iota \qquad \varphi \in Q_1.$$

This relation implies that

$$\tfrac{1}{2}(\varphi \circ \psi + \psi \circ \varphi) = -(\varphi, \psi) \cdot \iota \qquad \varphi, \psi \in Q_1. \tag{11.37}$$

Next, observe that for $\varphi \in Q_1$ and $\psi \in Q_1$

$$(\varphi \circ \psi - \psi \circ \varphi, \iota) = \tfrac{1}{2}(\operatorname{tr}(\varphi \circ \psi) - \operatorname{tr}(\psi \circ \varphi)) = 0$$

whence $\varphi \circ \psi - \psi \circ \varphi \in Q_1$. Thus a bilinear map, $\times$, is defined in Q_1 by

$$\varphi \times \psi = \tfrac{1}{2}(\varphi \circ \psi - \psi \circ \varphi). \tag{11.38}$$

It satisfies

$$(\varphi \times \psi, \varphi) = 0 \quad \text{and} \quad (\varphi \times \psi, \psi) = 0. \tag{11.39}$$

Moreover, equations (11.37) and (11.38) yield

$$\varphi \circ \psi = -(\varphi, \psi)\,\iota + \varphi \times \psi \qquad \varphi, \psi \in Q_1. \tag{11.40}$$

It follows that

$$|\varphi|^2 \cdot |\psi|^2 = (\varphi, \psi)^2 + |\varphi \times \psi|^2. \tag{11.41}$$

Finally, let Δ be the trilinear function in Q_1 given by

$$\Delta(\varphi, \psi, \chi) = (\varphi \circ \psi, \chi).$$

In view of (11.40) we can write

$$\Delta(\varphi, \psi, \chi) = (\varphi \times \psi, \chi). \tag{11.42}$$

This relation together with (11.39) implies that Δ is skew symmetric and hence a determinant function in Q_1. Moreover, setting $\chi = \varphi \times \psi$ yields, in view of (11.41),

$$\Delta(\varphi, \psi, \varphi \times \psi) = |\varphi|^2 \cdot |\psi|^2 - (\varphi, \psi)^2 \qquad \varphi, \psi \in Q_1.$$

Thus Δ is a *normed* determinant function in the Euclidean space Q_1. Hence it follows from (11.42) that the bilinear function $\times$ is the cross product in Q_1 if Q_1 is given the orientation determined by Δ (cf. sec. 7.16). Now equation (11.40) shows that Q is isomorphic to the algebra $\mathbb{H}$ of quaternions (cf. sec. 7.23).

Remark: In view of equation (11.34), Q is an associative division algebra of dimension 4 over $\mathbb{R}$. Therefore it must be isomorphic to $\mathbb{H}$ (cf. sec. 7.26). The above argument makes this isomorphism explicit.

11.15. The multiplication in C. Select a unit-vector a in C. Then to every vector $z \in C$ there exists a unique mapping $\varphi_z \in Q$ such that $\varphi_z a = z$. This mapping is determined by the equations

$$\begin{aligned} \varphi_z a &= \alpha a + \beta b \\ \varphi_z b &= -\bar{\beta} a + \bar{\alpha} b \end{aligned}$$

where b is a unit-vector orthogonal to a and

$$z = \alpha a + \beta b .$$

The correspondence $z \to \varphi_z$ obviously satisfies the relation

$$\varphi_{\lambda z_1 + \mu z_2} = \lambda \varphi_{z_1} + \mu \varphi_{z_2}$$

for any two real numbers λ and μ. Hence, it defines a linear mapping of C onto the linear space Q, if C is considered as a 4-dimensional real linear space. This suggests defining a multiplication among the vectors $z \in C$ in the following way:

$$z_1 z_2 = \varphi_{z_2} z_1 . \tag{11.43}$$

Then $\varphi_a = \iota$ and

$$\varphi_{z_1 z_2} = \varphi_{z_2} \circ \varphi_{z_1} \qquad z_1 z_2 \in C . \tag{11.44}$$

In fact, the two mappings $\varphi_{z_1 z_2}$ and $\varphi_{z_2} \circ \varphi_{z_1}$ are both contained in Q and send a into the same vector. Relation (11.44) shows that the correspondence $z \to \varphi_z$ preserves products. Consequently, the space C becomes a (real) division-algebra under the multiplication (11.43) and this algebra is isomorphic to the algebra of quaternions.

Equation (11.31) implies that

$$z + z^{-1} = 2(\varphi_z, \iota) a \tag{11.45}$$

for every unit-vector z.

In fact, if z is a unit-vector then φ_z is a unitary mapping with determinant 1 and thus (11.31) and (11.36) yield

$$z + z^{-1} = \varphi_z a + (\varphi_z a)^{-1} = \varphi_z a + \varphi_z^{-1} a = a \operatorname{tr} \varphi_z = 2a(\varphi_z, \iota) .$$

Finally, it will be shown that the inner products in C and Q are connected by the relation

$$\operatorname{Re}(z_1, z_2) = (\varphi_{z_1}, \varphi_{z_2}) . \tag{11.46}$$

To prove this we may again assume that z_1 and z_2 are unit-vectors. Then φ_{z_1} and φ_{z_2} are unitary mappings and we can write

$$(z_1, z_2) = (\varphi_{z_1} a, \varphi_{z_2} a) = (\varphi_{z_2}^{-1} \varphi_{z_1} a, a) = (\varphi_{z_1 z_2^{-1}} a, a). \qquad (11.47)$$

Since $\varphi_{z_1 z_2^{-1}}$ is also unitary formula (11.32) yields

$$\begin{aligned} \operatorname{Re}(\varphi_{z_1 z_2^{-1}} a, a) &= \tfrac{1}{2} \operatorname{tr} \varphi_{z_1 z_2^{-1}} = \tfrac{1}{2} \operatorname{tr}(\varphi_{z_2}^{-1} \circ \varphi_{z_1}) = \tfrac{1}{2} \operatorname{tr}(\widetilde{\varphi_{z_2}} \circ \varphi_{z_1}) \\ &= \tfrac{1}{2} \operatorname{tr}(\varphi_{z_1} \circ \widetilde{\varphi_{z_2}}) = (\varphi_{z_1}, \varphi_{z_2}). \end{aligned} \qquad (11.48)$$

Relations (11.47) and (11.48) imply (11.46).

Problems

1. Assume that an orthonormal basis is chosen in C. Prove that the transformations which correspond to the matrices

$$\begin{pmatrix} 1 & 0 \\ 0 & 1 \end{pmatrix} \quad \begin{pmatrix} i & 0 \\ 0 & -i \end{pmatrix} \quad \begin{pmatrix} 0 & -1 \\ 1 & 0 \end{pmatrix} \quad \begin{pmatrix} 0 & -i \\ -i & 0 \end{pmatrix}$$

form an orthonormal basis of Q.

2. Show that a transformation $\varphi \in Q$ is skew if and only if $\operatorname{tr} \varphi = 0$.

3. Prove that a transformation $\varphi \in Q$ satisfies the equation

$$\varphi^2 = -\iota$$

if and only if

$$\det \varphi = 1 \quad \text{and} \quad \operatorname{tr} \varphi = 0.$$

4. Verify the formula

$$(z_1 z, z_2 z) = (z_1, z_2)|z|^2 \qquad z_1, z_2, z \in C.$$

5. Let E be the underlying oriented 4-dimensional vector space of the complex plane C and define a positive definite inner product in E by

$$(x, y)_E = \operatorname{Re}(x, y).$$

Let a, b be an orthonormal basis of C such that a is the unit element for the multiplication defined in sec. 11.15. Show that the vectors

$$e = a, \quad e_1 = i a, \quad e_2 = b, \quad e_3 = i b$$

form an orthonormal basis of E and that, if Δ is the normed positive determinant function in E,

$$\Delta(e, e_1, e_2, e_3) = -1.$$

6. Let E_1 denote the orthogonal complement of e in E (cf. problem 5). Let ψ be a skew Hermitian transformation of C with $\operatorname{tr}\psi=0$.

a) Show that there is a unique vector $p\in E_1$ such that $\psi x=p\cdot x$, $x\in E$.

b) If

$$\begin{pmatrix} i\alpha & \beta+i\gamma \\ -\beta+i\gamma & -i\alpha \end{pmatrix}$$

is the matrix of ψ with respect to the orthonormal basis a, b in problem 5, show that

$$p=\alpha e_1+\beta e_2+\gamma e_3 .$$

§ 5.* Application to Lorentz-transformations

11.16. Selfadjoint linear transformations of the complex plane. Consider the set S of all selfadjoint mappings σ of the complex plane C. S is a real 4-dimensional linear space. In this space introduce a real inner product by

$$\langle\sigma,\tau\rangle=\tfrac{1}{2}(\operatorname{tr}(\sigma\circ\tau)-\operatorname{tr}\sigma\cdot\operatorname{tr}\tau). \tag{11.49}$$

This inner product is indefinite and has index 3. To prove this we note first that

$$\langle\sigma,\sigma\rangle=\tfrac{1}{2}(\operatorname{tr}\sigma^2-(\operatorname{tr}\sigma)^2)=-\det\sigma$$

and

$$\langle\sigma,\iota\rangle=-\tfrac{1}{2}\operatorname{tr}\sigma .$$

Now select an orthonormal basis z_1, z_2 of C and consider the transformations $\sigma_j (j=1,2,3)$ which correspond to the *Pauli-matrices*

$$\sigma_1:\begin{pmatrix}1&0\\0&-1\end{pmatrix}\quad \sigma_2:\begin{pmatrix}0&1\\1&0\end{pmatrix}\quad \sigma_3:\begin{pmatrix}0&-i\\i&0\end{pmatrix}.$$

Then it follows from (11.69) that

$$\langle\sigma_i,\sigma_j\rangle=\delta_{ij}$$

and

$$\langle\sigma_i,\iota\rangle=0,\quad \langle\iota,\iota\rangle=-1 .$$

These equations show that the mappings ι, σ_1, σ_2, σ_3 form an orthonormal basis of S with respect to the inner product (11.49) and that this inner product has index 3. Thus S becomes a Minkowski space (cf. sec. 9.27).

The orthogonal complement of the identity-map consists of all selfadjoint transformations with the trace zero.

11.17. The relation between the spaces Q and S. Recall from sec. 11.14 the definition of the 4-dimensional Euclidean space Q. Introduce a new inner product in Q by setting

$$\langle\varphi,\psi\rangle = -\tfrac{1}{2}\operatorname{tr}(\varphi\circ\psi) \qquad \varphi,\psi\in Q.$$

We shall define a linear isomorphism $\Omega: Q\xrightarrow{\cong} S$ such that

$$\langle\Omega\varphi,\Omega\psi\rangle=\langle\varphi,\psi\rangle \qquad \varphi,\psi\in Q.$$

In fact, set

$$\Omega\varphi = \frac{1-i}{2}\,\iota\operatorname{tr}\varphi + i\varphi \qquad \varphi\in Q.$$

Then

$$\widetilde{\Omega\varphi} - \Omega\varphi = i\cdot\iota\cdot\operatorname{tr}\varphi - i(\varphi+\tilde{\varphi}).$$

Since $\varphi\in Q$, we have $\varphi+\tilde{\varphi}=\iota\cdot\operatorname{tr}\varphi$ and so it follows that

$$\widetilde{\Omega\varphi} - \Omega\varphi = 0;$$

i.e., $\Omega\varphi$ is selfadjoint. The map $\Psi: Q\leftarrow S$ given by

$$\Psi(\sigma) = \frac{1+i}{2}\,\iota\cdot\operatorname{tr}\sigma - i\sigma \qquad \sigma\in S$$

is easily shown to be inverse to Ω and thus Ω is a linear isomorphism.

Finally, since

$$\Omega\varphi\circ\Omega\psi = \frac{(1-i)^2}{4}\,\iota\operatorname{tr}\varphi\cdot\operatorname{tr}\psi + \frac{1-i}{2}\,i(\psi\operatorname{tr}\varphi + \varphi\operatorname{tr}\psi) - \varphi\circ\psi$$

it follows that

$$\operatorname{tr}\cdot(\Omega(\varphi)\circ\Omega(\psi)) = \operatorname{tr}\varphi\cdot\operatorname{tr}\psi - \operatorname{tr}(\varphi\circ\psi)$$

whence

$$\begin{aligned}\langle\Omega\varphi,\Omega\psi\rangle &= \tfrac{1}{2}\{\operatorname{tr}((\Omega\varphi)\circ(\Omega\psi)) - \operatorname{tr}\Omega\varphi\cdot\operatorname{tr}\Omega\psi\}\\ &= \tfrac{1}{2}\{\operatorname{tr}\varphi\cdot\operatorname{tr}\psi - \operatorname{tr}\varphi\circ\psi - \operatorname{tr}\varphi\cdot\operatorname{tr}\psi\}\\ &= -\tfrac{1}{2}\operatorname{tr}(\varphi\circ\psi) = \langle\varphi,\psi\rangle.\end{aligned}$$

Note as well that, by definition,

$$\Omega(\psi\circ\varphi\circ\psi^{-1}) = \psi\circ\Omega\varphi\circ\psi^{-1} \qquad \varphi,\psi\in Q.$$

11.18. The transformations T_α. Now consider an arbitrary linear transformation α of the complex plane C such that $\det\alpha=1$. Then a linear transformation T_α: $S\to S$ is defined by

$$T_\alpha\sigma = \alpha\circ\sigma\circ\tilde{\alpha} \qquad \sigma\in S.$$

In fact, the equation

$$\widetilde{T_\alpha \sigma} = \alpha \circ \tilde{\sigma} \circ \tilde{\alpha} = \alpha \circ \sigma \circ \tilde{\alpha} = T_\alpha \sigma$$

shows that the mapping $T_\alpha \sigma$ is again selfadjoint. The transformation T_α preserves the inner product (11.49) and hence it is a Lorentz-transformation (cf. sec. 9.27):

$$\begin{aligned}\langle T_\alpha \sigma, T_\alpha \sigma \rangle &= -\det T_\alpha \sigma = -\det(\alpha \circ \sigma \circ \tilde{\alpha}) \\ &= -\det \sigma \, |\det \alpha|^2 = -\det \sigma = \langle \sigma, \sigma \rangle .\end{aligned}$$

Every Lorentz-transformation obtained in this way is proper. To prove this let $\alpha(t) (0 \leqq t \leqq 1)$ be a continuous family of linear transformations of C such that

$$\alpha(0) = \iota \qquad \alpha(1) = \alpha \quad \text{and} \quad \det \alpha(t) = 1 \qquad (0 \leqq t \leqq 1).$$

It follows from the result of sec. 4.36 that such a family exists. The continuous function

$$\det T_{\alpha(t)} \qquad (0 \leqq t \leqq 1)$$

is equal to ± 1 for every t. In particular

$$\det T_{\alpha(0)} = \det T_\iota = 1 .$$

This implies that

$$\det T_{\alpha(t)} = 1 \qquad (0 \leqq t \leqq 1)$$

whence

$$\det T_\alpha = 1 .$$

The transformations T_α are orthochronous. To prove this, observe that

$$T_\alpha \iota = \alpha \circ \tilde{\alpha}$$

whence

$$\langle \iota, T_\alpha \iota \rangle = \langle \iota, \alpha \circ \tilde{\alpha} \rangle = -\tfrac{1}{2} \operatorname{tr}(\alpha \circ \tilde{\alpha}) < 0 .$$

This relation shows that the time-like vectors ι and $T_\alpha \iota$ are contained in the same cone (cf. sec. (9.22)).

11.19. In this way every transformation α with determinant 1 defines a proper Lorentz-transformation T_α. Obviously,

$$T_{\alpha \circ \beta} = T_\alpha \circ T_\beta . \tag{11.50}$$

Now it will be shown that two transformations T_α and T_β coincide only if $\beta = \pm \alpha$. In view of (11.50) it is sufficient to prove that T_α is the identity operator only if $\alpha = \pm \iota$. If T_α is the identity, then

$$\alpha \circ \sigma \circ \tilde{\alpha} = \sigma \quad \text{for every} \quad \sigma \in S . \tag{11.51}$$

Inserting $\sigma=\iota$ we find that $\alpha\circ\tilde{\alpha}=\iota$ whence $\alpha=\tilde{\alpha}^{-1}$. Now equation (11.51) implies that

$$\alpha\circ\sigma=\sigma\circ\alpha \quad \text{for every} \quad \sigma\in S. \tag{11.52}$$

To show that $\alpha=\pm\iota$ select an arbitrary unit-vector $e\in C$ and define a selfadjoint mapping σ by

$$\sigma z=(z,e)e \qquad z\in C.$$

Then

$$(\sigma\circ\alpha)e=(\alpha e,e)e \quad \text{and} \quad (\alpha\circ\sigma)e=\alpha e.$$

Employing (11.52) we find that

$$\alpha e=(\alpha e,e)e.$$

In other words, every vector αz is a multiple of z. Now it follows from the linearity that $\alpha=\lambda\iota$ where λ is a complex constant. Observing that $\det\alpha=1$ we finally see that $\lambda=\pm 1$.

11.20. In this section it will be shown conversely that every proper orthochronous Lorentz-transformation T can be represented in the form T_α. Consider first the case that ι is invariant under T,

$$T\iota=\iota.$$

Employing the isomorphism $\Omega:Q\to S$ (cf. sec. 11.17) we introduce the transformation

$$T'=\Omega^{-1}\circ T\circ\Omega \tag{11.53}$$

of Q. Obviously,

$$\langle T'\varphi,T'\psi\rangle=\langle\varphi,\psi\rangle \qquad \varphi,\psi\in Q \tag{11.54}$$

and

$$T'\iota=\iota. \tag{11.55}$$

Besides the inner product of sec. 11.17 we have in Q the positive inner product defined in sec. 11.14. Comparing these two inner products we see that

$$(\varphi,\psi)=\langle\varphi,\psi\rangle+\tfrac{1}{2}\operatorname{tr}\varphi\operatorname{tr}\psi=\langle\varphi,\psi\rangle-2\langle\varphi,\iota\rangle\langle\psi,\iota\rangle. \tag{11.56}$$

Now formulae (11.54), (11.56) and (11.55) yield

$$(T'\varphi,T'\psi)=(\varphi,\psi) \qquad \varphi,\psi\in Q$$

showing that T' is also an isometry with respect to the positive definite inner product. Hence, by Proposition I, sec. 8.24, there exists a unit-vector $\beta\in Q$ such that

$$T'\varphi=\beta\circ\varphi\circ\beta^{-1}. \tag{11.57}$$

Using formulae (11.53), (11.57) and sec. 11.17 we thus obtain

$$T\sigma = (\Omega \circ T' \circ \Omega^{-1})\sigma = \Omega(\beta \circ \Omega^{-1}\sigma \circ \beta^{-1}) = \beta \circ \sigma \circ \beta^{-1}.$$

Since $\beta^{-1} = \tilde{\beta}$ (cf. sec. 11.14). This equation can be written in the form

$$T\sigma = \beta \circ \sigma \circ \tilde{\beta} = T_\beta \sigma.$$

Finally,

$$\det \beta = (\beta, \beta) = 1.$$

Now consider the case $T\iota \neq \iota$. Then, since T is a proper orthochronous Lorentz-transformation, $T\iota$ and ι are linearly independent. Consider the plane F generated by the vectors ι and $T\iota$. Let ω be a vector of F such that

$$\langle \iota, \omega \rangle = 0 \quad \text{and} \quad \langle \omega, \omega \rangle = 1. \tag{11.58}$$

In view of sec. 11.16 these conditions are equivalent to

$$\operatorname{tr} \omega = 0 \quad \text{and} \quad \det \omega = -1.$$

Therefore

$$\omega \circ \omega = \iota.$$

By hypothesis, T preserves fore-cone and past-cone. Hence $T\iota$ can be written as (cf. sec. 9.26)

$$T\iota = \iota \cosh \theta + \omega \sinh \theta. \tag{11.59}$$

Let α be the selfadjoint transformation defined by

$$\alpha = \iota \cosh \frac{\theta}{2} + \omega \sinh \frac{\theta}{2}. \tag{11.60}$$

Then

$$\begin{aligned} T_\alpha \iota = \alpha \circ \tilde{\alpha} &= \iota \cosh^2 \frac{\theta}{2} + 2\omega \cosh \frac{\theta}{2} \sinh \frac{\theta}{2} + \omega \circ \omega \sinh^2 \frac{\theta}{2} \\ &= \iota \cosh \theta + \omega \sinh \theta. \end{aligned} \tag{11.61}$$

Comparing (11.59) and (11.61) we see that

$$T\iota = T_\alpha \iota.$$

This equation shows that the transformation $T_\alpha^{-1} \circ T$ leaves vector ι invariant. As it has been shown already there exists a linear transformation $\beta \in Q$ of determinant 1 such that

$$T_\alpha^{-1} \circ T = T_\beta.$$

Hence,

$$T = T_\alpha \circ T_\beta = T_{\alpha \circ \beta}.$$

It remains to be proved that α has determinant 1. But this follows from the equation

$$\det \alpha = -\langle \alpha, \alpha \rangle = -\langle \iota, \iota \rangle \cosh^2 \frac{\theta}{2} - \langle \omega, \omega \rangle \sinh^2 \frac{\theta}{2} = 1.$$

Problems

1) Let α be the linear transformation of a complex plane defined by the matrix

$$\begin{pmatrix} 1 & 2i \\ -i & 3 \end{pmatrix}$$

Find the real 4×4 matrix which corresponds to the Lorentz-transformation T_α with respect to the basis $\iota, \sigma_1, \sigma_2, \sigma_3$ (cf. sec. 11.16).

2) A *fractional linear transformation* of the complex plane $\mathbb{C}$ is a transformation of the form

$$T(z) = \frac{az+b}{cz+d} \qquad ad - bc = 1$$

where a, b, c, d are complex numbers.

(i) Show that the fractional linear transformations form a group.

(ii) Show that this group is isomorphic to the group of proper orthochronous Lorentz transformations of a suitable Minkowski space.

Chapter XII

Polynomial Algebras

In this chapter Γ denotes a field of characteristic zero.

§ 1. Basic properties

In this paragraph we shall define the polynomial algebra over Γ and establish its elementary properties. Some of the work done here is simply a specialization of the more general results of Chapter VII, Volume II, and is included here so that the results of the following chapter will be accessible to the reader who has not seen the volume on multilinear algebra.

12.1. Polynomials over Γ. A *polynomial* over Γ is an infinite sequence

$$f = (\alpha_0, \alpha_1, \dots, \alpha_n, \dots) \qquad \alpha_\nu \in \Gamma$$

such that only finitely many α_ν are different from zero. The sum and the product of two polynomials

$$f = (\alpha_0, \alpha_1, \dots) \quad \text{and} \quad g = (\beta_0, \beta_1, \dots)$$

are defined by

$$f + g = (\alpha_0 + \beta_0, \alpha_1 + \beta_1, \dots)$$
$$fg = (\gamma_0, \gamma_1, \dots)$$

where

$$\gamma_k = \sum_{i+j=k} \alpha_i \beta_j .$$

These operations make the set of all polynomials over Γ into a commutative associative algebra. It is called the *polynomial algebra* over Γ and is denoted by $\Gamma[t]$. The unit element, 1, of $\Gamma[t]$ is the sequence $(1, 0 \dots)$. It is easy to check that the map $i: \Gamma \to \Gamma[t]$ given by

$$i(\alpha) = (\alpha, 0 \dots 0 \dots)$$

is an injective homomorphism of algebras.

Thus we may identify Γ with its image under i. The polynomial $(0, 1, 0 \dots)$ will be denoted by t

$$t = (0, 1, 0 \dots).$$

It follows that

$$t^k = (\underbrace{0, \ldots, 0}_{k}, 1, 0 \ldots) \qquad k = 0, 1 \ldots \qquad (t^\circ = 1)$$

Thus every polynomial f can be written in the form

$$f = \sum_{\nu=0}^{\infty} \alpha_\nu t^\nu \qquad \alpha_\nu \in \Gamma \tag{12.1}$$

where only finitely many α_ν are different from zero. Since the elements t^k ($k=0, 1, \ldots$) are linearly independent, the representation (12.1) is unique. Thus these elements form a basis of the vector space $\Gamma[t]$.

Now let

$$f = \sum_{\nu=0}^{n} \alpha_\nu t^\nu \qquad \alpha_n \neq 0$$

be a non-zero polynomial. Then α_n is called the *leading coefficient* of f. The leading coefficient of fg ($f \neq 0$, $g \neq 0$) is the product of the leading coefficients of f and g. This implies that $fg \neq 0$ whenever $f \neq 0$ and $g \neq 0$. A polynomial with leading coefficient 1 is called *monic*.

On the other hand, for every polynomial $f = \sum_\nu \alpha_\nu t^\nu$ the element $\alpha_0 \in \Gamma$ is called the *scalar term*. It is easily checked that the map $\varrho: \Gamma[t] \to \Gamma$ given by $f \mapsto \alpha_0$ is a surjective homomorphism.

Consider a non-zero polynomial

$$f = \sum_{k=0}^{n} \alpha_k t^k \qquad \alpha_n \neq 0 .$$

The number n is called the *degree* of f. If g is a second non zero polynomial, then clearly

$$\deg(f + g) \leqq \max(\deg f, \deg g)$$

and

$$\deg(f\, g) = \deg f + \deg g . \tag{12.2}$$

A polynomial of the form $\alpha_n t^n$ ($\alpha_n \neq 0$) is called a *monomial* of degree n.

Let $\Gamma_n[t]$ be the space of the monomials of degree n together with the zero element. Then clearly

$$\Gamma[t] = \sum_{n=0}^{\infty} \Gamma_n[t]$$

and by assigning the degree n to the elements of $\Gamma_n[t]$ we make $\Gamma[t]$ into a graded algebra as follows from (12.2). The homogeneous elements of degree n with respect to this gradation are precisely the monomials of degree n.

However, the structure of $\Gamma[t]$ as a graded algebra does not play a role in the subsequent theory. Consequently we shall consider simply its structure as an algebra.

12.2. The homomorphism $\Gamma[t]\to A$. Let A be any associative algebra with unit element e and choose a fixed element $a\in A$. Then the map

$$1\to e, t\to a$$

can be extended in a unique way to an algebra homomorphism

$$\Phi:\Gamma[t]\to A.$$

The uniqueness follows immediately from the fact that the elements 1 and t generate the algebra $\Gamma[t]$. To prove existence, simply set

$$\Phi\Big(\sum_k \alpha_k t^k\Big)=\sum_k \alpha_k a^k.$$

It follows easily that Φ is an algebra homomorphism. The image of $\Gamma[t]$ under Φ will be denoted by $\Gamma(a)$. It is clearly the subalgebra of A generated by e and a. Its elements are called *polynomials in a*, and are denoted by $f(a)$.

The homomorphism $\Phi: f\to f(a)$ induces a monomorphism

$$\bar{\Phi}:\Gamma[t]/\ker\Phi\to A.$$

In particular, if A is generated by e and a then Φ (and hence $\bar{\Phi}$) are surjective and so $\bar{\Phi}$ is an isomorphism

$$\bar{\Phi}:\Gamma[t]/\ker\Phi\xrightarrow{\cong} A$$

in this case.

Example: Let $A\in\Gamma[t]$ and let $g=\sum_\mu \beta_\mu t^\mu$ be a polynomial. Then we have

$$f(g)=\sum_\nu \alpha_\nu\Big(\sum_\mu \beta_\mu t^\mu\Big)^\nu.$$

In particular, the scalar term of $f(g)$ is given by

$$\gamma_0=\sum_\nu \alpha_\nu \beta_0^\nu=f(\beta_0).$$

Thus we can write

$$\varrho(f(g))=f(\varrho(g))\qquad f,g\in\Gamma[t]$$

where $\varrho:\Gamma[t]\to\Gamma$ is the homomorphism defined in sec. 12.1.

12.3. Differentiation. Consider the linear mapping

$$d:\Gamma[t]\to\Gamma[t]$$

defined by

$$d\,t^p = p\,t^{p-1} \qquad p\geqq 1$$
$$d1 = 0.$$

Then we have for $p,q\geqq 1$

$$\begin{aligned} d(t^p\cdot t^q) &= d(t^{p+q}) \\ &= (p+q)\,t^{p+q-1} \\ &= p\,t^{p-1}\,t^q + t^p\,q\,t^{q-1} \\ &= d\,t^p\cdot t^q + t^p\cdot d\,t^q \end{aligned}$$

i.e.

$$d(t^p\cdot t^q) = d\,t^p\cdot t^q + t^p\,d\,t^q. \tag{12.3}$$

It is easily checked that (12.3) continues to hold for $p=0$ or $q=0$. Since the polynomials t^p form a basis of $\Gamma[t]$ (cf. sec. 12.1) it follows from (12.3) that the mapping d is a derivation in the algebra $\Gamma[t]$.

d is called the *differentiation map* in $\Gamma[t]$, and is the unique derivation which maps t into 1. It follows from the definition of d that d lowers the degree of a polynomial by 1. In particular we have

$$\ker d = \Gamma$$

and so the relations

$$df = dg$$

and

$$f - g = \alpha, \qquad \alpha\in\Gamma$$

are equivalent. The polynomial df will be denoted by f'.

The *chain rule* states that for any two polynomials f and g,

$$(f(g))' = f'(g)\cdot g'.$$

For the proof we comment first that

$$\begin{aligned} d\,g^k &= k\,g^{k-1}\,d\,g \qquad k\geqq 1 \\ d\,g^0 &= 0 \end{aligned} \tag{12.4}$$

which follows easily from an induction argument. Now let

$$f = \sum_k \alpha_k t^k.$$

Then

$$f(g) = \sum_k \alpha_k g^k$$

and hence formula (12.4) yields

$$\begin{aligned}(f(g))' &= \sum_k \alpha_k k g^{k-1} \cdot dg \\ &= \sum_k k \alpha_k g^{k-1} \cdot g' \\ &= f'(g) \cdot g'.\end{aligned}$$

The polynomial $d^r f$ $(r \geqq 1)$ is called the r-th derivative of f and is usually denoted by $f^{(r)}$. We extend the notation to the case $r=0$ by setting $f^{(0)}=f$. It follows from the definition that $f^{(r)}=0$ if and only if r exceeds the degree of f.

12.4. Taylor's formula. Let A be an associative commutative algebra over Γ with unit element e. Recall from sec. 12.2 that if $f \in \Gamma[t]$ and $a \in A$ then an element $f(a) \in A$ is determined by

$$f(a) = \sum_{\nu=0}^{n} \alpha_\nu a^\nu.$$

Taylor's formula states that for $a \in A$, $b \in A$

$$f(a+b) = \sum_{p=0}^{n} \frac{f^{(p)}(a)}{p!} \cdot b^p. \tag{12.5}$$

Since the relation (12.5) is linear in f it is sufficient to consider the case $f=t^n$. Then we have by the binomial formula

$$f(a+b) = (a+b)^n = \sum_{p=0}^{n} \binom{n}{p} a^{n-p} \cdot b^p.$$

On the other hand,

$$f^{(p)}(t) = n(n-1)\dots(n-p+1)\, t^{n-p}$$

whence

$$f^{(p)}(a) = n(n-1)\dots(n-p+1)\, a^{n-p}$$

and so we obtain (12.5).

Problems

1. Consider the mapping $\Gamma[t] \times \Gamma[t] \to \Gamma[t]$ defined by $(f,g) \to f(g)$.

a) Show that this mapping does *not* make the space $\Gamma[t]$ into an algebra.

b) Show that the mapping is associative and has a left and right identity.

c) Show that the mapping is not commutative.

d) Prove that the mapping obeys the left cancellation law but not the right cancellation law; i.e., $f_1(g)=f_2(g)$ implies that $f_1=f_2$ but $f(g_1)=f(g_2)$ does not imply that $g_1=g_2$.

2. Construct a linear mapping
$$\int : \Gamma[t] \to \Gamma[t]$$
such that
$$d \circ \int = \iota .$$
Prove that if $\int_1$ and $\int_2$ are two such mappings, then there is a fixed scalar, α, such that
$$(\int_1 - \int_2) f = \alpha .$$
In particular, prove that there is a unique homogeneous linear mapping $\int$ of the graded space $\Gamma[t]$ into itself and calculate its degree. $\int$ is called the *integration operator*.

3. Consider the homomorphism $\varrho : \Gamma[t] \to \Gamma$ defined by
$$\varrho \sum_k \alpha_k t^k = \alpha_0$$
(cf. sec. 12.10). Show that
$$\varrho f = f(0) .$$
Prove that if $\int$ is the integration operator in $\Gamma[t]$, then
$$\int \circ d = \iota - \varrho .$$
Use this relation to obtain the formula for integration by parts:
$$\int f g' = f g - \varrho(f g) - \int g f' .$$

4. What is the Poincaré series for the graded space $\Gamma[t]$?

5. Show that if $\partial : \Gamma[t] \to \Gamma[t]$ is a non-trivial homogeneous antiderivation in the graded algebra $\Gamma[t]$ with respect to the canonical involution, and $\partial^2 = 0$ then
$$H_p(\Gamma[t]) = 0, \qquad p \geqq 1 .$$

6. Calculate Taylor's expansion
$$f(g+h) = f(g) + h f'(g) + \cdots$$
for the following cases and so verify that it holds in each case:

a) $f=t^2-t+1, \quad g=t^3+2t, \quad h=t-5$
b) $f=t^2+1, \quad g=t^3+t-1, \quad h=-t+1$
c) $f=3t^2+2t+5, \quad g=1, \quad h=-1$
d) $f=t^3-t^2+t-1, \quad g=t, \quad h=t^2-t+1$

7. For the polynomials in problem 6 verify the chain rule

$$[f(g)]' = f'(g) \cdot g'$$

explicitly. Express the polynomial $f(g(h))'$ in terms of the derivatives of f, g and h and calculate $[f(g(h))]'$ explicitly for the polynomials of problem 6.

§ 2. Ideals and divisibility

12.5. Ideals in $\Gamma[t]$. In this section it will be shown that every ideal in the algebra $\Gamma[t]$ is a principal ideal (cf. sec. 5.3). We first prove the following

Lemma I: (Euclid algorithm): Let $f \neq 0$ and $g \neq 0$ be two polynomials. Then there exist polynomials q and r such that

$$f = gq + r$$

and $\deg r < \deg g$ or $r = 0$.

Proof: Let $\deg f = n$ and $\deg g = m$. If $m > n$ we write

$$f = g \cdot 0 + f$$

and the lemma is proved. Now consider the case $n \geqq m$. Without loss of generality we may assume that f and g are monic polynomials. Then we have

$$f = t^{n-m} g + f_1, \quad \deg f_1 < n \quad \text{or} \quad f_1 = 0. \tag{12.6}$$

If $f_1 \neq 0$ assume (by induction on n) that the lemma holds for f_1. Then

$$f_1 = g q_1 + r_1 \tag{12.7}$$

where $\deg r_1 < \deg g$ or $r_1 = 0$. Combining (12.6) and (12.7) we obtain

$$f = (t^{n-m} + q_1) g + r_1$$

and so the lemma follows by induction.

Proposition 1: Every ideal in $\Gamma[t]$ is principal.

Proof: Let I be the given ideal. We may assume that $I \neq 0$. Let h be a monic polynomial of minimum degree in I. It will be shown that $I = I_h$ (cf. sec. 5.3). Clearly, $I_h \subset I$. Conversely, let $f \in I$ be an arbitrary polynomial. Then by the lemma,

$$f = hq + r$$

where

$$\deg r < \deg h \quad \text{or} \quad r = 0. \tag{12.8}$$

Since $f\in I$ and $h\in I$ we have

$$r=f-hq\in I$$

and hence if $r\neq 0$ $\deg r\geqq \deg h$. Now (12.8) implies that $r=0$; i.e. $f=hq$ and so $f\in I_h$.

The monic polynomial h is uniquely determined by I. In fact, assume that k is a second polynomial such that $I=I_k$. Then there are monic polynomials g_1 and g_2 such that $k=g_1h$ and $h=g_2k$. It follows that $k=g_1g_2k$ whence $g_1g_2=1$. Since g_1 and g_2 are monic we obtain $g_1=g_2=1$ whence $k=h$.

12.6. Ideals and divisors in $\Gamma[t]$. Let f and g be non-zero polynomials. We say that g *divides* f or f is a *multiple* of g if there is a polynomial h such that

$$f=g\cdot h.$$

In this case we write g/f. Clearly, f is a multiple of g if and only if it is contained in the ideal, I_g, generated by g. If h/g and g/f, then h/f. Two monic polynomials which divide each other are equal.

Next, let f and g be monic polynomials and consider the ideal I_f+I_g. In view of Proposition I, sec. 12.5, there is a unique monic polynomial, $f\vee g$, such that

$$I_{f\vee g}=I_f+I_g.$$

It is called the *greatest common divisor* of f and g. Since $I_f\subset I_{f\vee g}$ and $I_g\subset I_{f\vee g}$, $f\vee g$ is indeed a common divisor of f and g. On the other hand, if h/f and h/g, then $I_f\subset I_h$ and $I_g\subset I_h$ whence $I_{f\vee g}\subset I_h$ and so $h/f\vee g$. This shows that every common divisor of f and g is a divisor of $f\vee g$. A polynomial f whose only divisors are scalars and scalar multiples of f is called *irreducible* ar *prime*.

In a similar way the greatest common divisor of a finite number of monic polynomials f_i $(i=1\dots r)$ is defined. It is denoted by $f_1\vee\cdots\vee f_r$ and is characterized by the relation

$$I_{f_1\vee\cdots\vee f_r}=I_{f_1}+\cdots+I_{f_r}. \tag{12.9}$$

If

$$f_1\vee\cdots\vee f_r=1$$

the polynomials f_i are called *relatively prime*.

12.7. Again let f and g be monic polynomials and consider the ideal $I_f\cap I_g$. By Proposition I, sec. 12.5, there is a unique monic polynomial, $f\wedge g$, such that

$$I_{f\wedge g}=I_f\cap I_g.$$

It is called the *least common multiple* of f and g. It follows from the definition that $f \wedge g$ is indeed a common multiple of f and g and every common multiple of f and g is a multiple of $f \wedge g$.

If f_i $(i=1 \dots r)$ are monic polynomials their least common multiple, $f_1 \wedge \cdots \wedge f_r$, is defined by the equation

$$I_{f_1 \wedge \cdots \wedge f_r} = I_{f_1} \cap \cdots \cap I_{f_r}. \tag{12.10}$$

Proposition II: If f is the greatest common divisor of the f_i $(i=1 \dots r)$ then there are polynomials g_i such that

$$f = \sum_{i=1}^{r} f_i g_i. \tag{12.11}$$

Proof: It follows from (12.9) that every element $h \in I_f$ can be written as

$$h = \sum_{i=1}^{r} f_i g_i.$$

Since $f \in I_f$, f must be of the form (12.11).

Corollary: If the polynomials f_i are relatively prime there exist polynomials g_i $(i=1 \dots r)$ such that

$$\sum_{i=1}^{r} f_i g_i = 1. \tag{12.12}$$

Conversely if there exists a relation of the form (12.12) then the f_i are relatively prime.

Proof: The first part follows immediately from the proposition. Now assume that there is a relation of the form (12.12). Then every common divisor of the f_i divides 1 and hence is a scalar.

Proposition III. Let f be the greatest common divisor of the monic polynomials f_1, f_2 and write

$$f_i = f h_i \qquad (i = 1, 2).$$

Then the polynomials h_1, h_2 are relatively prime and the least common multiple of the polynomials f_1, f_2 is given by $f \cdot h_1 \cdot h_2$.

Proof: In view of Proposition II we can write

$$f = \sum_i f_i g_i. \tag{12.13}$$

It follows that

$$f = \sum_i f h_i g_i$$

whence

$$\sum_i h_i g_i = 1.$$

This shows that h_1 and h_2 are relatively prime.

Clearly, the polynomial $f h_1 h_2$ is a common multiple of f_1 and f_2. Now let g be any common multiple of f_1 and f_2 and write

$$g = f_1 p_1, \qquad g = f_2 p_2.$$

Then (12.13) yields

$$\begin{aligned} f p_1 &= g_1 f_1 p_1 + g_2 f_2 p_1 = g_1 g + g_2 f_2 p_1 = f_2 (g_1 p_2 + g_2 p_1) \\ &= f h_2 (g_1 p_2 + g_2 p_1). \end{aligned}$$

This implies that

$$p_1 = h_2 (g_1 p_2 + g_2 p_1)$$

whence

$$g = f_1 p_1 = f h_1 h_2 (g_1 p_2 + g_2 p_1);$$

i.e., g is a multiple of $f h_1 h_2$.

Corollary: If the monic polynomials f_i $(i=1 \dots r)$ are relatively prime, then

$$f_1 \wedge \cdots \wedge f_r = f_1 \dots f_r.$$

Proof: If $r=2$ this follows immediately from the proposition. If $r>2$ a simple induction argument is required.

12.8. The lattice of ideals in $\Gamma[t]$. Let $\mathscr{I}$ denote the set of all ideals in $\Gamma[t]$. Recall from sec. 5.9 that $\mathscr{I}$ is a lattice with respect to the partial order given by inclusion.

On the other hand, let $\mathscr{P}$ denote the set of all monic polynomials in $\Gamma[t]$ together with the zero polynomial. Define a partial order in $\mathscr{P}$ by setting

$$f \leqq g \quad \text{if } g/f \qquad (f \neq 0,\ g \neq 0);$$

$$g \leqq 0 \quad \text{for every } g \in \Gamma[t].$$

Then, in view of sec. 12.6 and 12.7, $\mathscr{P}$ becomes a lattice as well. Now let $\Phi: \mathscr{P} \to \mathscr{I}$ be the map given by $\Phi: f \mapsto I_f$. Then $f \leqq g$ implies that $I_f \leqq I_g$ and so Φ is a lattice homomorphism. Moreover, since Φ is bijective, it is a lattice isomorphism (cf. Proposition I, sec. 12.5).

12.9. The decomposition of a polynomial into prime factors.

Theorem I: Every monic polynomial can be written

$$f = f_1^{k_1} \dots f_r^{k_r} \tag{12.14}$$

where the f_i are distinct irreducible monic polynomials and $\deg f_i \geqq 1$. The decomposition is unique up to the ordering of the prime factors.

Proof: The existence of the decomposition (12.14) is proved by induction on the degree of f. If $\deg f = 0$ then $f = 1$ and the decomposition is trivial. Suppose that the decomposition (12.14) exists for polynomials of degree $< n$ and let f be of degree n. Then either f is irreducible in which case we have nothing to prove; or f is a product

$$f = gh \qquad \deg g \geqq 1, \deg h \geqq 1 .$$

Since $\deg g < \deg f$ and $\deg h < \deg f$ it follows by induction that

$$g = g_1^{i_1} \dots g_s^{i_s}$$

and

$$h = h_1^{j_1} \dots h_t^{j_t}$$

whence

$$f = g_1^{i_1} \dots g_s^{i_s} h_1^{j_1} \dots h_t^{j_t} .$$

Collecting the powers of the same prime polynomials we obtain the decomposition (12.14).

The uniqueness part follows (with the aid of a similar induction argument) from

Lemma II: Let f, g, h be monic polynomials and assume that h is irreducible. Let m be a positive integer. Then h^m divides fg if and only if there is an integer p $(1 \leqq p \leqq m)$ such that

$$h^p/f \quad \text{and} \quad h^{m-p}/g .$$

Proof. The "if" part of the statement is trivial. Now suppose that h^m/fg. Let $p \geqq 0$ be the largest integer such that h^p/f. If $p = m$ there is nothing to prove. If $p < m$, write

$$f = h^p f_1 .$$

Then

$$fg = h^p f_1 g .$$

On the other hand, by hypothesis,

$$fg = h^m k$$

for some polynomial k.

These relations yield

$$f_1 g = h^{m-p} k. \tag{12.15}$$

By the definition of p, f_1 is not divisible by h. Since h is irreducible, it follows that h and f_1 are relatively prime. Thus there are polynomials u and v such that (cf. the corollary to Proposition II, sec. 12.7)

$$u h + v f_1 = 1.$$

Set

$$g_1 = u g + v h^{m-p-1} k.$$

Then we have

$$h g_1 = h u g + v h^{m-p} k = h u g + f_1 v g = (h u + f_1 v) g = g;$$

i.e., $h g_1 = g$.

Now equation (12.15) implies that

$$f_1 g_1 = h^{m-p-1} k.$$

Continuing this process we see that h^{m-p} divides g. This establishes the lemma and so the proof of Theorem I is complete.

Corollary. The monic polynomials which divide the polynomial

$$f = f_1^{k_1} \dots f_r^{k_r}$$

are precisely the polynomials

$$g = f_1^{j_1} \dots f_r^{j_r} \qquad j_\nu \le k_\nu \quad (\nu = 1, \dots, r).$$

Now let (12.14) be the decomposition of the monic polynomial f and set

$$q_i = f_1^{k_1} \dots \hat{f}_i^{k_i} \dots f_r^{k_r} \quad (i = 1 \dots r).$$

It will be shown that the q_i are relatively prime and that for every i, f is the least common multiple of q_i and the polynomial $\bigvee_{j \neq i} q_j$,

$$\bigvee_{i=1}^{r} q_i = 1 \tag{12.16}$$

$$q_i \wedge (\bigvee_{j \neq i} q_j) = f. \tag{12.17}$$

Let g be a monic polynomial which divides q_i. Then Theorem I implies that g has the form

$$g = f_1^{j_1} \dots \hat{f}_i^{k_i} \dots f_r^{j_r} \qquad j_\nu \le k_\nu.$$

Hence, if g divides all polynomials q_i, it follows that $g = 1$, whence (12.16).

A similar argument shows that $\bigvee_{j \neq i} q_j = f_i^{k_i}$ and now formula (12.17) follows from the relation

$$q_i \wedge \left(\bigvee_{j \neq i} q_j\right) = q_i\, f_i^{k_i} = f$$

and the fact that q_i and $f_i^{k_i}$ are relatively prime.

Proposition IV: Suppose g is a product of relatively prime irreducible polynomials and suppose f is a polynomial such that g/f^m for some $m \geqq 1$. Then g/f.

Proof: Let

$$f = f_1^{k_1} \dots f_s^{k_s}$$

be the decomposition of f into its prime factors. Then the corollary of Theorem I implies that g is of the form

$$g = f_1^{j_1} \dots f_s^{j_s}.$$

Since, by hypothesis, g is a product of relatively prime irreducible polynomials it follows that $j_\nu \leqq 1$ $(\nu = 1 \dots s)$ whence

$$g/f_1 \dots f_s$$

and so g/f.

Proposition V: A polynomial f is the product of relatively prime irreducible polynomials if and only if f and f' are relatively prime.

Proof: Let

$$f = f_1^{k_1} \dots f_r^{k_r}$$

be the decomposition of f into prime factors. Suppose that $k_i > 1$ for some i $(1 \leqq i \leqq r)$.

Then writing

$$f = h\, f_i^2 \qquad h \in \Gamma[t]$$

we obtain

$$f' = h' f_i^2 + 2h f_i f_i' = f_i(h' f_i + 2h f_i')$$

and so f_i divides f and f'. Consequently, f and f' are not relatively prime.

Conversely, assume that $k_i = 1$ $(i = 1 \dots r)$. Then, if f and f' have a common factor, the corollary to Theorem I, implies that, for some i, f_i divides f'.

Since

$$f' = \sum_{j=1}^{r} f_1 \dots f_j' \dots f_r = \sum_{j \neq i} f_1 \dots f_j' \dots f_r + f_1 \dots f_i' \dots f_r$$

we obtain

$$f_i / f_1 \cdots f_i' \cdots f_r .$$

Since f_i is irreducible and the polynomials $f_1, \ldots, f_r$ are relatively prime, it follows that f_i / f_i'. But this is impossible since $\deg f_i' < \deg f_i$. This contradiction shows that f and f' are relatively prime.

12.10. Polynomial functions. Let $(\Gamma; \Gamma)$ be the space of all set maps $\Gamma \to \Gamma$ furnished with the linear structure defined in sec. 1.2, Example 3. Then every polynomial $f = \sum_{i=0}^{r} \alpha_i t^i$ determines an element $\tilde{f}$ of $(\Gamma; \Gamma)$ defined by

$$\tilde{f}(\xi) = \sum_{i=1}^{n} \alpha_i \xi^i = f(\xi), \qquad \xi \in \Gamma \ (\xi^0 = 1).$$

The functions $\tilde{f}$ are called *polynomial functions.*

If $f(\lambda) = 0$ for some $\lambda \in \Gamma$, then λ is called a *root* of f. λ is a root of f if and only if $t - \lambda$ divides f. In fact, if

$$f = (t - \lambda) g$$

it follows that $f(\lambda) = 0$. Conversely, if $t - \lambda$ does not divide f, then $t - \lambda$ and f are relatively prime and hence there exist polynomials q and s such that

$$f q + (t - \lambda) s = 1 .$$

This implies that

$$f(\lambda) q(\lambda) = 1$$

whence $f(\lambda) \neq 0$. It follows from the above remark that a polynomial of degree n has at most n roots.

Proposition VI: The mapping $f \to \tilde{f}$ is injective.

Proof: Suppose $\tilde{f} = 0$. Then $\tilde{f}(\xi) = 0$ for every $\xi \in \Gamma$. Since Γ has characteristic zero it contains infinitely many elements and hence it follows that $f = 0$.

In view of the above proposition we may denote the polynomial function $\tilde{f}$ simply by f.

Problems

1. Let f be a polynomial such that $f'(0) \neq 0$. Consider two polynomials g_1 and g_2 such that $g_1 \neq g_2$ and $f(g_1) = f(g_2)$. Prove that g_1 and g_2 are relatively prime.

2. Consider the set of all pairs (f,g) where $g \neq 0$. Define an equivalence relation in this set by

$$(f,g) \sim (\tilde{f},\tilde{g}) \quad \text{if and only if} \quad f\tilde{g} = \tilde{f}g.$$

Show that this is indeed an equivalence relation. Denote the equivalence classes by $\overline{(f,g)}$. Prove that the operations

$$\overline{(f_1,g_1)} + \overline{(f_2,g_2)} = \overline{(f_1 g_2 + f_2 g_1, g_1 g_2)}$$

and

$$\overline{(f_1,g_1)}\,\overline{(f_2,g_2)} = \overline{(f_1 f_2, g_1 g_2)}$$

are well defined.

Show that with these operations the set of equivalence classes becomes a field, denoted by $\mathbb{Q}_\Gamma[t]$.

Prove that the mapping

$$f \to \overline{(f,1)}$$

is a monomorphism of the algebra $\Gamma[t]$ into the algebra $\mathbb{Q}_\Gamma[t]$.

3. Extend the derivation d to a derivation in $\mathbb{Q}_\Gamma[t]$ and show that this extension is unique. Show that the integration operator $\int$ (cf. problem 2, § 1) cannot be extended to $\mathbb{Q}_\Gamma[t]$.

4. Show that any ideal in $\Gamma[t]$ is contained in only finitely many ideals.

5. Consider the mapping $\mathbb{R}[t] \times \mathbb{R}[t] \to \mathbb{R}[t]$ defined by

$$\left(\sum_\nu \alpha_\nu t^\nu, \sum_\nu \beta_\nu t^\nu\right) \to \sum_\nu \alpha_\nu \beta_\nu .$$

Show that this mapping makes $\mathbb{R}[t]$ into an inner product space. Prove that the induced topology makes $\mathbb{R}[t]$ into a topological algebra (addition, scalar multiplication, multiplication and division are continuous).

6. Let $\mathbb{R}[t]$ have the inner product of problem 5. Let I be any ideal. Calculate I explicitly. Under what conditions do either of the equations

$$\mathbb{R}[t] = I \oplus I^\perp$$
$$(I^\perp)^\perp = I$$

hold? Show that

$$((I^\perp)^\perp)^\perp = I^\perp.$$

7. Let f and g be any two non-zero polynomials and assume that $\deg f \geqq \deg g$. Write

$$f = p_1 g + g_1$$

where $g_1 = 0$ or $\deg g_1 < \deg g$. Prove that the greatest common divisor of f and g coincides with the greatest common divisor of g and g_1, unless

g divides f. If $g_1 \neq 0$ write

$$g = p_2 g_1 + g_2, \quad g_2 = 0 \quad \text{or} \quad \deg g_2 < \deg g_1 .$$

Show that the repeated application of this process yields an explicit calculation of the greatest common divisor of f and g. (This method is called the *Euclidean algorithm*).

8. Calculate the greatest common divisors and the least common multiples of the following polynomials over $\mathbb{R}[t]$:

a) $t^5 + t^4 + t^3 + t^2 + t + 1, \quad t^6 - 1$
b) $t^3 + 3t^2 + 1, \quad t^4 - t + 7, \quad 7t^2 + 16$
c) $5t^4 - \frac{1}{6}t^3 + t^2 - 3t + 7, \quad 4t^4 - 17t^3 + 16, \quad \frac{1}{10}t^5 - \frac{1}{8}t^2 + 2t + 1$
d) $8t^8 + \sqrt{6}t^4 - \sqrt{2}t^2 - 72 + 2\sqrt{6}, \quad 2t^8 + t^5 + 5t^4 - 6t^2 - 3t - 15$
e) $3t^4 + 50t^3 - 9t^2 + 84t + 5, \quad t^4 + 15t^3 - 29t^2 - 64t + 4$.

9. If f, g are two polynomials and d is their greatest common divisor use the Euclidean algorithm to construct polynomials r, s such that

$$f\,r + g\,s = d .$$

10. Construct the polynomials r, s explicitly for the polynomials of problem 8, (in parts b) and c) it will be necessary to construct three polynomials).

11. Decide whether the following polynomials are products of relatively prime irreducible polynomials:
a) $t^7 - t^5 + t^3 - t$;
b) $t^4 + 2t^3 + 2t^2 + \frac{1}{4}$;
c) the polynomials of problem 6, §1;
d) the polynomials of problem 8.

12. Let f, g_1, g_2 be non-zero polynomials such that $g_1 \neq g_2$. Show that $g_1 - g_2$ divides $f(g_1) - f(g_2)$.

§ 3. Factor algebras

12.11. Minimum polynomial. Let A be a finite dimensional associative commutative algebra with unit element e. Fix an element $a \in A$ and consider the homomorphism $\Phi: \Gamma[t] \to A$ given by

$$\Phi(f) = f(a) \qquad f \in \Gamma[t]$$

(cf. sec. 12.2). Its image is the subalgebra of A generated by e and a. It will be denoted by $\Gamma(a)$.

The kernel, K of Φ is an ideal in $\Gamma[t]$. Since A has finite dimension, $K \neq 0$. By Proposition I, (Sec. 12.5) there is a unique monic polynomial, μ, such that $K = I_\mu$. μ is called the *minimum polynomial* of a.

If $A \neq 0$, μ must have positive degree. To see this assume that $\mu = 1$. Then $I_\mu = \Gamma[t]$ and so $\Phi = 0$. It follows that $e = \Phi(1) = 0$ whence $A = 0$.

The homomorphism Φ factors over the canonical projection to induce an isomorphism

$$\Psi: \Gamma[t]/I_\mu \xrightarrow{\cong} \Gamma(a)$$

such that the diagram

$$\begin{array}{ccc} \Gamma[t] & \xrightarrow{\Phi} & \Gamma(a) \\ \pi\downarrow & \nearrow_{\Psi} \cong & \\ \Gamma[t]/I_\mu & & \end{array}$$

commutes.

Example I: If $a = 0$, then K consists of all polynomials whose scalar term is zero and so $\mu = t$.

Example II: If $a = e$, then K consists of all polynomials $f = \sum_\nu \alpha_\nu t^\nu$ satisfying $\sum_\nu \alpha_\nu = 0$. In this case we have $\mu = t - 1$.

Example III: Set $A = \Gamma[t]/I_h$ (where h is a fixed monic polynomial) and $a = \bar{t}$ where $\bar{t} = \pi(t)$. Then, for every polynomial $f = \sum_\nu \alpha_\nu t^\nu$,

$$\Phi(f) = \sum_\nu \alpha_\nu \pi(t^\nu) = \sum_\nu \alpha_\nu \pi(t)^\nu = \sum_\nu \alpha_\nu \bar{t}^\nu = \pi(f)$$

and so Φ coincides with the canonical projection. It follows that the minimum polynomial of $\bar{t}$ is the polynomial h.

Proposition I: The dimension of $\Gamma(a)$ is equal to the degree of the minimum polynomial of a.

Proof: Let r denote the degree of μ. Then it is easily checked that the elements $1, \bar{t}, \ldots, \bar{t}^{r-1}$ form a basis of $\Gamma[t]/I_\mu$. Thus we have, in view of the isomorphism Ψ, $\dim \Gamma(a) = \dim \Gamma[t]/I_\mu = r$.

Proposition II: Every ideal in $\Gamma(a)$ is principal.

Proof: Let I_A be an ideal in $\Gamma(a)$ and set $I = \Phi^{-1}(I_A)$. Then I is an ideal in $\Gamma[t]$. Thus, by Proposition I, sec. 12.5, there is an element $f \in \Gamma[t]$ such that $I = I_f$. It follows that

$$I_A = \Phi(I) = \Phi(I_f) = I_{f(a)}.$$

12.12. Nilpotent elements in $\Gamma(a)$. Suppose that $f(a)$ is a nilpotent element of $\Gamma(a)$. Then, for some $m \geqq 1$, $f(a)^m = 0$. It follows that

$$\Phi(f^m) = (\Phi(f))^m = f(a)^m = 0$$

and so the minimum polynomial, μ, of a divides f^m.

Now decompose μ into its prime factors

$$\mu = f_1^{k_1} \dots f_r^{k_r}$$

and set

$$g = f_1 \dots f_r.$$

Then we have $g/\mu/f^m$. Now Proposition IV, sec. 12.9, implies that g divides f.

Conversely, assume that g/f. Set $k = \max(k_1, \dots, k_r)$. Then μ/g^k and so we have

$$\mu/g^k/f^k.$$

It follows that $f(a)^k = 0$ and so $f(a)$ is nilpotent.

Thus $f(a)$ is nilpotent if and only if g divides f. In particular, if all the exponents in the decomposition of μ are 1, then $\mu = g$ and so $f(a)$ is nilpotent if and only if μ/f; i.e., if and only if $f(a) = 0$. Hence, in this case there are no non-zero nilpotent elements in $\Gamma(a)$.

12.13. Factor algebras of an irreducible polynomial.

Theorem I: Let f be a polynomial. Then the factor algebra $\Gamma(\bar{t})$ is a field if and only if f is irreducible.

Proof: Suppose $f = gh$, where $\deg g \geqq 1$ and $\deg h \geqq 1$. Then $g \notin I_f$, $h \notin I_f$ and so $\bar{g} \neq 0$, $\bar{h} \neq 0$. On the other hand, $\bar{g} \cdot \bar{h} = \overline{gh} = \bar{f} = 0$ and so $\Gamma(\bar{t})$ has zero divisors. Consequently, it is not a field.

Conversely, suppose f is irreducible. $\Gamma(\bar{t})$ is an associative commutative algebra with identity $\bar{1}$. To prove that $\Gamma(\bar{t})$ is a field we need only show that every non-zero element $\bar{g}$ has a multiplicative inverse. Let $g \in \bar{g}$ be any representing polynomial. Then since $\bar{g} \neq 0$, it follows that g is not divisible by f. Since f is irreducible, f and g are relatively prime and so there exist polynomials h and k such that

$$gh + fk = 1$$

whence

$$\bar{g}\bar{h} + \bar{f}\bar{k} = \bar{1}.$$

But $\bar{f} = 0$ and so

$$\bar{g}\bar{h} = \bar{1}.$$

Hence $\bar{h}$ is an inverse of $\bar{g}$.

Corollary: If f is irreducible, then $\Gamma(\bar{t})$ is an extension field of Γ.

Proof: Consider the homomorphism $\varphi:\Gamma\to\Gamma(\bar{t})$ defined by

$$\varphi:\alpha\to\bar{\alpha}.$$

It is clear that φ is a monomorphism and so the corollary follows.

Problems

1. Consider the irreducible polynomial $f=t^2+5t+1$ as an element of $\mathbb{Q}[t]$ (where $\mathbb{Q}$ is the field of rational numbers). Let

$$\pi: Q[t]\to Q[t]/I_f$$

be the canonical projection.

a) Decide whether the polynomials of problem 6, § 1, problem 8, § 2 (except for part d) and problems 11a) and b), § 2 are in the kernel of π.

b) For each polynomial p of part a) such that $\pi p\neq 0$ construct a polynomial $g\in\mathbb{Q}[t]$ such that

$$\pi q=(\pi p)^{-1}.$$

2. Let $f\in\Gamma[t]$ be any polynomial. Consider an arbitrary element $x\in\Gamma[t]/I_f$. Prove that the minimum polynomial of x has degree $\leqq \deg f$.

3. Suppose $f\in\Gamma[t]$ is an irreducible polynomial, and consider the polynomial algebra $\Gamma[t]/I_f[t]$ denoted by $\Gamma_f[t]$.

a) Show that $\Gamma[t]$ may be identified in a natural way with a subalgebra of $\Gamma_f[t]$.

b) Prove, that if two polynomials in $\Gamma[t]$ are relatively prime, then they are relatively prime when considered as polynomials in $\Gamma_f[t]$.

c) Construct an example to prove that an irreducible polynomial in $\Gamma[t]$ is not necessarily irreducible in $\Gamma_f[t]$.

d) Suppose that f has degree 2, and that g is an irreducible polynomial of degree 3 in $\Gamma[t]$. Prove that g is irreducible in $\Gamma_f[t]$.

§ 4.* The structure of factor algebras

In this paragraph f will denote a fixed monic polynomial, and $\Gamma(\bar{t})$ will denote the factor algebra $\Gamma[t]/I_f$.

12.14. The lattice of ideals in $\Gamma(\bar{t})$. Consider the set of all monic polynomials which divide f. These polynomials form a sublattice $\mathscr{P}_f$ of $\mathscr{P}$ (cf. sec. 12.8). In fact, if $f_1\dots f_r$ is any finite set in $\mathscr{P}_f$ then the greatest common divisor and the least common multiple of the f_i is again contained in $\mathscr{P}_f$. f is a lower bound and 1 is an upper bound of $\mathscr{P}_f$.

On the other hand, consider the lattice $\mathscr{I}_f$ of ideals in the algebra $\Gamma(\bar{t})=\Gamma[t]/I_f$. The remarks of sec. 12.11 establish a bijection

$$\Phi:\mathscr{P}_f \to \mathscr{I}_f$$

defined by

$$\Phi g = I_{\bar{g}}$$

where $I_{\bar{g}}$ denotes the ideal in $\Gamma(\bar{t})$ generated by $\bar{g}$. The reader can easily check that Φ and Φ^{-1} are order preserving and so Φ is a lattice isomorphism; i.e.,

$$\Phi(\bigvee_i f_i) = \sum_i I_{\bar{f}_i}$$

and

$$\Phi(\bigwedge_i f_i) = \bigcap_i I_{\bar{f}_i}.$$

In particular,

$$\Phi(1) = \Gamma(\bar{t})$$

and

$$\Phi(f) = 0.$$

12.15. Decomposition of $\Gamma(\bar{t})$ into irreducible ideals. Let $f=f_1 \dots f_m$ and let I_j denote the ideal in $\Gamma(\bar{t})$ generated by $\bar{f}_j$. Consider the ideal

$$I = \sum_j I_j. \tag{12.18}$$

Proposition I: $I=\Gamma(\bar{t})$ if and only if the polynomials f_j are relatively prime. The sum (12.18) is direct if and only if, for each j, the polynomials f_j and $\bigvee_{i \neq j} f_i$ have f as least common multiple.

Proof: To prove the first part of the proposition we notice that

$$\Gamma(\bar{t}) = \sum_i I_i$$

is equivalent to

$$\Phi(1) = \Phi(\bigvee_i f_i)$$

which in turn is equivalent to

$$1 = \bigvee_i f_i.$$

But according to sec. 12.6, this holds if and only if the f_i are relatively prime.

For the second part we observe that the sum is direct if and only if

$$I_j \cap \sum_{i \neq j} I_i = 0 \qquad (j = 1 \ldots m).$$

Since

$$I_j \cap \sum_{i \neq j} I_i = \Phi(f_j \wedge (\bigvee_{i \neq j} f_i))$$

this is equivalent to

$$f_j \wedge (\bigvee_{i \neq j} f_i) = f \qquad (j = 1 \ldots m).$$

Theorem I: Let

$$f = f_1^{k_1} \ldots f_r^{k_r}$$

be the decomposition of f into prime polynomials and let the polynomials q_i be defined by

$$q_i = f_1^{k_1} \ldots \hat{f}_i^{k_i} \ldots f_r^{k_2}.$$

Then

$$\Gamma(\bar{t}) = \sum_{i=1}^{r} I_i \tag{12.19}$$

where I_i denotes the ideal generated by $\bar{q}_i$. Moreover let

$$\bar{1} = \bar{e}_1 + \cdots + \bar{e}_r, \qquad \bar{e}_i \in I_i \tag{12.20}$$

and

$$\bar{t} = \bar{t}_1 + \cdots + \bar{t}_r, \qquad \bar{t}_i \in I_i \tag{12.21}$$

be the decompositions determined by (12.19). Then $\bar{e}_i$ is an identity in I_i and for every $\bar{q} \in \Gamma(\bar{t})$

$$\bar{q} = \sum_{i=1}^{r} q(\bar{t}_i). \tag{12.22}$$

Finally, if I is any ideal in $\Gamma(\bar{t})$, then

$$I = \sum_{i=1}^{r} I \cap I_i. \tag{12.23}$$

Proof: The relation (12.19) is an immediate consequence of Proposition I, and formulae (12.16) and (12.17). To show that $\bar{e}_i$ is an identity in I_i let $\bar{q} \in I_i$ be arbitrary. Then

$$\bar{q} = \bar{1}\,\bar{q} = \sum_j \bar{e}_j \bar{q}.$$

Since for $j \neq i$

$$\bar{e}_j \bar{q} \in I_j \cap I_i = 0$$

it follows that

$$\bar{e}_i \bar{q} = \bar{q}.$$

Now let $\bar{q}$ be an arbitrary element of $\Gamma(\bar{t})$ and let $q \in \bar{q}$ be any representative. Writing

$$q = \sum_k \alpha_k t^k$$

we obtain

$$\bar{q} = \alpha_0 (\bar{e}_1 + \cdots + \bar{e}_r) + \sum_{k=1}^{m} \alpha_k (\bar{t}_1 + \cdots + \bar{t}_r)^k.$$

But

$$\bar{t}_i \bar{t}_j \in I_i \cap I_j = 0$$

and so

$$(\bar{t}_1 + \cdots + \bar{t}_r)^k = \bar{t}_1^k + \cdots + \bar{t}_r^k.$$

It follows that

$$\bar{q} = \sum_{k=0}^{r} \alpha_k (\bar{t}_1^k + \cdots + \bar{t}_r^k) = \sum_{i=1}^{r} q(\bar{t}_i) \qquad (\bar{t}_i^0 = \bar{e}_i).$$

Finally, let I be any ideal in $\Gamma(\bar{t})$. Then clearly

$$I \cap I_i = \bar{e}_i I$$

and so (12.23) is an immediate consequence of (12.20).

Theorem II. With the notation of Theorem I let

$$\varphi_i : \Gamma[t] \to I_i$$

be the homomorphism defined by

$$\varphi_i(1) = \bar{e}_i; \ \varphi_i(t) = \bar{t}_i.$$

Then φ_i is an epimorphism and $\ker \varphi_i = I_{f_i^{k_i}}$.
Thus φ_i induces an isomorphism

$$\Gamma[t]/I_{f_i^{k_i}} \xrightarrow{\cong} I_i. \tag{12.24}$$

In particular, the minimum polynomial of $\bar{t}_i$ is $f_i^{k_i}$.

Proof: (12.22) shows that φ_i is an epimorphism. Next we prove that $I_{f_i^{k_i}} = \ker \varphi_i$. We have

$$\sum_i f(\bar{t}_i) = f(\bar{t}) = 0.$$

Since $f(\bar{t}_i) \in I_i$ and the sum (12.19) is direct it follows that

$$f(\bar{t}_i) = 0 \qquad (i = 1 \cdots r);$$

i.e., $\varphi_i(f)=0$. Now consider the induced map

$$\bar{\varphi}: \Gamma[t]/I_f \to I_i.$$

Then

$$\bar{\varphi}_i(\bar{q}) = q(\bar{t}_i) \qquad q \in \Gamma[t]$$

and, in view of (12.22),

$$\ker \bar{\varphi}_i = \sum_{j \neq i} I_i.$$

But $\sum_{j \neq i} I_j$ is the ideal generated by $f_i^{k_i}$ and thus

$$\ker \varphi_i = I_{f_i^{k_i}}.$$

This completes the proof.

Corollary I: An element $\bar{q} \in \Gamma(\bar{t})$ is contained in I_i if and only if

$$q(\bar{t}_i) = \bar{q} \quad \text{and} \quad q(\bar{t}_j) = 0 \qquad j \neq i.$$

Theorem III: The ideals I_i are irreducible and (12.19) is the unique decomposition of $\Gamma(\bar{t})$ into a direct sum of irreducible ideals.

Proof: Let $f_i^{k_i} = g_i$. In view of the isomorphism (12.24) it is sufficient to prove that the algebra $\Gamma[t]/I_{g_i}$ is irreducible. According to Proposition I, $\Gamma[t]/I_{g_i}$ is the direct sum of two ideals I_1 and I_2 only if

$$I_1 = I_{\bar{q}_1} \quad \text{and} \quad I_2 = I_{\bar{q}_2}$$

where q_1 and q_2 are relatively prime divisors of g_i. But this can only happen if either $q_1 = 1$ or $q_2 = 1$. If $q_1 = 1$, say, then I_1 is the full algebra and so $I_2 = 0$. It follows that $\Gamma(t)/I_{g_i}$ is irreducible.

Now suppose that I is an irreducible ideal in $\Gamma(\bar{t})$. Then Theorem I above gives

$$I = \sum_{i=1}^{r} I \cap I_i.$$

Since I is irreducible, it follows that $I \subset I_i$ for some i. Thus if

$$\Gamma(\bar{t}) = I \oplus J$$

for some ideal J, then intersection with I_i gives

$$I_i = I \oplus (J \cap I_i).$$

Since I_i is irreducible, it follows that $J \cap I_i = 0$, whence $I_i = I$. This completes the proof.

Corollary I: The irreducible factor algebras $\Gamma(\bar{f})$ are precisely those for which f is a power of an irreducible polynomial.

Corollary II: Suppose the ideal I is a direct summand in $\Gamma(\bar{f})$. Then I is a direct sum of the I_i.

Proof: Let J be an ideal such that $I \oplus J = \Gamma(\bar{f})$. It is obvious that in a finite dimensional algebra any ideal is a direct sum of irreducible ideals. Since I and J are ideals, $I \cdot J = 0$, and so any ideal in $I(J)$ is an ideal in $\Gamma(\bar{f})$. Now the result follows from Theorem III.

12.16. Semisimple elements. Let $A \neq 0$ be a finite dimensional associative commutative algebra with unit element e and let $a \in A$. Recall from sec. 12.11 that the minimum polynomial, f, of a has positive degree. The element a is called *semisimple*, if f is the product of relatively prime irreducible polynomials.

Lemma I: a is semisimple if and only if the element $f'(a)$ is invertible in the algebra $\Gamma(a)$.

Proof: If a is semisimple, then the polynomials f and f' are relatively prime (cf. Proposition V, sec. 12.9). Thus, by Corollary I to Proposition II, sec. 12.7, there are polynomials p and q such that

$$pf + qf' = 1.$$

Since $f(a) = 0$ it follows that $q(a)f'(a) = e$ and so $f'(a)$ is invertible. Conversely, assume that $f'(a)$ is invertible in $\Gamma(a)$. Then there is a polynomial g such that $f'(a)g(a) = e$. Set

$$h = f'g - 1.$$

Then

$$h(a) = f'(a)g(a) - e = 0.$$

Since f is the minimum polynomial of a it follows that f/h. Thus we can write

$$f'g - 1 = f \cdot q$$

where q is some polynomial. It follows that f and f' are relatively prime.

Lemma II: Assume that a is semisimple and let h be a polynomial such that $h(a)^m = 0$ for some $m \geqq 1$. Then $h(a) = 0$.

Proof: Let f be the minimum polynomial of a. Then the hypothesis implies that f/h^m. Since a is semisimple, f is a product of relatively prime irreducible polynomials and it follows that f/h (cf. Proposition IV, sec. 12.9). Thus $h(a) = 0$.

Theorem IV: Let A be a finite dimensional associative commutative algebra. Let $x \in A$ and let

$$f = f_1^{k_1} \cdots f_r^{k_r}$$

be the decomposition of the minimum polynomial of x into prime factors. Set

$$g = f_1 \ldots f_r$$

and

$$k = \max k_i \qquad (i = 1 \ldots r).$$

Then there are unique elements $x_S \in A$ and $x_N \in A$ such that x_S is semisimple, x_N is nilpotent and

$$x = x_S + x_N.$$

The minimum polynomials of x_S and x_N are given by

$$\mu_S = g \quad \text{and} \quad \mu_N = t^k.$$

Proof: 1. Existence: Identify the subalgebra $\Gamma(x)$ with the factor algebra $\Gamma(\bar{t}) = \Gamma[t]/I_f$ via $x = \bar{t}$. Lemma IV below (cf. sec. 12.17) yields elements $u \in \Gamma[t]$ and $\omega \in \Gamma[t]$ with the following properties:

(i) $g^k / g(u)$,
(ii) $u + \omega = t$,
(iii) g divides ω.

Projection onto the quotient algebra yields the relations

$$g(\bar{u}) = 0 \tag{12.25}$$

$$\bar{\omega}^k = 0, \tag{12.26}$$

and

$$\bar{u} + \bar{\omega} = \bar{t}. \tag{12.27}$$

Now set

$$x_S = \bar{u} \quad \text{and} \quad x_N = \bar{\omega}.$$

Then $g(x_S) = 0$.

Thus the minimum polynomial of x_S divides g and hence it is a product of relatively prime irreducible polynomials. Thus, by definition, x_S is semisimple. Relation (12.26) shows that x_N is nilpotent and from (12.27) we obtain the relation

$$x_N + x_S = x.$$

Finally, x_S and x_N are polynomials in x.

2. *Minimum polynomials:* Next we show that g is the minimum polynomial of x_S. Let h be the minimum polynomial of x_S. Then h divides g. On the other hand, Taylor's expansion gives

$$h(x) = h(x_S + x_N) = h(x_S) + q \cdot x_N = q \cdot x_N$$

where $q \in A$ is some element. This shows that $h(x)$ is nilpotent. Now Sec. 12.12 implies that g/h. Thus $g = h$.

Now consider the minimum polynomial, μ_N, of x_N. Since g/ω we have g^k/ω^k. Thus $f/g^k/\omega^k$. It follows that

$$x_N^k = \bar{\omega}^k = 0.$$

Thus $\mu_N = t^l$ with $l \leqq k$.

To show that $k \leqq l$ we may assume that $k \geqq 2$. Then $\omega = g \cdot v$ and v is relatively prime to g (cf. Lemma III and IV, sec. 12.17). Hence v is relatively prime to f. It follows that $v(x)$ is invertible. Now

$$x_N = \bar{\omega} = \bar{g} \cdot \bar{v} = g(x) v(x)$$

and so

$$g(x)^l \cdot v(x)^l = x_N^l = \mu_N(x) = 0.$$

Since $v(x)$ is invertible we obtain $g(x)^l = 0$ whence $g^l(x) = 0$. This implies that f/g^l whence $k \leqq l$. Thus $k = l$; i.e., $\mu_N = t^k$.

3. *Uniqueness:* Assume a decomposition

$$x = y_S + y_N$$

where y_S is semisimple and y_N is nilpotent. Set $y_S - x_S = z$. Then

$$z = x_N - y_N$$

and so z is nilpotent. We must show that $z = 0$. Taylor expansion yields

$$g(y_S) = g(x_S + z) = g(x_S) + \sum_{j=1}^{\infty} \frac{g^{(j)}(x_S)}{j!} z^j = \sum_{j=0}^{\infty} \frac{g^{(j)}(x_S)}{j!} z^j$$

showing that $g(y_S)$ is nilpotent.

Hence, for some m,

$$(g(y_S))^m = 0.$$

Since y_S is semisimple it follows that $g(y_S) = 0$ (cf. Lemma II) sec. 12.16 and so

$$\sum_{j=1}^{\infty} \frac{g^{(j)}(x_S)}{j!} z^j = 0. \tag{12.28}$$

Let $l \geqq 1$ denote the degree of nilpotency of z. We show that $l=1$. In fact, assume that $l \geqq 2$. Then, multiplying the equation above by z^{l-2} we obtain

$$g'(x_S)\, z^{l-1} = 0.$$

In view of Lemma I (applied with $A = \Gamma(x_S)$), $g'(x_S)$ is invertible in $\Gamma(x_S)$ and hence invertible in A. It follows that $z^{l-1}=0$, in contradiction to the choice of l. Thus $l=1$; i.e., $z=0$. Thus $y_S = x_S$ and $y_N = x_N$ and the proof is complete.

Definition: The elements x_N and x_S are called the *semisimple* and *nilpotent parts of* x.

12.17. *Lemma III:* Let g be a polynomial such that g and g' are relatively prime. Then for each integer $k \geqq 1$ there are polynomials u and v with the following properties:

(i) $g^k/g(u)$,

(ii) $u + g v = t$.

(iii) If $k \geqq 2$, then v is relatively prime to g.

Proof: For $k=1$ set $u=t$ and $v=0$. Next consider the case $k=2$. Since g and g' are relatively prime there are polynomials p and q such that

$$1 + pg' = qg.$$

The Taylor expansion (cf. sec. 12.4) gives, in view of the relation above,

$$\begin{aligned} g(t+pg) &= \sum_{i=0}^{\infty} \frac{g^{(i)}}{i!}(pg)^i = g + g'pg + \cdots + \frac{g^{(n)}}{n!}(pg)^n + \cdots \\ &= g(1+g'p) + g^2 \cdot l \qquad (12.29) \\ &= g^2 q + g^2 l = g^2(q+l). \end{aligned}$$

Now set

$$u = t + gp \quad \text{and} \quad v = -p.$$

Then we have

$$u + gv = t \qquad (12.30)$$

and, in view of (12.29),

$$g(u) = g^2(q+l)$$

which achieves the result for $k=2$.

Finally, suppose the proposition holds for $k-1$ ($k \geqq 3$) and define u_k by

$$u_k = u_2(u_{k-1}).$$

Replacing t by u_{k-1} in (12.30) we obtain

$$u_2(u_{k-1}) + g(u_{k-1})\,v_2(u_{k-1}) = u_{k-1}. \tag{12.31}$$

By induction hypothesis

$$g^{k-1}/g(u_{k-1})$$

whence $g/g(u_{k-1})$. Thus we can write

$$g(u_{k-1}) = g \cdot q$$

where q is some polynomial. Now (12.31) can be written as

$$u_k + g \cdot q \cdot v_2(u_{k-1}) = u_{k-1}. \tag{12.32}$$

Finally, (ii) yields in view of the induction hypothesis

$$u_{k-1} = t - g\,v_{k-1}. \tag{12.33}$$

Relations (12.32) and (12.33) imply that

$$u_k + g \cdot (q \cdot v_2(u_{k-1}) + v_{k-1}) = t.$$

Setting $v_k = q\,v_2(u_{k-1}) + v_{k-1}$ we obtain

$$u_k + g\,v_k = t.$$

It remains to be shown that g^k divides $g(u_k)$. But

$$g(u_k) = g(u_2(u_{k-1})) = (g(u_2))(u_{k-1}).$$

Now the lemma, applied for $k=2$, shows that $g(u_{k-1})^2/g(u_k)$. Since $g^{k-1}/g(u_{k-1})$ it follows that $g^k/g(u_k)$ and so the induction is closed.

Thus property (i) and (ii) are established.

(iii): Let $k \geqq 2$. Suppose that h is a common divisor of g and v,

$$g = p \cdot h, \qquad v = q \cdot h.$$

Then, by (ii)

$$g(u) = g(t - v\,g) = g(t - p\,q\,h^2)$$

and so Taylor's expansion yields

$$g(u) = g - h^2 \cdot l$$

where l is some polynomial.

Since $k \geqq 2$, (i) implies that $g^2/g(u)$ and so $h^2/g(u)$. Now the equation above shows that h^2 divides g

$$g = h^2 \cdot m.$$

It follows that

$$g' = 2hh'm + h^2 m'$$

and so h is a common divisor of g and g'. Since g and g' are relatively prime, h must be a scalar. Hence g and v are relatively prime. This completes the proof.

Lemma IV: Let g be a polynomial such that g and g' are relatively prime. Then for each $k \geqq 1$ there are polynomials u and ω with the following properties:

(*i*) *g^k divides $g(u)$*,

(*ii*) $u+\omega=t$,

(*iii*) *g divides ω*.

Proof: Define ω by $\omega=g \cdot v$ where v is the polynomial obtained in Lemma III.

12.18. Decomposition of $\Gamma(\bar{t})$ into the radical and a direct sum of fields. Let

$$f = f_1^{k_1} \dots f_r^{k_r}$$

be the decomposition of f into its prime factors and set

$$g = f_1 \dots f_r .$$

Consider the factor algebra $\Gamma(\bar{t})=\Gamma[t]/I_f$.

By Theorem IV there is a unique decomposition

$$\bar{t} = \bar{t}_S + \bar{t}_N$$

where $\bar{t}_S$ is semisimple and $\bar{t}_N$ is nilpotent. Moreover, g is the minimum polynomial of $\bar{t}_S$.

Now let A be the subalgebra of $\Gamma(\bar{t})$ generated by $\bar{1}$ and $\bar{t}_S$. Let I_i be the ideal in $\Gamma[\bar{t}]$ generated by $\bar{f}_i$ $(i=1 \dots r)$.

Theorem V: 1. A is the (unique) direct sum of irreducible ideals I_i $(i=1 \dots r)$,

$$A = I_1 \oplus \dots \oplus I_r$$

and

$$I_i \cong \Gamma[t]/I_{f_i} \qquad (i = 1 \dots r).$$

In particular, each I_i is a field and so A contains no non-zero nilpotent elements.

2. The radical of $\Gamma(\bar{t})$ consists precisely of the nilpotent elements in $\Gamma(\bar{t})$, and is generated as an ideal by $\bar{t}_N$.

3. The vector space $\Gamma(\bar{t})$ is the direct sum of the subalgebra A and the ideal rad $\Gamma(\bar{t})$,

$$\Gamma(\bar{t}) = A \oplus \operatorname{rad} \Gamma(\bar{t}).$$

4. The subalgebra A consists precisely of the semisimple elements in $\Gamma(\bar{t})$.

5. A is the only subalgebra of $\Gamma(\bar{t})$ which complements the radical.

Proof: *1*) Consider the surjective homomorphism $\Gamma(t) \to A$ given by $t \mapsto \bar{t}_S$. Since g is the minimum polynomial of $\bar{t}_S$, it induces an isomorphism

$$\Gamma[t]/I_g \xrightarrow{\cong} A.$$

According to Theorem III (applied to g) $\Gamma[t]/I_g$ is the unique direct sum of irreducible ideals I_i where $I_i \cong \Gamma[t]/I_{f_i}$ $(i = 1 \dots r)$. Let I_i also denote the corresponding ideal in A. Then

$$A = I_1 \oplus \cdots \oplus I_r$$

and the I_i are irreducible ideals.

Finally, since each f_i is irreducible, Theorem I, sec. 12.13, implies that $\Gamma[t]/I_{f_i}$ is a field.

2) *and 3*): Let I_N denote the ideal generated by $\bar{t}_N$ and let I be the ideal of nilpotent elements in $\Gamma(\bar{t})$. Since $\bar{t}_N$ is nilpotent we have

$$I_N \subset I \subset \operatorname{rad} \Gamma(\bar{t}). \tag{12.34}$$

Next, since $\bar{t} = \bar{t}_S + \bar{t}_N$, it follows that

$$\Gamma(\bar{t}) = A + I_N. \tag{12.35}$$

Moreover, since A is the direct sum of the fields I_i,

$$\operatorname{rad} \Gamma(\bar{t}) \cap A \subset \operatorname{rad} A = \sum_{j=1}^{r} \operatorname{rad} I_i = 0. \tag{12.36}$$

Relations (12.34) and (12.36) imply that the decomposition (12.35) is direct,

$$\Gamma(\bar{t}) = A \oplus I_N.$$

This yields

$$\operatorname{rad} \Gamma(\bar{t}) = \operatorname{rad} \Gamma(\bar{t}) \cap A + I_N = I_N$$

whence

$$I_N = I = \operatorname{rad} \Gamma(\bar{t}).$$

It follows that

$$\Gamma(\bar{t}) = A \oplus \operatorname{rad} \Gamma(\bar{t}). \tag{12.37}$$

4) We show first that every element in A is semisimple. In fact, let $\bar{x} \in A$. Then, by Theorem IV, sec. 12.16, there are elements $\bar{x}_S \in \Gamma(\bar{x})$, $\bar{x}_N \in \Gamma(\bar{x})$ such that $\bar{x}_S$ is semisimple, $\bar{x}_N$ is nilpotent and

$$\bar{x} = \bar{x}_S + \bar{x}_N.$$

In particular, $\bar{x}_N$ is nilpotent in A. Thus, by 1), $\bar{x}_N = 0$ and so $\bar{x}$ is semisimple.

Conversely, let $\bar{x} \in \Gamma(\bar{t})$ be semisimple. In view of 3) we can write

$$\bar{x} = \bar{x}_A + \bar{x}_R \qquad \bar{x}_A \in A, \; \bar{x}_R \in \operatorname{rad} \Gamma(\bar{t}). \tag{12.38}$$

Then $\bar{x}_A$ is semisimple and $\bar{x}_R$ is nilpotent. Thus (12.38) must be the unique decomposition of $\bar{x}$ into its semisimple and nilpotent parts. Since $\bar{x}$ is semisimple it follows that $\bar{x}_R = 0$ whence $\bar{x} \in A$.

5) Let B be any subalgebra of $\Gamma(\bar{t})$ which complements the radical. Then dividing out by the radical we obtain an algebra isomorphism $A \xrightarrow{\cong} B$. Thus, by 4), the elements of B are semisimple. Hence, again by 4), $B \subset A$. It follows that $B = A$.

Corollary: Let $x \in \Gamma(\bar{t})$. Then the decomposition

$$x = x_N + x_R$$

obtained from the decomposition 3) is the decomposition of x into its semisimple and nilpotent parts.

The results of this paragraph yield at once

Theorem VI: Let

$$\Gamma(\bar{t}) = I_1 \oplus \cdots \oplus I_r \tag{12.39}$$

be the decomposition of the algebra $\Gamma(\bar{t})$ into irreducible ideals. Then every ideal I_i is a direct sum

$$I_i = A_i \oplus \operatorname{rad} I_i \tag{12.40}$$

where A_i is a field isomorphic to $\Gamma[t]/I_{f_i}$ $(i = 1 \ldots r)$. Moreover, the decompositions (12.39) and (12.40) are connected by

$$\operatorname{rad} \Gamma(\bar{t}) = \sum_{i=1}^{r} \operatorname{rad} I_i$$

and

$$A = \sum_{i=1}^{r} A_i.$$

Problems

1. Consider the Polynomials

a) $f = t^3 - 6t^2 + 11t - 16$,

b) $f = t^2 + t + 7$,

c) $f = t^2 - 5$.

Prove that in each case f is the product of relatively prime irreducible polynomials. For $k = 2, 3$, construct polynomials u and v which satisfy the conditions of Lemma III, sec. 12.17.

2. Show that if $\tilde{u}$ and $\tilde{v}$ are any two polynomials satisfying the conditions of Lemma III (for some fixed k), then g^k divides $\tilde{u} - u$ and g^{k-1} divides $\tilde{v} - v$.

Chapter XIII

Theory of a linear transformation

In this chapter E will denote a finite-dimensional non-trivial vector space defined over an arbitrary field Γ of characteristic 0, and $\varphi: E \to E$ will denote a linear transformation.

§ 1. Polynomials in a linear transformation

13.1. Minimum polynomial of a linear transformation. Consider the algebra $A(E; E)$ of linear transformations and fix an element $\varphi \in A(E; E)$. Then a homomorphism $\Phi: \Gamma[t] \to A(E; E)$ is defined by

$$\Phi: f \mapsto f(\varphi)$$

(cf. sec. 12.11). Let μ be the minimum polynomial of φ. Since $A(E; E)$ is non-trivial and has finite dimension, it follows that $\deg \mu \geqq 1$ (cf. sec. 12.11). The minimum polynomial of the zero transformation is t whereas the minimum polynomial of the identity map is $t-1$.

Proposition I, sec. 12.11, shows that

$$\dim \Gamma(\varphi) = \deg \mu$$

where

$$\Gamma(\varphi) = \operatorname{Im} \Phi.$$

13.2. The space K(f). Fix a polynomial f and denote by $K(f)$ the kernel of the linear transformation $f(\varphi)$. Then the subspace $K(f) \subset E$ is stable under φ. In fact, if $x \in K(f)$, then $f(\varphi)x = 0$ and so

$$f(\varphi)\varphi x = \varphi f(\varphi) x = 0;$$

i.e.,

$$\varphi x \in K(f).$$

In particular we have

$$K(1) = 0, \quad K(t) = \ker \varphi \quad \text{and} \quad K(\mu) = E,$$

where μ denotes the minimum polynomial of φ.

Let $F \subset E$ be any subspace stable under φ and let $\varphi_F: F \to F$ be the induced linear transformation. Then $\mu(\varphi_F)=0$ and so μ_F/μ where μ_F denotes the minimum polynomial of φ_F.

Now let g be a second polynomial and assume that g/f. Then

$$K(g) \subset K(f). \tag{13.1}$$

In fact, writing $f = gg_1$ we obtain that for every vector $x \in K(g)$

$$f(\varphi)x = g_1(\varphi)g(\varphi)x = 0$$

whence $x \in K(f)$. This proves (13.1).

Proposition I: Let f and g be any two non-zero polynomials, and let d be their greatest common divisor. Then

$$K(d) = K(f) \cap K(g).$$

Proof: Since d/f and d/g it follows that

$$K(d) \subset K(f) \quad \text{and} \quad K(d) \subset K(g)$$

whence

$$K(d) \subset K(f) \cap K(g). \tag{13.2}$$

On the other hand, since d is the greatest common divisor of f and g, there exist polynomials f_1 and g_1 such that

$$d = f_1 f + g_1 g.$$

Thus if $x \in K(f) \cap K(g)$ we have

$$d(\varphi)x = f_1(\varphi)f(\varphi)x + g_1(\varphi)g(\varphi)x = 0$$

and hence $x \in K(d)$. It follows that

$$K(d) \supset K(f) \cap K(g) \tag{13.3}$$

which, together with (13.2) establishes the proposition.

Corollary I: Let f be any polynomial and let d be the greatest common divisor of f and μ. Then

$$K(f) = K(d).$$

Proof: Since $K(\mu) = E$, it follows from the proposition that

$$K(d) = K(f) \cap E = K(f).$$

Proposition II: Let f and g be any two non-zero polynomials, and let v be their least common multiple. Then

$$K(v) = K(f) + K(g). \tag{13.4}$$

Proof: Since $f|v$ and $g|v$ it follows that $K(f) \subset K(v)$ and $K(g) \subset K(v)$; whence

$$K(v) \supset K(f) + K(g). \tag{13.5}$$

On the other hand, since v is the least common multiple of f and g, there are polynomials f_1 and g_1 such that

$$f_1 f = v = g_1 g$$

and f_1 and g_1 are relatively prime. Choose polynomials f_2 and g_2 so that $f_2 f_1 + g_2 g_1 = 1$; then

$$f_2(\varphi) f_1(\varphi) + g_2(\varphi) g_1(\varphi) = \iota.$$

Consequently each $x \in E$ can be written as

$$x = x_1 + x_2$$

where

$$x_1 = f_2(\varphi) f_1(\varphi) x \quad \text{and} \quad x_2 = g_2(\varphi) g_1(\varphi) x. \tag{13.6}$$

Now suppose that $x \in K(v)$. Then

$$f_1(\varphi) f(\varphi) x = g_1(\varphi) g(\varphi) x = v(\varphi) x = 0$$

and so (13.6) implies that

$$f(\varphi) x_1 = 0 = g(\varphi) x_2.$$

Hence $x_1 \in K(f)$, and $x_2 \in K(g)$, so that

$$x \in K(f) + K(g)$$

that is,

$$K(v) \subset K(f) + K(g). \tag{13.7}$$

(13.4) follows from (13.5) and (13.7).

Corollary I: If f and g are relatively prime, then

$$K(fg) = K(f) \oplus K(g). \tag{13.8}$$

Proof: Since f and g are relatively prime, their least common multiple is fg and it follows from Proposition II that

$$K(fg) = K(f) + K(g). \tag{13.9}$$

On the other hand the greatest common divisor of f and g is 1, and so Proposition I yields that

$$K(f) \cap K(g) = K(1) = 0. \tag{13.10}$$

Now (13.9) and (13.10) imply (13.8).

Corollary II: Suppose

$$f = f_1 \dots f_r$$

is a decomposition of the polynomial f into relatively prime factors. Then

$$K(f) = K(f_1) \oplus \dots \oplus K(f_r).$$

Proof: This is an immediate consequence of Corollary I with the aid of an induction argument on r.

Let f and g be any two non-zero polynomials such that g is a proper divisor of f. Then $K(g) \subset K(f)$ but the inclusion need not be proper. In fact, let $g = \mu$ and $f = h\mu$ where h is any polynomial with $\deg h \geqq 1$. Then $g | f$ (properly) but $K(g) = E = K(f)$.

Proposition III: Let f and g be non-zero polynomials such that

(i) $$f | \mu$$

and

(ii) $$g | f \quad \text{(properly)}$$

Then

$$K(g) \subset K(f) \qquad \text{(properly)}.$$

Proof: (i) and (ii) imply that there are polynomials f_1 and g_1 such that

$$\mu = f f_1 \quad \text{and} \quad f = g g_1 \qquad \deg g_1 > 0. \tag{13.11}$$

Setting $g_2 = g f_1$ we have $\deg g_2 < \deg \mu$ and so μ is not a divisor of g_2. It follows that g_2 cannot annihilate all the vectors of E; i.e., there is a vector $x \in E$ such that

$$g_2(\varphi) x \neq 0. \tag{13.12}$$

Let $y = f_1(\varphi)x$. Then we obtain from (13.11) and (13.12) that

$$f(\varphi) y = f(\varphi) f_1(\varphi) x = \mu(\varphi) x = 0$$

while

$$g(\varphi) y = g(\varphi) f_1(\varphi) x = g_2(\varphi) x \neq 0.$$

Thus $y \in K(f)$, but $y \notin K(g)$ and so $K(g)$ is a proper subspace of $K(f)$.

Corollary I: Let f be a non-zero polynomial. Then

$$K(f) = 0 \tag{13.13}$$

if and only if f and μ are relatively prime.

Proof: If f and μ are relatively prime, then Corollary I to Proposition II gives

$$K(f) = K(f) \cap E = K(f) \cap K(\mu) = 0 .$$

Conversely suppose (13.13) holds, and let d be the greatest common divisor of f and μ. Then

$$1/d \quad \text{and} \quad d/\mu$$

but

$$K(d) = K(f) \cap K(\mu) = 0 = K(1) .$$

It follows from Proposition III that 1 cannot be a proper divisor of d; hence $d = 1$ and f and μ are relatively prime.

Corollary II: Let f be any non-zero monic polynomial that divides μ, and let φ_f denote the restriction of φ to $K(f)$. Let μ_f denote the minimum polynomial of φ_f. Then

$$\mu_f = f . \tag{13.14}$$

Proof: We have from the definitions that $f(\varphi_f) = 0$, and hence μ_f / f. It follows that $K(\mu_f) \subset K(f)$ and since $K(\mu_f) \supset K(f)$ we obtain

$$K(\mu_f) = K(f) .$$

On the other hand, $f | \mu$ and $\mu_f | f$. Now Proposition III implies that μ_f cannot be a proper divisor of f, which yields (13.14).

Proposition IV: Let

$$\mu = f_1 \dots f_r$$

be a decomposition of μ into relatively prime factors. Then

$$E = K(f_1) \oplus \cdots \oplus K(f_r) .$$

Moreover, if φ_i denotes the restriction of φ to $K(f_i)$ and μ_i is the minimum polynomial of φ_i, then

$$\mu_i = f_i .$$

Proof: The proposition is an immediate consequence of Corollary II to Proposition II and Corollary II to Proposition III.

13.3. Eigenvalues. Recall that an eigenvalue of φ is a scalar $\lambda \in \Gamma$ such that

$$\varphi x = \lambda x \tag{13.15}$$

for some non-zero vector $x \in E$, and that x is called an eigenvector corresponding to the eigenvalue λ. (13.15) is clearly equivalent to

$$K(f) \neq 0 \tag{13.16}$$

where f is the polynomial $f = t - \lambda$.

In view of Corollary I to Proposition III, (13.16) is equivalent to requiring that f and μ have a non-scalar common divisor. Since $\deg f = 1$, this is the same as requiring that $f \mid \mu$. Thus the eigenvalues of φ are precisely the distinct roots of μ.

Now consider the *characteristic polynomial* of φ,

$$\chi = \sum_{\nu} \alpha_\nu t^{n-\nu}$$

where the α_ν are the characteristic coefficients of φ defined in sec. 4.19. The corresponding polynomial function is then given by

$$\chi(\lambda) = \det(\varphi - \lambda\iota);$$

it follows from the definition that

$$\dim E = \deg \chi .$$

Recall that the distinct roots of χ are precisely the eigenvalues of φ (cf. sec. 4.20). Hence the distinct roots of the characteristic polynomial coincide with those of the minimum polynominal. In sec. 13.20 it will be shown that the minimum polynomial is even a divisor of the characterisic polynomial.

Problems

1. Calculate the minimum polynomials for the following linear transformations:

a) $\varphi = \lambda\iota$

b) φ is a projection operator

c) φ is an involution

d) φ is a differential operator

e) φ is a (proper or improper) rotation of a Euclidean plane or of Euclidean 3-space

2. Given an example of linear transformations $\varphi, \psi: E \to E$ such that $\psi \circ \varphi$ and $\varphi \circ \psi$ do not have the same minimum polynomial.

3. Suppose $E = E_1 \oplus E_2$ and $\varphi = \varphi_1 \oplus \varphi_2$ where $\varphi_i: E_i \to E_i$ $(i = 1,2)$ are linear transformations. Let μ, μ_1, μ_2 be the minimum polynomials of φ, φ_1 and φ_2. Prove that μ is the least common multiple of μ_1 and μ_2.

4. More generally, suppose $E_1, E_2 \subset E$ are stable under φ and $E = E_1 + E_2$. Let $\varphi_1: E_1 \to E_1$, $\varphi_2: E_2 \to E_2$ and $\varphi_{12}: E_1 \cap E_2 \to E_1 \cap E_2$ be the restrictions of φ and suppose that they have minimum polynomials μ_1, μ_2, μ_{12}. Show that

a) μ is the least common multiple of μ_1 and μ_2.

b) $\mu_{12} | \nu$ where ν is the greatest common divisor of μ_1 and μ_2.

c) Give an example showing that in general $\mu_{12} \neq \nu$.

5. Suppose $E_1 \subset E$ is a subspace stable under φ. Let μ, μ_1 and $\bar{\mu}$ be the minimum polynomials of $\varphi: E \to E$, $\varphi_1: E_1 \to E_1$ and $\bar{\varphi}: E/E_1 \to E/E_1$ and let ν be the least common multiple of μ_1 and $\bar{\mu}$. Prove that $\nu | \mu | \bar{\mu}\mu_1$. Construct an example where $\nu = \mu \neq \bar{\mu}\mu_1$ and an example where $\nu \neq \mu = \bar{\mu}\mu_1$. Finally construct an example where $\nu \neq \mu \neq \bar{\mu}\mu_1$.

6. Show that the minimal polynomial μ of a linear transformation φ can be constructed in the following way: Select an arbitrary vector $x_1 \in E$ and determine the smallest integer k_1, such that the vectors $\varphi^\nu x_1$ $(\nu = 0 \ldots k_1)$ are linearly dependent,

$$\sum_{\nu=0}^{k_1} \lambda_\nu \varphi^\nu x_1 = 0.$$

Define a polynomial f_1 by

$$f_1 = \sum_{\nu=0}^{k_1} \lambda_\nu t^\nu.$$

If the vectors $\varphi^\nu x_1$ $(\nu = 0 \ldots k_1)$ do not generate the space E select a vector x_2 which is not a linear combination of these vectors and apply the same construction to x_2. Let f_2 be the corresponding polynomial. Continue this procedure until the whole space E is exhausted. Then μ is the least common multiple of the polynomials f_ϱ.

7. Construct the minimum and characteristic polynomials for the following linear transformations of $\mathbb{R}^4$. Verify in each case that μ divides χ.

a) $\varphi(\xi^1, \xi^2, \xi^3, \xi^4) = (\xi^1 - \xi^2 + \xi^3, \xi^1, \xi^2 + \xi^4, 0)$
b) $\varphi(\xi^1, \xi^2, \xi^3, \xi^4) = (\xi^3 + 3\xi^2 + 2\xi^4, 2\xi^2, \xi^1 - 3\xi^2 - 4\xi^4, 2\xi^4)$
c) $\varphi(\xi^1, \xi^2, \xi^3, \xi^4) = (\xi^1 + \xi^3, \xi^2 + \xi^4, \xi^2 + \xi^3, \xi^4)$
d) $\varphi(\xi^1, \xi^2, \xi^3, \xi^4) = (\xi^1 - \xi^2 + \xi^3 - \xi^4, \xi^1 - \xi^2 + \xi^3 - \xi^4, \xi^1 - \xi^2, \xi^1)$.

8. Let φ be a rotation of an inner product space. Prove that the coefficients α_ν of the minimum polynomial satisfy the relations

$$\alpha_\nu = \varepsilon\,\alpha_{k-\nu} \qquad k = \deg \mu,\ \nu = 0 \ldots k$$

where $\varepsilon = \pm 1$ depending on whether the rotation is proper or improper.

9. Show that the minimum polynomial of a selfadjoint transformation of a unitary space has real coefficients.

10. Assume that a conjugation $z \to \bar{z}$ is defined in the complex vector space E (cf. sec. 11.7). Let $\varphi: E \to E$ be a linear transformation such that $\overline{\varphi z} = \varphi \bar{z}$. Prove that the minimum polynomial of φ has real coefficients.

11. Show that the set of stable subspaces of E under a linear transformation φ is a lattice with respect to inclusion. Establish a lattice homomorphism of this lattice onto the lattice of ideals in $\Gamma(\varphi)$.

12. Given a regular linear transformation φ show that φ^{-1} is a polynomial in φ.

13. Suppose $\varphi \in L(E, E)$ is regular. Assume that for every $\psi \in L(E; E)$ $\varphi\psi = \lambda\psi\varphi$ (some $\lambda \in \Gamma$). Prove that $\lambda^k = 1$ for some k. If k is the least integer such that $\lambda^k = 1$, prove that the minimum polynomial, μ, of φ can be written

$$\mu = \sum_\nu \alpha_\nu t^{k\nu}$$

§ 2. Generalized eigenspaces

13.4. Generalized eigenspaces. Let

$$\mu = f_1^{k_1} \ldots f_r^{k_r} \qquad f_i \text{ irreducible} \tag{13.17}$$

be the decomposition of μ into its prime factors (cf. sec. 12.9). Then the spaces

$$E_i = K(f_i^{k_i}) \qquad i = 1, \ldots, r$$

are called the *generalized eigenspaces of* φ. It follows from sec. 13.2 that the E_i are stable under φ. Moreover, Proposition IV, sec. 13.2 implies that

$$E = E_1 \oplus \cdots \oplus E_r \tag{13.18}$$

and

$$\mu_i = f_i^{k_i}$$

where μ_i denotes the minimum polynomial of the restriction φ_i of φ to E_i. In particular, $\dim E_i > 0$.

Now suppose λ is an eigenvalue for φ. Then $t-\lambda|\mu$, and so for some i $(1 \le i \le r)$

$$f_i = t - \lambda .$$

Hence the eigenspaces of φ are precisely the spaces

$$K(f_i)$$

where the f_i are the those polynomials in the decomposition (13.17) which are linear.

13.5. The projection operators. Let the projection operators in E associated with the decomposition (13.18) be denoted by π_i. It will be shown that the mappings π_i are polynomials in φ,

$$\pi_i \in \Gamma(\varphi) \qquad i = 1, \ldots, r .$$

If $r=1$, $\pi_1 = \iota$ and the assertion is trivial. Suppose now that $r>1$ and define polynomials g_i by

$$g_i = f_1^{k_1} \ldots \hat{f}_i^{k_i} \ldots f_r^{k_r} .$$

Then, according to sec. 12.9, the g_i are relatively prime, and hence there exist polynomials h_i such that

$$\sum_i g_i h_i = 1 . \tag{13.19}$$

On the other hand, it follows from Corollary II, Proposition II, sec. 13.2 that

$$K(g_i) = \sum_{j \neq i} E_j$$

and so, in particular,

$$h_i(\varphi) g_i(\varphi) x = 0 \qquad x \in \sum_{j \neq i} E_j . \tag{13.20}$$

Now let $x \in E$ be an arbitrary vector, and let

$$x = x_1 + \cdots + x_r \qquad x_i \in E_i$$

be the decomposition of x determined by (13.18). Then (13.19) and (13.20) yield the relation

$$\sum_i x_i = x = \sum_i h_i(\varphi)g_i(\varphi)x = \sum_{i,j} h_i(\varphi)g_i(\varphi)x_j$$
$$= \sum_i h_i(\varphi)g_i(\varphi)x_i$$

whence

$$x_i = h_i(\varphi)g_i(\varphi)x_i \qquad i = 1, \ldots, r. \tag{13.21}$$

It follows at once from (13.20) and (13.21) that

$$\pi_i = h_i(\varphi)g_i(\varphi) \qquad i = 1, \ldots, r$$

which completes the proof.

13.6. Arbitrary stable subspaces. Let $F \subset E$ be any stable subspace. Then

$$F = \sum_i F \cap E_i \tag{13.22}$$

where the E_i are the generalized eigenspaces of φ. In fact, since the projection operators π_i are polynomials in φ, it follows that F is stable under each π_i.

Now we have for each $x \in F$ that

$$x = \iota x = \sum_i \pi_i x$$

and

$$\pi_i x \in F \cap E_i .$$

It follows that $x \in \sum_i F \cap E_i$, whence

$$F \subset \sum_i F \cap E_i .$$

Since inclusion in the other direction is obvious, (13.22) is established.

13.7. The Fitting decomposition. Suppose F_0 is the generalized eigenspace of φ corresponding to the irreducible polynomial t (if t does not divide μ, then of course $F_0 = 0$). Let F_1 be the direct sum of the remaining generalized eigenspaces. The decomposition

$$E = F_0 \oplus F_1$$

is called the *Fitting decomposition* of E with respect to φ. F_0 and F_1 are called respectively the Fitting-null component and the Fitting-one component of E.

Clearly F_0 and F_1 are stable subspaces. Moreover it follows from the definitions that if φ_0 and φ_1 denote the restrictions of φ to F_0 and F_1, then φ_0 is *nilpotent*; i.e.,

$$\varphi_0^l = 0 \qquad \text{for some } l > 0$$

while φ_1 is a linear isomorphism. Finally, we remark that the corresponding projection operators are polynomials in φ, since they are sums of the projection operators π_i defined in sec. 13.5.

13.8. Dual mappings. Let E^* be a space dual to E and let

$$\varphi^* : E^* \leftarrow E^*$$

be the linear transformation dual to φ. Then if f is any polynomial, it follows from sec. 2.25 that

$$f(\varphi^*) = [f(\varphi)]^* .$$

This implies that $f(\varphi^*)=0$ if and only if $f(\varphi)=0$. In particular, the minimum polynomials of φ and φ^* coincide.

Now suppose that F is any stable subspace of E. Then $F^\perp$ is stable under φ^*. In fact, if $y \in F$ and $y^* \in F^\perp$ are arbitrarily chosen, we have

$$\langle \varphi^* y^*, y \rangle = \langle y^*, \varphi y \rangle = 0$$

whence $\varphi^* y^* \in F^\perp$. This proves that $F^\perp$ is stable. Thus φ^* induces a linear transformation

$$(\varphi^*)^\perp : E^*/F^\perp \leftarrow E^*/F^\perp .$$

On the other hand, let

$$\varphi_F : F \to F$$

be the restriction of φ to F. It will now be shown that φ_F and $(\varphi^*)^\perp$ are dual with respect to the induced scalar product between F and $E^*/F^\perp$ (cf. sec. 2.23). In fact, if $y \in F$ is any vector and y^* is a representative of an arbitrary vector $\bar{y}^* \in E^*/F^\perp$, then

$$\begin{aligned} \langle (\varphi^*)^\perp \bar{y}^*, y \rangle &= \langle \varphi^* y^*, y \rangle = \langle y^*, \varphi y \rangle \\ &= \langle y^*, \varphi_F y \rangle = \langle \bar{y}^*, \varphi_F y \rangle \end{aligned}$$

which proves the duality of φ_F and $(\varphi^*)^\perp$. Thus we may write $(\varphi^*)^\perp = (\varphi_F)^*$.

Suppose next that

$$E = F_1 \oplus F_2$$

is a decomposition of E into two stable subspaces. Then it follows that

$$E^* = F_2^\perp \oplus F_1^\perp$$

is a decomposition of E^* into stable subspaces (under φ^*). Moreover, the pairs $F_1, F_2^\perp$ and $F_2, F_1^\perp$ are dual,

$$F_1^* = F_2^\perp \quad \text{and} \quad F_2^* = F_1^\perp$$

(cf. sec. 2.30), and it is easily checked that φ and φ^* induce dual mappings in each pair.

Conversely, assume that $F_1 \subset E$ and $F_1^* \subset E^*$ are two dual subspaces stable under φ and φ^* respectively. Then we have the direct decompositions

$$E = F_1 \oplus (F_1^*)^\perp$$

and

$$E^* = F_1^* \oplus F_1^\perp$$

(cf. sec. 2.30). Clearly the subspaces $(F_1^*)^\perp$ and $F_1^\perp$ are again stable.

More generally, a direct decomposition

$$E = F_1 \oplus \cdots \oplus F_r$$

of E into several stable subspaces determines a direct decomposition of E^* into stable subspaces,

$$E^* = F_1^* \oplus \cdots \oplus F_r^* \qquad F_i^* = \Big(\sum_{j \neq i} F_j\Big)^\perp$$

as follows by an argument similar to that used above in the case $r=2$. Each pair F_i, F_i^*, is dual and the restrictions φ_i, φ_i^* of φ and φ^* to F_i and F_i^* are dual mappings.

Proposition I: Let

$$\mu = f_1^{k_1} \dots f_r^{k_r}$$

be the decomposition of the common minimum polynomial, μ, of φ and φ^*. Consider the direct decompositions

$$E = E_1 \oplus \cdots \oplus E_r$$

and

$$E^* = E_1^* \oplus \cdots \oplus E_r^* \tag{13.23}$$

of E and E^* into the generalized eigenspaces of φ and φ^*. Then

$$E_i^* = \Big(\sum_{j \neq i} E_j\Big)^\perp \qquad i = 1, \dots, r. \tag{13.24}$$

Proof: Consider the subspaces $F_i^* \subset E^*$ defined by

$$F_i^* = \Big(\sum_{j \neq i} E_j\Big)^\perp \qquad i = 1, \dots, r.$$

Then, as shown above, the F_i^* are stable under φ^* and

$$E^* = F_1^* \oplus \cdots \oplus F_r^* . \tag{13.25}$$

It will now be shown that

$$F_i^* \subset E_i^* . \tag{13.26}$$

Let $y^* \in F_i^*$ be arbitrarily chosen. Then for each $x \in E_i$ we have

$$\langle f_i^{k_i}(\varphi^*) y^*, x \rangle = \langle y^*, f_i^{k_i}(\varphi) x \rangle = \langle y^*, 0 \rangle = 0 .$$

In view of the duality between E_i and F_i^*, this implies that $f_i^{k_i}(\varphi^*) y^* = 0$; i.e.,

$$y^* \in E_i^* .$$

This establishes (13.26). Now a comparison of the decompositions (13.23) and (13.25) yields (13.24).

Problems

1. Show that the minimum polynomial of φ is completely reducible (i.e. all prime factors are of degree 1) if and only if every stable subspace contains an eigenvector.

2. Suppose that the minimum polynomial μ of φ is completely reducible. Construct a basis of E with respect to which the matrix of φ is lower triangular; i.e., the matrix has the form

$$\begin{pmatrix} \lambda_1 & & 0 \\ & \ddots & \\ * & & \lambda_n \end{pmatrix} .$$

Hint: Use problem 1.

3. Let E be an n-dimensional real vector space and $\varphi: E \to E$ be a regular linear transformation. Show that φ can be written $\varphi = \varphi_1 \varphi_2$ where every eigenvalue of φ_1 is positive and every eigenvalue of φ_2 is negative.

4. Use problem 3 to derive a simple proof of the basis deformation theorem of sec. 4.32.

5. Let $\varphi: E \to E$ be a linear transformation and consider the subspaces F_0 and F_1 defined by

$$F_0 = \sum_{j \geqq 1} \ker \varphi^j \quad \text{and} \quad F_1 = \bigcap_{j \geqq 1} \operatorname{Im} \varphi^j$$

a) Show that $F_0 = \bigcup_{j \geq 1} \ker \varphi^j$.

b) Show that $E = F_0 \oplus F_1$.

c) Prove that F_0 and F_1 are stable under φ and that the restrictions $\varphi_0: F_0 \to F_0$ and $\varphi_1: F_1 \to F_1$ are respectively nilpotent and regular.

d) Prove that c) characterizes the decomposition $E = F_0 \oplus F_1$ and conclude that F_0 and F_1 are respectively the Fitting null and the Fitting 1-component of E.

6. Consider the linear transformations of problem 7, § 1. For each transformation

a) Construct the decomposition of $\mathbb{R}^4$ into the generalized eigenspaces.

b) Determine the eigenspaces.

c) Calculate explicitly polynomials g_i such that the $g_i(\varphi)$ are the projection operators in E corresponding to the generalized eigenspaces. Verify by explicit consideration of the vectors $g_i(\varphi)x$ that the $g_i(\varphi)$ are in fact the projection operators.

d) Determine the Fitting decomposition of E.

7. Let $E = \sum_i E_i$ be the decomposition of E into generalized eigenspaces of φ, and let π_i be the corresponding projection operators. Show that there exist unique polynomials g_i such that

$$g_i(\varphi) = \pi_i \quad \text{and} \quad \deg g_i \leq \deg \mu .$$

Conclude that the polynomials g_i depend only on μ.

8. Let E^* be dual to E and $\varphi^*: E^* \leftarrow E^*$ be dual to φ. If $E^* = \sum_i E_i^*$ is the decomposition of E^* into generalized eigenspaces of φ^* prove that

$$\pi_i^* = g_i(\varphi^*)$$

where the π_i^* are the corresponding projection operators and the g_i are defined in problem 7.

Use this result to show that π_i and π_i^* are dual and to obtain formula (13.24).

9. Let $F \subset E$ be stable under φ and consider the induced mappings $\varphi_F: F \to F$ and $\bar{\varphi}: E/F \to E/F$. Let $E = \sum_i E_i$ be the decomposition of E into generalized eigenspaces of φ. Let $j: F \to E$ be the canonical injection and $\varrho: E \to E/F$ be the canonical projection.

a) Show that the decomposition of F into generalized eigenspaces is given by

$$F = \sum_i F_i \quad \text{where} \quad F_i = F \cap E_i .$$

b) Show that the decomposition of E/F into generalized eigenspaces of $\bar{\varphi}$ is given by

$$E/F = \sum_i (E/F)_i \quad \text{where} \quad (E/F)_i = \varrho(E_i).$$

Conclude that ϱ determines a linear isomorphism

$$E_i/F_i \to (E/F)_i .$$

c) If π_i, π_i^F, $\bar{\pi}_i$ denote the projection operators in E, F and E/F associated with the decompositions, prove that the diagrams

$$\begin{array}{ccc} E & \xrightarrow{\pi_i} & E \\ \varrho\downarrow & & \downarrow\varrho \\ E/F & \xrightarrow{\bar{\pi}_i} & E/F \end{array} \quad \text{and} \quad \begin{array}{ccc} E & \xrightarrow{\pi_i} & E \\ j\uparrow & & \uparrow j \\ F & \xrightarrow{\pi_i^F} & F \end{array}$$

are commutative, and that $\bar{\pi}_i$, π_i^F are the unique linear mappings for which this is the case. Conclude that if g_i are the polynomials of problem 7, then

$$\bar{\pi}_i = g_i(\bar{\varphi}) \quad \text{and} \quad \pi_i^F = g_i(\varphi_F).$$

10. Suppose that the minimum polynomial μ of φ is completely reducible.

a) By considering first the case $\mu = (t-\lambda)^k$ prove that

$$\deg \mu \leqq \dim E .$$

b) With the aid of a) prove that $\mu | \chi$, χ the characteristic polynomial of φ.

§ 3. Cyclic spaces

13.9. The map σ_a. Fix a vector $a \in E$ and consider the linear map $\sigma_a: \Gamma[t] \to E$ given by

$$\sigma_a(f) = f(\varphi)\, a \qquad f \in \Gamma[t].$$

Let K_a denote the kernel of σ_a. It follows from the definition that $f \in K_a$ if and only if $a \in K(f)$. K_a is an ideal in $\Gamma[t]$. Clearly $I_\mu \subset K_a$, where μ is the minimum polynomial of φ.

Proposition I: There exists a vector $a \in E$ such that $K_a = I_\mu$.
Proof: Consider first the case that μ is of the form

$$\mu = f^k, \qquad k \geqq 1, \qquad f \text{ irreducible}.$$

Then there is a vector $a\in E$ such that

$$f^{k-1}(\varphi)(a)\neq 0. \tag{13.27}$$

Suppose now that $h\in K_a$ and let g be the greatest common divisor of h and μ. Since $a\in K(h)$, Corollary I to Proposition I, sec. 13.2 yields

$$a\in K(g). \tag{13.28}$$

Since g/μ it follows that $g=f^l$ where $l\leqq k$. Hence $f^l(\varphi)\,a=0$ and relation (13.27) implies that $l=k$. Thus $g=\mu$ and hence $h\in I_\mu$.

In the general case decompose μ in the form

$$\mu=f_1^{k_1}\dots f_r^{k_r}, \qquad f_i \text{ irreducible}$$

and let

$$E=E_1\oplus\cdots\oplus E_r$$

be the corresponding decomposition of E. Let $\varphi_i\colon E_i\to E_i$ $(i=1\dots r)$ be the induced transformation. Then the minimum polynomial of φ_i is given by (cf. sec. 13.4)

$$\mu_i=f_i^{k_i} \qquad (i=1\dots r).$$

Thus, by the first part of the proof, there are vectors $a_i\in E_i$ such that

$$K_{a_i}=I_{\mu_i} \qquad (i=1\dots r).$$

Now set

$$a=a_1+\cdots+a_r.$$

Assume that $f\in K_a$. Then $f(\varphi)\,a=0$ whence

$$\sum_{i=1}^r f(\varphi_i)\,a_i=0.$$

Since $f(\varphi_i)\,a_i\in E_i$, it follows that

$$f(\varphi_i)\,a_i=0 \qquad (i=1,\dots,r)$$

whence $f\in K_{a_i}$ $(i=1\dots r)$. Thus

$$f\in I_{\mu_1}\cap\cdots\cap I_{\mu_r}=I_\mu$$

and so $f\in I_\mu$. This shows that $K_a\subset I_\mu$ whence $K_a=I_\mu$.

13.10. Cyclic spaces. The vector space E is called *cyclic* (with respect to the linear transformation φ) if there exists a vector $a\in E$ such that the map σ_a is surjective. Every such vector is called a *generator* of E. If a is a generator of E, then $K_a=I_\mu$. In fact, let $f\in K_a$ and let $x\in E$. Then, for some $g \in \Gamma[t]$, $x = g(\varphi)a$. It follows that

$$f(\varphi)x = f(\varphi)g(\varphi)a = g(\varphi)f(\varphi)a = g(\varphi)0 = 0$$

whence $f(\varphi) = 0$ and so $f \in I_\mu$.

Proposition II: If E is cyclic, then

$$\deg \mu = \dim E .$$

Proof. Let a be a generator of E. Then, since $K_a = I_\mu$, σ_a induces an isomorphism

$$\Gamma[t]/I_\mu \xrightarrow{\cong} E .$$

It follows that (cf. Proposition I, sec. 12.11)

$$\dim E = \dim \Gamma[t]/I_\mu = \deg \mu .$$

Proposition III: Let a be a generator of E and let $\deg \mu = m$. Then the vectors

$$a, \varphi(a), \ldots, \varphi^{m-1}(a)$$

form a basis of E.

Proof: Let l be the largest integer such that the vectors

$$a, \varphi(a), \ldots, \varphi^{l-1}(a) \tag{13.29}$$

are linearly independent. Then these vectors generate E. In fact, every vector $x \in E$ can be written in the form $x = f(\varphi)a$ where $f = \sum_{\nu=0}^{k} \alpha_\nu t^\nu$ is a polynomial. It follows that

$$x = \sum_{\nu=0}^{k} \alpha_\nu \varphi^\nu(a) = \sum_{j=0}^{l-1} \mu_j \varphi^j(a).$$

Thus the vectors (13.29) form a basis of E. Now Proposition II implies that $l = m$.

Proposition IV: The space E is cyclic if and only if there exists a basis a_ν $(\nu = 0 \ldots n-1)$ of E such that

$$\varphi(a_\nu) = a_{\nu+1} \qquad (\nu = 0 \ldots n-2). \tag{13.30}$$

Proof: If E is cyclic with generator a set $a_0 = a$ and apply Proposition III. Conversely, assume that a_ν $(\nu = 0 \ldots n-1)$ is a basis satisfying the conditions above. Let $x = \sum_{\nu=0}^{n-1} \xi_\nu a_\nu$ be an arbitrary vector and define $f_x \in \Gamma[t]$ by

$$f_x(t) = \sum_{\nu=0}^{n-1} \xi_\nu t^\nu.$$

Then

$$f_x(\varphi)a_0 = x$$

as is easily checked and so E is cyclic.

Corollary: Let a_ν ($\nu = 0, \ldots, n-1$) be a basis as in the Proposition above. Then the minimum polynomial of φ is given by

$$\mu = t^n - \sum_{\nu=0}^{n-1} \alpha_\nu t^\nu, \tag{13.31}$$

where the α_ν are determined by $\varphi(a_{n-1}) = \sum_{\nu=0}^{n-1} \alpha_\nu a_\nu$.

Proof: It is easily checked that $\mu(\varphi) = 0$ and so μ is a multiple of the minimum polynomial of φ. On the other hand, by Proposition II, the minimum polynomial of φ has degree n and thus it must coincide with μ.

13.11. Cyclic subspaces. A stable subspace $F \subset E$ is called a *cyclic subspace* if it is cyclic with respect to the induced transformation. Every vector $a \in E$ determines a cyclic subspace, namely the subspace

$$E_a = \operatorname{Im} \sigma_a.$$

Proposition V: There exists a cyclic subspace whose dimension is equal to $\deg \mu$.

Proof: In view of Proposition I there is a vector $a \in E$ such that $\ker \sigma_a = I_\mu$. Then σ_a induces an isomorphism

$$\Gamma[t]/I_\mu \xrightarrow{\cong} E_a.$$

It follows that

$$\dim E_a = \dim \Gamma[t]/I_\mu = \deg \mu.$$

Theorem I: The degree of the minimum polynomial satisfies

$$\deg \mu \leqq \dim E.$$

Equality holds if and only if E is cyclic. Moreover, if F is any cyclic subspace of E, then

$$\dim F \leqq \deg \mu.$$

Proof: Proposition V implies that $\deg \mu \leqq \dim E$. If E is cyclic, equality holds (cf. Proposition III). Conversely, assume that $\deg \mu = \dim E$. By Proposition V there exists a cyclic subspace $F \subset E$ with $\dim F = \deg \mu$. It follows that $F = E$ and so E is cyclic.

Finally, let $F \subset E$ be any cyclic subspace and let ν denote the minimum polynomial of the induced transformation. Then, as we have seen above, $\dim F = \deg \nu$. But ν/μ (cf. sec. 13.2) and so we obtain $\dim F \leqq \deg \mu$.

Corollary: Let $F \subset E$ be any cyclic subspace, and let ν denote the minimum polynomial of the linear transformation induced in F by φ. Then

$$\nu = \mu$$

if and only if

$$\dim F = \deg \mu .$$

Proof: From the theorem we have

$$\dim F = \deg \nu$$

while according to sec. 13.2 ν divides μ.

Hence $\nu = \mu$ if and only if $\deg \nu = \deg \mu$; i.e., if and only if

$$\dim F = \deg \mu .$$

13.12. Decomposition of E into cyclic subspaces. *Theorem II:* There exists a decomposition of E into a direct sum of cyclic subspaces.

Proof: The theorem is an immediate consequence (with the aid of an induction argument on the dimension of E) of the following lemma.

Lemma I: Let E_a be a cyclic subspace of E such that

$$\dim E_a = \deg \mu .$$

Then there is a complementary stable subspace, $F \subset E$,

$$E = E_a \oplus F .$$

Proof: Let

$$\varphi_a : E_a \to E_a$$

denote the restriction of φ to E_a, and let

$$\varphi^* : E^* \leftarrow E^*$$

be the linear transformation in E^* dual to φ. Then (cf. sec. 13.8) $E_a^{\perp}$ is stable under φ^*, and the induced linear transformation

$$\varphi_a^* : E^*/E_a^{\perp} \leftarrow E^*/E_a^{\perp}$$

is dual to φ_a with respect to the induced scalar product between E_a and $E^*/E_a^\perp$.

The corollary to Theorem I, sec. 13.11 implies that the minimum polynomial of φ_a is again μ. Hence (cf. sec. 13.8) the minimum polynomial of φ_a^* is μ. But E_a and $E^*/E_a^\perp$ are dual, so that

$$\dim E^*/E_a^\perp = \dim E_a = \deg \mu .$$

Thus Theorem I, sec. 13.11 implies that $E^*/E_a^\perp$ is cyclic with respect to φ_a^*. Now let

$$\pi : E^* \to E^*/E_a^\perp$$

be the projection and choose $a^* \in E^*$ so that the element $\bar{a}^* = \pi(a^*)$ generates $E^*/E_a^\perp$. Then, by Proposition III, the vectors $\bar{a}_\mu^* = (\varphi_a^*)^\mu \bar{a}^*$ $(\mu = 0 \ldots m-1)$ form a basis of $E^*/E_a^\perp$. Hence the vectors $a_\mu^* = (\varphi^*)^\mu a^*$ $(\mu = 1 \ldots m-1)$ are linearly independent.

Now consider the cyclic subspace $E_{a^*}^*$. Since $a_\mu^* \in E_{a^*}^*$ $(\mu = 0 \ldots m-1)$ it follows that $\dim E_{a^*}^* \geqq m$. On the other hand, Theorem I, sec. 13.11, implies that $\dim E_{a^*}^* \leqq m$. Hence

$$\dim E_{a^*}^* = m .$$

Finally, since $\pi^* a_\mu^* = \bar{a}_\mu^*$ $(\mu = 0 \ldots m-1)$ it follows that the restriction of π to $E_{a^*}^*$ is injective. Thus $E_{a^*}^* \cap E_a^\perp = 0$. But

$$\dim E_{a^*}^* + \dim E_a^\perp = m + (n - m) = n \qquad (n = \dim E)$$

and thus we have the direct decomposition

$$E^* = E_{a^*}^* \oplus E_a^\perp$$

of E^* into stable subspaces. Taking orthogonal complements we obtain the direct decomposition

$$E = E_a \oplus (E_{a^*}^*)^\perp \tag{13.32}$$

of E into stable subspaces which completes the proof.

§ 4. Irreducible spaces

13.13. Definition. A vector space E is called *indecomposable* or *irreducible* with respect to a linear transformation φ, if it can not be expressed as a direct sum of two proper stable subspaces. A stable subspace $F \subset E$ is called *irreducible* if it is irreducible with respect to the linear transformation induced by φ.

Proposition I: E is the direct sum of *irreducible* subspaces.

Proof: Let

$$E = \sum_{j=1}^{s} F_j \qquad \dim F_j > 0$$

be a decomposition of E into stable subspaces such that s is maximized (this is clearly possible, since for all decompositions we have $s \le \dim E$). Then the spaces F_i are irreducible. In fact, assume that for some i,

$$F_i = F_i' \oplus F_i'' \qquad \dim F_i' > 0, \dim F_i'' > 0$$

is a decomposition of F_i into stable subspaces. Then

$$E = \sum_{j \neq i} F_j \oplus F_i' \oplus F_i''$$

is a decomposition of E into $(s+1)$ stable subspaces, which contradicts the maximality of s.

An irreducible space E is always cyclic. In fact, by Theorem II, sec. 13.12, E is the direct sum of cyclic subspaces.

$$E = \sum_{j=1}^{r} F_r.$$

If E is indecomposable, it follows that $j=1$ and so E is cyclic.

On the other hand, a cyclic space is not necessarily indecomposable. In fact, let E be a 2-dimensional vector space with basis a, b and define $\varphi: E \to E$ by setting $\varphi a = b$ and $\varphi b = a$. Then φ is cyclic (cf. Proposition IV, sec. 13.10). On the other hand, E is the direct-sum of the stable subspaces generated by $a+b$ and $a-b$.

Theorem I: E is irreducible if and only if

i) $\mu = f^k$, f irreducible;

ii) E is cyclic.

Proof: Suppose E is irreducible. Then, by the remark above, E is cyclic. Moreover, if $E = E_1 \oplus \cdots \oplus E_r$ is the decomposition of E into the generalized eigenspaces (cf. sec. 13.4), it follows that $r=1$ and so $\mu = f^k$ (f irreducible).

Conversely, suppose that (i) and (ii) hold. Let

$$E = E_1 \oplus E_2$$

be any decomposition of E into stable subspaces. Denote by φ_1 and φ_2 the linear transformations induced in E_1 and E_2 by φ, and let μ_1 and μ_2 be the minimum polynomials of φ_1 and φ_2. Then sec. 13.2 implies that $\mu_1|\mu$ and $\mu_2|\mu$. Hence, we obtain from (i) that

$$\mu_1 = f^{k_1}, \mu_2 = f^{k_2} \qquad k_1, k_2 \leq k. \tag{13.33}$$

Without loss of generality we may assume that $k_1 \geq k_2$. Then

$$f^{k_1}(\varphi)x = 0 \qquad x \in E_1 \quad \text{or} \quad x \in E_2$$

and so

$$f^{k_1}(\varphi) = 0.$$

It follows that $\mu | f^{k_i}$ whence $k_1 \geq k$. In view of (13.33) we obtain $k_1 = k$ i.e.,

$$\mu = \mu_1.$$

Now Theorem I, sec. 13.11 yields that

$$\dim E_1 \geq \deg \mu. \tag{13.34}$$

On the other hand, since E is cyclic, the same Theorem implies that

$$\dim E = \deg \mu. \tag{13.35}$$

Relations (13.34) and (13.35) give that

$$\dim E = \dim E_1.$$

Thus $E = E_1$ and $E_2 = 0$. It follows that E is irreducible.

Corollary I: Any decomposition of E into a direct sum of irreducible subspaces is simultaneously a decomposition into cyclic subspaces.

Corollary II: Suppose that $\mu = f^k$, f irreducible. Then a stable subspace of E is cyclic if and only if it is irreducible.

13.14. The Jordan canonical matrix. Suppose that E is irreducible with respect to φ. Then it follows from Theorem I sec. 13.13 that E is cyclic and that the minimum polynomial of φ has the form

$$\mu = f^k \qquad k \geq 1 \tag{13.36}$$

where f is an irreducible polynomial. Let e be a generator of E and consider the vectors

$$a_{ij} = f(\varphi)^{i-1}\varphi^{j-1}e \qquad \begin{matrix} i = 1,\dots,k \\ j = 1,\dots,p \end{matrix} \qquad p = \deg f\,. \tag{13.37}$$

It will be shown that these form a basis of E.

Since

$$\dim E = \deg \mu = p\,k$$

it is sufficient to show that the vectors (13.37) generate E. Let $F \subset E$ be the subspace generated by the vectors (13.37).

Writing

$$f = \sum_{\nu=0}^{p} \alpha_\nu t^\nu \qquad \alpha_p = 1$$

we obtain that

$$\varphi\, a_{ij} = a_{ij+1} \qquad i = 1,\dots,k \quad j = 1,\dots,p-1$$

$$\varphi\, a_{ip} = \varphi f(\varphi)^{i-1}\varphi^{p-1}e = f(\varphi)^{i-1}\varphi^{p}e = f(\varphi)^{i}e - \sum_{\nu=0}^{p-1}\alpha_\nu f^{i-1}(\varphi)\varphi^\nu e$$

$$= a_{i+11} - \sum_{\nu=0}^{p-1}\alpha_\nu a_{i\,\nu+1} \qquad i = 1,\dots,k-1$$

$$\varphi\, a_{kp} = -\sum_{\nu=0}^{p-1}\alpha_\nu a_{k\,\nu+1}\,.$$

These equations show that the subspace F is stable under φ. Moreover, $e = a_{11} \in F$. On the other hand, since E is cyclic, E is the *smallest* stable subspace containing e. It follows that $F = E$.

Now consider the basis

$$a_{11},\dots,a_{1p};\,a_{21}\dots a_{2p};\dots;\,a_{k1}\dots a_{kp}$$

of E. The matrix of φ relative to this basis has the form

$$\begin{pmatrix} \boxed{A_1} & 1 & & & 0 \\ & \boxed{A_2} & 1 & & \\ & & \ddots & \ddots & \\ & & & & 1 \\ 0 & & & & \boxed{A_k} \end{pmatrix} \tag{13.38}$$

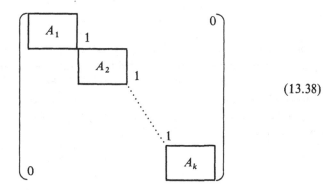

where the submatrices A_j are all equal, and given by

$$A_j = \begin{pmatrix} 0 & 1 & & & 0 \\ & 0 & 1 & & \\ & & 0 & \ddots & \\ & & & \ddots & \\ 0 & & & & 1 \\ -\alpha_0 & -\alpha_1 & \dots & & -\alpha_{p-1} \end{pmatrix} \qquad j = 1, \dots, k.$$

The matrix (13.38) is called a *Jordan canonical matrix* of the irreducible transformation φ.

Next let φ be an arbitrary linear transformation. In view of sec. 13.12 there exists a decomposition of E into irreducible subspaces. Choose a basis in every subspace relative to which the induced transformation has the Jordan canonical form. Combining these bases we obtain a basis of E. In this basis the matrix of φ consists of submatrices of the form (13.38) following each other along the main diagonal. This matrix is called a *Jordan canonical matrix of* φ.

13.15. Completely reducible minimum polynomials. Suppose now that E is an irreducible space and that the minimum polynomial is completely reducible ($p=1$); i.e. that

$$\mu = (t - \lambda)^k.$$

It follows that the A_j are (1×1)-matrices given by $A_j = (\lambda)$. Hence the Jordan canonical matrix of φ is given by

$$\begin{pmatrix} \lambda & 1 & & & 0 \\ & \lambda & 1 & & \\ & & \ddots & \ddots & \\ & & & \ddots & 1 \\ 0 & & & & \lambda \end{pmatrix} \tag{13.39}$$

In particular, if E is a complex vector space (or more generally a vector space over an algebraically closed field) which is irreducible with respect to φ, then the Jordan canonical matrix of φ has the form (13.39).

13.16. Real vector spaces. Next, let E be a real vector space which is irreducible with respect to φ. Then the polynomial f in (13.36) has one of the two forms

$$f = t - \lambda \qquad \lambda \in \mathbb{R}$$

or

$$f = t^2 + \alpha t + \beta \qquad \alpha, \beta \in \mathbb{R}, \quad \alpha^2 - 4\beta < 0.$$

In the first case the Jordan canonical matrix of φ has the form (13.39). In the second case the A_j are 2×2-matrices given by

$$A_j = \begin{pmatrix} 0 & 1 \\ -\beta & -\alpha \end{pmatrix}.$$

Hence the Jordan canonical matrix of φ has the form

$$\left(\begin{array}{cc|c|cc|c|c|cc}
0 & 1 & & & & & & & \\
-\beta & -\alpha & 1 & & & & & & \\
 & & & 0 & 1 & & & & \\
 & & & -\beta & -\alpha & 1 & & & \\
 & & & & & \ddots & & & \\
 & & & & & & 1 & & \\
 & & & & & & & 0 & 1 \\
0 & & & & & & & -\beta & -\alpha
\end{array}\right).$$

13.17. The number of irreducible subspaces. It is clear from the construction in sec. 13.12 that a vector space can be decomposed in several ways into irreducible subspaces. However, the number of irreducible subspaces of any given dimension is uniquely determined by φ, as will be shown in this section.

Consider first the case that the minimum polynomial of φ has the form

$$\mu = f^k \qquad \deg f = p$$

where f is irreducible. Assume that a decomposition of E into irreducible subspaces is given. The dimension of every such subspace is of the form $p\kappa\,(1 \le \kappa \le k)$, as follows from sec. 13.12. Denote by F_κ the direct sum of the irreducible subspaces E_κ^λ of dimension $p\kappa$ and denote by N_κ the number of the subspaces E_κ^λ.

Then we have

$$E = \sum_{\kappa=1}^{k} F_\kappa. \tag{13.40}$$

Comparing the dimensions in (13.40) we obtain the equation

$$n = \sum_{\kappa=1}^{k} \dim F_\kappa = p \sum_{\kappa=1}^{k} \kappa N_\kappa$$

where $\dim E = n$.

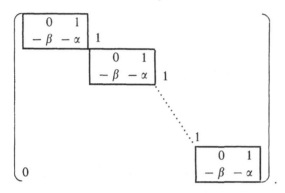

Now consider the transformation

$$\psi = f(\varphi).$$

Since the subspaces F_κ are stable under ψ it follows from (13.40) that

$$\psi E = \sum_{\kappa=1}^{k} \psi F_\kappa . \tag{13.41}$$

By the definition of F_κ we have

$$F_\kappa = \sum_{\lambda=1}^{N_\kappa} E_\kappa^\lambda \qquad \dim E_\kappa^\lambda = p\kappa$$

whence

$$\psi F_\kappa = \sum_{\lambda=1}^{N_\kappa} \psi E_\kappa^\lambda .$$

Since the dimension of each E_κ^λ decreases by p under ψ (cf. sec. 13.14) it follows that

$$\dim \psi F_\kappa = p(\kappa - 1) N_\kappa . \tag{13.42}$$

Equations (13.41) and (13.42) yield ($r(\psi) = \operatorname{rank} \psi$)

$$r(\psi) = p \sum_{\kappa=2}^{k} (\kappa - 1) N_\kappa .$$

Repeating the above argument we obtain the equations

$$r(\psi^j) = p \sum_{\kappa=j+1}^{k} (\kappa - j) N_\kappa \qquad j = 1, \dots, k .$$

Replacing j by $j+1$ and $j-1$ respectively we find that

$$r(\psi^{j+1}) = p \sum_{\kappa=j+2}^{k} (\kappa - j - 1) N_\kappa = p \sum_{\kappa=j+2}^{k} (\kappa - j) N_\kappa - p \sum_{\kappa=j+2}^{k} N_\kappa \tag{13.43}$$

and

$$r(\psi^{j-1}) = p \sum_{\kappa=j}^{k} (\kappa - j + 1) N_\kappa = p \sum_{\kappa=j}^{k} (\kappa - j) N_\kappa + p \sum_{\kappa=j}^{k} N_\kappa . \tag{13.44}$$

Adding (13.43) and (13.44) we obtain

$$\begin{aligned} r(\psi^{j+1}) + r(\psi^{j-1}) &= 2p \sum_{\kappa=j+2}^{k} (\kappa - j) N_\kappa + p N_{j+1} + p(N_j + N_{j+1}) \\ &= 2p \sum_{\kappa=j+1}^{k} (\kappa - j) N_\kappa + p N_j = 2r(\psi^j) + p N_j \end{aligned}$$

whence

$$N_j = \frac{1}{p}[r(\psi^{j+1}) + r(\psi^{j-1}) - 2r(\psi^j)] \qquad j = 1, \ldots, k.$$

This equation shows that the numbers N_j are uniquely determined by the ranks of the transformations ψ^j $(j = 1, \ldots, k)$.
In particular,

$$N_k = \frac{1}{p} r(\psi^{k-1}) \geqq 1$$

and $N_j = 0$ if $j > k$. Thus the degree of μ is pk where k is the largest integer k such that $N_k > 0$.

In the general case consider the decomposition of E into the generalized eigenspaces E_i $(i = 1, \ldots, r)$ and suppose that

$$E = \sum_\lambda F_\lambda$$

is a decomposition of E as a direct sum of irreducible subspaces. Then every irreducible subspace F_λ is contained in some E_i (cf. Theorem I, sec. 13.13). Hence the decomposition determines a decomposition of each E_i into irreducible subspaces. Moreover, it is clear that these subspaces are irreducible with respect to the induced transformation $\varphi_i: E_i \to E_i$. Hence the number of irreducible subspaces in E_i of a given dimension is determined by φ_i and thus by φ. It follows that the number of spaces F_λ of a given dimension depends only on φ.

13.18. Conjugate linear transformations. Let E and F be n-dimensional vector spaces. Two linear transformations $\varphi: E \to E$ and $\psi: F \to F$ are called *conjugate* if there is a linear isomorphism $\alpha: E \xrightarrow{\cong} F$ such that

$$\psi = \alpha \circ \varphi \circ \alpha^{-1}.$$

Proposition II: Two linear transformations φ and ψ are conjugate if and only if they satisfy the following conditions:

(i) The minimum polynomials of φ and ψ have the same prime factors $f_1, \ldots, f_r$.

(ii) $r(f_i(\varphi)^j) = r(f_i(\psi)^j)$ $i = 1 \ldots r, j \geqq 1$.

Proof: If φ and ψ are conjugate the conditions above are clearly satisfied.

To prove the converse we distinguish two cases.

Case I: The minimum polynomials of φ and ψ are of the form

$$\mu_\varphi = f^k \quad \text{and} \quad \mu_\psi = f^l, \qquad f \text{ irreducible.}$$

Decompose E and F into the irreducible subspaces

$$E = \sum_{i=1}^{k} \sum_{j=1}^{N_i(\varphi)} E_i^j, \quad F = \sum_{i=1}^{l} \sum_{j=1}^{N_i(\psi)} F_i^j.$$

The numbers $N_i(\varphi)$ and $N_i(\psi)$ are given by

$$N_i(\varphi) = \frac{1}{p}\left[r(f(\varphi)^{i+1}) + r(f(\varphi)^{i-1}) - 2r(f(\varphi)^i)\right]$$

and $\qquad p = \deg f$

$$N_i(\psi) = \frac{1}{p}\left[r(f(\psi)^{i+1}) + r(f(\psi)^{i-1}) - 2r(f(\psi)^i)\right]$$

(cf. sec. 13.17). Thus (ii) implies that

$$N_i(\varphi) = N_i(\psi) \qquad i \geqq 1.$$

Since

$$N_k(\varphi) \neq 0, \quad N_i(\varphi) = 0 \qquad i > k$$

and

$$N_l(\psi) \neq 0, \quad N_i(\psi) = 0 \qquad i > l$$

it follows that $k = l$.

In view of Theorem I, sec. 13.13, the spaces E_i^j and F_i^j are cyclic. Thus Lemma I below yields an isomorphism $\alpha_i^j: E_i^j \xrightarrow{\cong} F_i^j$ such that

$$\psi = \alpha_i^j \circ \varphi \circ (\alpha_i^j)^{-1}.$$

These isomorphisms determine an isomorphism $\alpha: E \xrightarrow{\cong} F$ such that

$$\psi = \alpha \circ \varphi \circ \alpha^{-1}.$$

Case II: φ and ψ are arbitrary. Decompose E and F into the generalized eigenspaces

$$E = E_1 \oplus \cdots \oplus E_r, \quad F = F_1 \oplus \cdots \oplus F_r$$

and set

$$p_i = \dim E_i, \quad q_i = \dim F_i \qquad (i = 1 \ldots r).$$

Then $f_i(\varphi)$ restricts to a linear isomorphism in the subspace

$$E_1 \oplus \cdots \oplus \hat{E}_i \oplus \cdots \oplus E_r.$$

Moreover, for $p \geqq p_i$, $f_i(\varphi)^p$ is zero in E_i. Thus

$$\dim E - \dim E_i = r\left(f_i(\varphi)^{p_i+q_i}\right) \qquad i = 1 \dots r.$$

Similarly,

$$\dim F - \dim F_i = r\left(f_i(\psi)^{p_i+q_i}\right) \qquad i = 1 \dots r.$$

Now (ii) implies that

$$\dim E_i = \dim F_i \qquad i = 1 \dots r.$$

Let $\varphi_i: E_i \to E_i$ and $\psi_i: F_i \to F_i$ denote the restrictions of φ and ψ respectively. The minimum polynomials of φ_i and ψ_i are of the form

$$\mu_i = f_i^{k_i} \qquad i = 1 \dots \mathrm{r}$$

and

$$\nu_i = f_i^{l_i} \qquad i = 1 \dots r.$$

Since

$$r\left(f_i(\varphi)^j\right) = r\left(f_i(\varphi_i)^j\right) + \dim E - \dim E_i$$

and $\qquad\qquad i = 1 \dots r,\ j \geqq 1$

$$r\left(f_i(\psi)^j\right) = r\left(f_i(\psi_i)^j\right) + \dim F - \dim F_i$$

it follows from (ii) that

$$r\left(f_i(\varphi_i)^j\right) = r\left(f_i(\psi_i)^j\right) \qquad \begin{array}{l} i = 1 \dots r \\ j \geqq 1. \end{array}$$

Thus the pair φ_i, ψ_i satisfies the hypotheses of the theorem and so we are reduced to case I.

Lemma I: Let $\varphi: E \to E$ and $\psi: F \to F$ ($\dim E = \dim F = n$) be linear transformations having the same minimum polynomial. Then, if E and F are cyclic, φ and ψ are conjugate.

Proof: Choose generators a and b of E and F. Then the vectors $a_j = \varphi^j(a)$ and $b_j = \psi^j(b)$ $(j = 0 \dots n-1)$ form a basis of E and F respectively. Now set $a_n = \varphi^n(a)$ and $b_n = \psi^n(b)$. Then, for some $\alpha_j, \beta_j \in \Gamma$,

$$a_n = \sum_{j=0}^{n-1} \alpha_j a_j \quad \text{and} \quad b_n = \sum_{j=0}^{n-1} \beta_j b_j.$$

Moreover, the minimum polynomials of φ and ψ are given by

$$\mu_\varphi = t^n - \sum_{j=0}^{n-1} \alpha_j t^j \quad \text{and} \quad \mu_\psi = t^n - \sum_{j=0}^{n-1} \beta_j t^j$$

(cf. the corollary of Proposition IV, sec. 13.10). Since $\mu_\varphi = \mu_\psi$, it follows that

$$\alpha_j = \beta_j \qquad (j = 0 \dots n-1).$$

Now define an isomorphism $\gamma: E \xrightarrow{\cong} F$ by setting

$$\gamma(a_j) = b_j \qquad j = 0 \dots n-1.$$

Then we have

$$\gamma(a_n) = \gamma \sum_{j=0}^{n-1} \alpha_j a_j = \sum_{j=0}^{n-1} \alpha_j b_j = b_n.$$

These equations imply that

$$\psi \gamma(a_j) = \psi b_j = \psi^{j+1}(b) = b_{j+1} = \gamma(a_{j+1}) = \gamma \varphi(a_j) \qquad j = 0 \dots n-1.$$

This shows that

$$\psi = \gamma \circ \varphi \circ \gamma^{-1}.$$

Anticipating the result of sec. 13.20 we also have

Corollary I: Two linear transformations are conjugate if and only if they have the same minimum polynomial and satisfy condition (ii) of Proposition II.

Corollary II: **Two linear transformations are conjugate if they have the same characteristic polynomial (cf. sec. 4.19) and satisfy (ii).**

Proof: Let f be the common characteristic polynomial of φ and ψ. Write

$$f = f_1^{m_1} \dots f_r^{m_r}$$

$$\mu_\varphi = f_1^{k_1} \dots f_r^{k_r} \quad (f_i \text{ irreducible}).$$

and

$$\mu_\psi = f_1^{l_1} \dots f_r^{l_r}$$

Set $\tilde{E}_i = \ker(f_i(\varphi)^{m_i})$, $\tilde{F}_i = \ker(f_i(\psi)^{m_i})$, $i = 1 \dots r$; let E_i, F_i $(i = 1 \dots r)$ denote the generalized eigenspaces for φ and ψ respectively. Then we have the direct decompositions

$$E = \sum_{i=1}^{r} \tilde{E}_i, \quad E = \sum_{i=1}^{r} E_i$$

and

$$F = \sum_{i=1}^{r} \tilde{F}_i, \quad F = \sum_{i=1}^{r} F_i.$$

Since $k_i \leqq m_i$, it follows that $E_i \subset \tilde{E}_i$ whence $E_i = \tilde{E}_i$ $(i = 1 \dots r)$. Similarly, $F_i = \tilde{F}_i$ $(i = 1 \dots r)$. Thus we may assume that f is of the form $f = g^m$ (g irreducible). Then $\mu_\varphi = g^k$ $(k \leqq m)$, $\mu_\psi = g^l$ $(l \leqq m)$ and the corollary follows immediately from the proposition.

Corollary III: Every linear transformation is conjugate to its dual.

Problems

1. Let $\Phi: A(E; E) \to A(E; E)$ be a non-zero algebra endomorphism.

(i) Show that for any projection operator π

$$r(\Phi(\pi)) = r(\pi).$$

Hint: Use problems 3 and 6, chapter IV, §7.

(ii) Use (i) to prove that $r(\Phi(\varphi)) = r(\varphi)$, $\varphi \in A(E; E)$.

2. Show that every non-zero endomorphism Φ of the algebra $A(E; E)$ is of the form $\Phi(\varphi) = \alpha \circ \varphi \circ \alpha^{-1}$, where α is a fixed isomorphism $E \to E$.

Hint: Use Problem 1 and Proposition II, sec. 13.18.

3. Construct a decomposition of $\mathbb{R}^4$ into *irreducible* subspaces with respect to the linear transformations of problem 7, § 1. Hence obtain the Jordan canonical matrices.

4. Which of the linear transformations of problem 7, § 1 make $\mathbb{R}^4$ into a cyclic space? For each such transformation find a generator. For which of these transformations is $\mathbb{R}^4$ irreducible?

5. Let $E_1 \subset E$ be a stable subspace under φ and consider the induced mappings $\varphi_1: E_1 \to E_1$ and $\bar{\varphi}: E/E_1 \to E/E_1$. Let μ, μ_1, $\bar{\mu}$ be the corresponding minimum polynomials.

a) Prove that E is cyclic if and only if

i) E_1 is cyclic

ii) E/E_1 is cyclic

iii) $\mu = \mu_1 \bar{\mu}$.

In particular conclude that every subspace of a cyclic space is again cyclic.

b) Construct examples in which conditions i), ii), iii) respectively fail, while the remaining two continue to hold.

Hint: Use problem 5, § 1.

6. Let $E = \sum_{j=1}^{s} F_j$ be a decomposition of E into subspaces and suppose $\varphi_j: F_j \to F_j$ are linear transformations with minimum polynomials μ_j.

Define a linear transformation φ of E by

$$\varphi = \varphi_1 \oplus \cdots \oplus \varphi_s$$

a) Prove that E is cyclic if and only if each φ_j is cyclic and the μ_j are relatively prime.

b) Conclude that if E is cyclic, then each F_j is a sum of generalized eigenspaces for φ.

c) Prove that if E is cyclic and

$$a = a_1 + \cdots + a_s \qquad a_j \in F_j$$

is any vector in E, then a generates E if and only if a_j generates F_j $(j = 1 \ldots s)$.

7. Suppose $F \subset E$ is stable under φ and let $\varphi_F : F \to F$ (minimum polynomial μ_F) and $\bar{\varphi} : E/F \to E/F$ (minimum polynomial $\bar{\mu}$) be the induced transformations. Show that E is irreducible if and only if

i) E/F is irreducible.

ii) F is irreducible.

iii) $\mu_F = f^k, \bar{\mu} = f^l, \mu = f^{k+l}$ where f is an irreducible polynomial.

8. Suppose that E is irreducible with respect to φ. Let f^k (f irreducible) be the minimum polynomial of φ.

a) Prove that the k subspaces $K(f), \ldots, K(f^{k-1})$ are the only non-trivial stable subspaces of E

b) Conclude that

$$\operatorname{Im} f(\varphi)^{\kappa} = K(f^{k-\kappa}) \qquad 0 \leqq \kappa \leqq k.$$

9. Find necessary and sufficient conditions that E have no non-trivial stable subspaces.

10. Let ∂ be a differential operator in E.

a) Show that in any decomposition of E into irreducible subspaces, each subspace has dimension 1 or 2.

b) Let N_j be the number of j-dimensional irreducible subspaces in the above decomposition ($j = 1, 2$). Show that

$$N_1 + 2N_2 = \dim E \quad \text{and} \quad N_1 = \dim H(E).$$

c) Using part b) prove that two differential operators in E are conjugate if and only if the corresponding homology spaces coincide.

Hint: Use Proposition II, sec. 13.18.

11. Show that two linear transformations of a 3-dimensional vector space are conjugate if and only if they have the same minimum polynomial.

12. Let $\varphi: E \to E$ be a linear transformation. Show that there exists a (not necessarily unique) multiplication in E such that

i) E is an associative commutative algebra

ii) E contains a subalgebra A isomorphic to $\Gamma(\varphi)$

iii) If $\Phi: \Gamma(\varphi) \xrightarrow{\cong} A$ is the isomorphism, then

$$\Phi(\psi) \cdot x = \psi x \qquad \psi \in \Gamma(\varphi), x \in E.$$

13. Let E be irreducible (and hence cyclic) with respect to φ. Show that the set S of generators of the cyclic space E is not a subspace. Construct a subspace F such that S is in $1-1$ correspondence with the non-zero elements of E/F.

14. Let φ be a linear transformation of a real vector space having distinct eigenvalues, all negative. Show that φ can not be written in the form $\varphi = \psi^2$.

§ 5. Applications of cyclic spaces

In this paragraph we shall apply the theory developed in the preceding paragraph to obtain three important, independent theorems.

13.19. Generalized eigenspaces. Direct sums of the generalized eigenspaces of φ are characterized by the following

Theorem I: Let

$$E = F_1 \oplus \cdots \oplus F_s \tag{13.45}$$

be any decomposition of E into a direct sum of stable subspaces. Then the following three conditions are equivalent:

(i) Each F_j is a direct sum of some of the generalized eigenspaces E_i of φ.

(ii) The projection operators ϱ_j in E associated with the decomposition (13.45) are polynomials in φ.

(iii) Every stable subspace $U \subset E$ satisfies

$$U = \sum_j U \cap F_j.$$

Proof: Suppose that (i) holds. Then the projection operators ϱ_j are sums of the projection operators associated with the decomposition of E into generalized eigenspaces, and so it follows from sec. 13.5 that they are polynomials in φ. Thus (i) implies (ii).

Now suppose that (ii) holds, and let $U\subset E$ be any stable subspace. Then since $\sum_j \varrho_j = \iota$, we have

$$U \subset \sum_j \varrho_j U.$$

Since U is stable under ϱ_j it follows that $\varrho_j U \subset U \cap F_j$; whence

$$U \subset \sum_j U \cap F_j.$$

The inclusion in the other direction is obvious. Thus (ii) implies (iii).

Finally, suppose that (iii) holds. To show that (i) must also hold we first prove

Lemma I: Suppose that (iii) holds, and let

$$E = E_1 \oplus \cdots \oplus E_r$$

be the decomposition of E into generalized eigenspaces. Then to every i, $(i=1,\ldots,r)$ there corresponds precisely one integer j, $(1 \leq j \leq s)$ such that

$$E_i \cap F_j \neq 0.$$

Proof of lemma I: Suppose first that $E_i \cap F_j = 0$ for a fixed i and for every j $(1 \leq j \leq s)$. Then from (iii) we obtain

$$E_i = \sum_{j=1}^{s} E_i \cap F_j = 0$$

which is clearly false (cf. sec. 13.4). Hence there is at least one j such that $E_i \cap F_j \neq 0$.

To prove that there is at most one j (for any fixed i) such that $E_i \cap F_j \neq 0$, we shall assume that for some i, j_1, j_2,

$$E_i \cap F_{j_1} \neq 0 \quad \text{and} \quad E_i \cap F_{j_2} \neq 0$$

and derive a contradiction. Without loss of generality we may assume that

$$E_1 \cap F_1 \neq 0 \quad \text{and} \quad E_1 \cap F_2 \neq 0.$$

Choose two non-zero vectors

$$y_1 \in E_1 \cap F_1 \quad \text{and} \quad y_2 \in E_1 \cap F_2.$$

Then since $y_1, y_2 \in E_1$ we have that

$$f_1^{k_1}(\varphi)\, y_1 = 0 = f_1^{k_1}(\varphi)\, y_2$$

where $\mu = f_1^{k_1} \dots f_r^{k_r}$, f_i irreducible, is the decomposition of μ. Let l_1 and l_2 be the least integers such that $f_1^{l_1}(\varphi)y_1 = 0$ and $f_1^{l_2}(\varphi)y_2 = 0$. We may assume that $l_1 = l_2$. In fact, if $l_1 > l_2$ we simply replace y_1 by the vector $f_1^{l_1 - l_2}(\varphi)y_1$. Then

$$f_1^{l_1}(\varphi) f_1^{l_1 - l_2}(\varphi) y_1 = 0 \quad \text{and} \quad f_1^k(\varphi) f_1^{l_1 - l_2}(\varphi) y_1 \neq 0 \quad \text{for} \quad k < l_2 .$$

Now set

$$y = y_1 + y_2$$

and let Y be the cyclic subspace generated by y. Clearly $Y \subset F_1 \oplus F_2$, and so in view of (iii) we obtain

$$Y = Y \cap F_1 \oplus Y \cap F_2 .$$

It will now be shown that $Y \cap F_1 = 0$.

Let $u \in Y \cap F_1$ be any vector. Since $u \in Y$ we have that

$$u = f(\varphi) y = f(\varphi) y_1 + f(\varphi) y_2$$

for some polynomial f. Since $u \in F_1$, it follows that

$$f(\varphi) y_2 = u - f(\varphi) y_1 \in F_1 \cap F_2 = 0$$

and hence $d(\varphi) y_2 = 0$ where d is the greatest common divisor of f and $f_1^{k_1}$. Thus we obtain $d = f_1^p$ where

$$k_1 \geqq p \geqq l_1 .$$

But $d \mid f$; whence $f_1^{l_1} \mid f$, and so $u = f(\varphi)y = f(\varphi)y_1 + f(\varphi)y_2 = 0$.
This proves that $Y \cap F_1 = 0$. A similar argument shows that $Y \cap F_2 = 0$ so that

$$Y = Y \cap F_1 \oplus Y \cap F_2 = 0 .$$

This is the desired contradiction, and it completes the proof of the lemma.

We now revert to the proof of the theorem. Recall that we assume that (iii) holds, and are required to prove (i). In view of the above lemma we can define a set mapping

$$\tau : (1, \dots, r) \to (1, \dots, s)$$

such that

$$E_i \cap F_{\tau(i)} \neq 0 \quad i = 1, \dots, r \quad \text{and} \quad E_i \cap F_j = 0 \qquad \begin{matrix} j \neq \tau(i) \\ i = 1, \dots, r . \end{matrix}$$

Then (iii) yields that

$$E_i = \sum_{j=1}^{s} E_i \cap F_j = E_i \cap F_{\tau(i)} .$$

Finally, the relation

$$E = \sum_{i=1}^{r} E_i = \sum_i E_i \cap F_{\tau(i)} \subset \sum_i F_{\tau(i)} \subset \sum_j F_j = E$$

implies that

$$\sum_i F_{\tau(i)} = \sum_j F_j \quad \text{and} \quad F_j = \sum_{i \in \tau^{-1}(j)} E_i .$$

Hence τ is a surjection, and for every integer j $(1 \le j \le s)$. F_j is a direct sum of some of the E_i and so (i) is proved. Thus (iii) implies (i), and the proof of the theorem is complete.

13.20. Cayley-Hamilton theorem. It is the purpose of this section to prove the

Theorem II: (Cayley-Hamilton) Let χ denote the characteristic polynomial of φ. Then

$$\mu | \chi$$

or, equivalently, φ satisfies its own characteristic equation.

Before proceeding to the proof of this theorem we establish some elementary results.

Suppose $\lambda \in \Gamma$ is any scalar, and let ν denote the minimum polynomial of $\varphi - \lambda\iota$. Assume further that ν has degree m, and let ν be given explicitly by

$$\nu = t^m + \sum_{j=0}^{m-1} \beta_j t^j . \tag{13.46}$$

Then

$$\begin{aligned} 0 &= (\varphi - \lambda\iota)^m + \sum_{j=0}^{m-1} \beta_j (\varphi - \lambda\iota)^j \\ &= \varphi^m + \sum_{j=0}^{m-1} \alpha_j \varphi^j \\ &= f(\varphi) \end{aligned} \tag{13.47}$$

where

$$f = t^m + \sum_{j=0}^{m-1} \alpha_j t^j .$$

It follows that μ / f, and so in particular

$$\deg \mu \le \deg f = \deg \nu . \tag{13.48}$$

On the other hand,

$$\varphi = (\varphi - \lambda\iota) + \lambda\iota$$

and thus a similar argument shows that

$$\deg \nu \leq \deg \mu .$$

This, together with (13.48) implies that

$$\deg \nu = \deg f = \deg \mu . \tag{13.49}$$

Since $f|\mu$ and f has leading coefficient 1, we obtain that

$$f = \mu .$$

In particular, since $f = \nu(g)$, where $g = t - \lambda$, we have

$$\mu(\lambda) = \nu(g(\lambda)) = \nu(0) = \beta_0 . \tag{13.50}$$

Lemma II: Suppose E is cyclic and $\dim E = m$, and let χ be the characteristic polynomial for φ. Then

$$\chi = (-1)^m \mu .$$

Proof: Let $\lambda \in \Gamma$ be any scalar, and let ν be the minimum polynomial for $\varphi - \lambda \iota$. Since E is cyclic (with respect to φ), Theorem I, sec. 13.11 implies that

$$\deg \mu = \dim E = m .$$

Now we obtain from (13.49) that $\deg \nu = m$, and so a second application of Theorem I, sec. 13.11 shows that E is cyclic with respect to $\varphi - \lambda \iota$.

Let a be any generator of E (with respect to $\varphi - \lambda \iota$). Then

$$a, (\varphi - \lambda \iota) a, \ldots, (\varphi - \lambda \iota)^{m-1} a$$

is a basis for E (cf. Proposition III, sec. 13.10).

Now suppose that Δ is a non-trivial determinant function for E. Then

$$\begin{aligned} \chi(\lambda) \cdot \Delta(a, (\varphi - \lambda \iota) a, \ldots, (\varphi - \lambda \iota)^{m-1} a) & \\ = \det(\varphi - \lambda \iota) \cdot \Delta(a, \ldots, (\varphi - \lambda \iota)^{m-1} a) & \\ = \Delta((\varphi - \lambda \iota) a, \ldots, (\varphi - \lambda \iota)^m a) & \qquad (13.51) \\ = (-1)^{m-1} \Delta((\varphi - \lambda \iota)^m a, (\varphi - \lambda \iota) a, \ldots, (\varphi - \lambda \iota)^{m-1} a) . & \end{aligned}$$

On the other hand, if (13.46) gives the minimum polynomial of $\varphi - \lambda \iota$, we obtain that

$$(\varphi - \lambda \iota)^m a = - \sum_{j=0}^{m-1} \beta_j (\varphi - \lambda \iota)^j a$$

and substitution in (13.51) yields the relation

$$\begin{aligned}\chi(\lambda)\cdot\Delta(a,\ldots,(\varphi-\lambda\iota)^{m-1}a) &= (-1)^m\sum_{j=0}^{m-1}\beta_j\Delta((\varphi-\lambda\iota)^j a,(\varphi-\lambda\iota)a,\ldots,(\varphi-\lambda\iota)^{m-1}a)\\ &= (-1)^m\beta_0\Delta(a,(\varphi-\lambda\iota)a,\ldots,(\varphi-\lambda\iota)^{m-1}a).\end{aligned}$$

It follows that

$$\chi(\lambda)=(-1)^m\beta_0$$

and in view of (13.50) we obtain that

$$\chi(\lambda)=(-1)^m\mu(\lambda)\qquad\lambda\in\Gamma. \tag{13.52}$$

Finally, since (13.52) holds for every $\lambda\in\Gamma$, we can conclude (cf. sec. 12.10) that

$$(-1)^m\mu=\chi.$$

Proof of Theorem II: According to Proposition V, sec. 13.11 there exists a cyclic subspace $E_a\subset E$ such that

$$\dim E_a=\deg\mu.$$

By Lemma I, sec. 13.12, E_a has a complementary stable subspace F,

$$E=E_a\oplus F.$$

Let χ_a and χ_F be the characteristic polynomials of the linear transformations induced in E_a and in F by φ. Then

$$\chi=\chi_a\chi_F$$

(cf. sec. 4.21).

On the other hand, the minimum polynomial of the linear transformation induced in E_a by φ is μ as follows from the corollary to Theorem I, sec. 13.11. Now the lemma implies that

$$\chi_a=\pm\mu.$$

Hence, $\mu|\chi$.

13.21.* The commutant of φ. The *commutant of* φ, $C(\varphi)$, is the subalgebra of $A(E;E)$ consisting of all the linear transformations that commute with φ.

Let f be any polynomial. Then $K(f)$ is stable under every $\psi\in C(\varphi)$. In fact if $y\in K(f)$ is any vector, then

$$f(\varphi)\psi y=\psi f(\varphi)y=0\qquad\psi\in C(\varphi)$$

and so $\psi y\in K(f)$.

Next suppose that $\psi \in C(\varphi)$ is any linear transformation. Consider the decompositions of E into generalized eigenspaces of φ and of ψ,

$$E = E_1 \oplus \cdots \oplus E_r \quad (\text{for } \varphi)$$

and

$$E = F_1 \oplus \cdots \oplus F_s \quad (\text{for } \psi)$$

and the corresponding projection operators in E, π_i and ϱ_j. Since the mappings π_i and ϱ_j are respectively polynomials in φ and ψ (cf. sec. 13.5) it follows that

$$\pi_i \circ \varrho_j = \varrho_j \circ \pi_i \qquad \begin{array}{l} i = 1, \dots, r \\ j = 1, \dots, s. \end{array}$$

Now define linear transformations τ_{ij} in E by

$$\tau_{ij} = \pi_i \circ \varrho_j .$$

Then we obtain that

$$\tau_{ij}^2 = \pi_i \circ \varrho_j \circ \pi_i \circ \varrho_j = \pi_i^2 \circ \varrho_j^2 = \pi_i \circ \varrho_j = \tau_{ij}$$

and hence the τ_{ij} are again projection operators in E.

Since

$$\operatorname{Im} \tau_{ij} \subset E_i \cap F_j$$

and

$$\sum_{i,j} \tau_{ij} = \Big(\sum_i \pi_i\Big) \circ \Big(\sum_j \varrho_j\Big) = \iota$$

it follows that

$$E = \sum_{i,j} \operatorname{Im} \tau_{ij} \subset \sum_{i,j} E_i \cap F_j \subset E$$

whence

$$\operatorname{Im} \tau_{ij} = E_i \cap F_j$$

and

$$E = \sum_{i,j} E_i \cap F_j .$$

Proposition I: Let $E = F_1 \oplus \cdots \oplus F_s$ be any decomposition of E as a direct sum of subspaces. Then the subspaces F_j are stable under φ if and only if the projection operators σ_j are contained in $C(\varphi)$.

Proof: Since

$$F_j = \bigcap_{l \neq j} \ker \sigma_l$$

it follows that the F_j are stable under φ if the $\sigma_j \in C(\varphi)$. Conversely, if the F_j are stable under φ we have for each $y \in F_j$ that $\varphi y \in F_j$, and hence

$$\sigma_j \varphi y = \varphi y = \varphi \sigma_j y$$

while

$$\sigma_l \varphi y = 0 = \varphi \sigma_l y \qquad l \neq j .$$

Thus the σ_l commute with φ.

13.22.* The bicommutant of φ. The *bicommutant*, $C^2(\varphi)$, of φ is the subalgebra of $A(E; E)$ consisting of all the linear transformations which commute with every linear transformation in $C(\varphi)$.

Theorem III: $C^2(\varphi)$ coincides with the linear transformations which are polynomials in φ; $C^2(\varphi) = \Gamma(\varphi)$.

Proof: Clearly

$$C^2(\varphi) \supset \Gamma(\varphi).$$

Conversely, suppose $\psi \in C^2(\varphi)$ is any linear transformation and let

$$E = F_1 \oplus \cdots \oplus F_s \qquad (13.53)$$

be a decomposition of E into cyclic subspaces with respect to φ. A decomposition (13.53) exists by Theorem II, sec. 13.12. Let $\{a_i\}$ be any fixed generators of the spaces F_i.

Denote by φ_i the linear transformation in F_i induced by φ, and let μ_i be the minimum polynomial of φ_i. Then (cf. sec. 13.2) $\mu_i | \mu$ so we can write

$$\mu = \mu_i \nu_i \qquad i = 1, \ldots, s .$$

In view of Proposition V and the corollary to Theorem I, sec. 13.11 we may (and do) assume that $\mu_1 = \mu$.

Now the F_i are stable subspaces of E (under φ), and so by Proposition I, sec. 13.21 the projection operators in E associated with (13.53) commute with φ. Hence they commute with ψ as well, and so a second application of Proposition I shows that the F_i are stable under ψ. In particular, $\psi a_i \in F_i$. Since F_i is cyclic with respect to φ we can write

$$\psi a_i = g_i(\varphi) a_i \qquad i = 1, \ldots, s .$$

Thus if $h(\varphi) a_i \in F_i$ is an arbitrary vector in F_i we obtain

$$\psi h(\varphi) a_i = h(\varphi) \psi a_i = h(\varphi) g_i(\varphi) a_i = g_i(\varphi) h(\varphi) a_i$$

since φ and ψ commute. It follows that

$$\psi_i = g_i(\varphi) \qquad i = 1, \ldots, s$$

where ψ_i denotes the restriction of ψ to F_i. In the following it will be shown that

$$\psi = g_1(\varphi)$$

thus proving the theorem.

Consider now linear transformations $\chi_i\,(2 \leqq i \leqq s)$ in E defined by

$$\chi_i x = x \qquad x \in F_j \quad j \neq i$$
$$\chi_i f(\varphi) a_i = f(\varphi) v_i(\varphi) a_1 .$$

To show that χ_i is well-defined it is clearly sufficient to prove that

$$f(\varphi) v_i(\varphi) a_1 = 0 \quad \text{whenever} \quad f(\varphi) a_i = 0 .$$

But if $f(\varphi)a_i = 0$, then $\mu_i | f$ and so $\mu = \mu_i v_i$ divides $v_i f$: whence

$$f(\varphi) v_i(\varphi) = 0 .$$

The relation

$$\chi_i \varphi f(\varphi) a_i = \varphi f(\varphi) v_i(\varphi) a_1 = \varphi \chi_i f(\varphi) a_i$$

shows that χ_i commutes with φ, and hence with ψ. On the other hand we have that

$$\chi_i \psi a_i = \chi_i g_i(\varphi) a_i = g_i(\varphi) v_i(\varphi) a_1 = v_i(\varphi) g_i(\varphi) a_1$$

and

$$\psi \chi_i a_i = \psi v_i(\varphi) a_1 = v_i(\varphi) \psi a_1 = v_i(\varphi) g_1(\varphi) a_1$$

whence

$$v_i(\varphi)[g_i(\varphi) - g_1(\varphi)] a_1 = 0 .$$

This relation implies that $\mu_1 | v_i(g_i - g_1)$. But $\mu_1 = \mu$ and so

$$\mu \,|\, v_i(g_i - g_1) .$$

Since $\mu = v_i \mu_i$, we obtain that

$$\mu_i \,|\, g_i - g_1 .$$

This last relation yields that for any vector $x \in F_i$,

$$\psi x = g_i(\varphi) x = g_1(\varphi) x \qquad i = 2, \ldots, s .$$

It follows that

$$\psi = g_1(\varphi)$$

which completes the proof.

Problems

1. Let

$$\mu = f_1^{k_1} \ldots f_r^{k_r}$$

be the decomposition of the minimum polynomial of φ, and let E_i be the generalized eigenspaces. If f_i has degree p_i denote by N_{ij} the number of irreducible subspaces of E_i of dimension $p_i j$ $(1 \leqq j \leqq k_i)$. Set

$$l_i = \sum_{j=1}^{k_i} j N_{ij}.$$

Show that the characteristic polynomial of φ is given by

$$\chi = f_1^{l_1} \dots f_r^{l_r}.$$

2. Prove that E is cyclic if and only if

$$\chi = \pm \mu.$$

3. Let φ be a linear transformation of E and assume that $E = \sum_j F_j$ is a decomposition of E as a direct sum of stable subspaces. If each F_j is a sum of generalized eigenspaces, prove that each F_j is stable under every $\psi \in C(\varphi)$. Conversely, assume that each F_j is stable under every $\psi \in C(\varphi)$ and prove that each F_j is a sum of generalized eigenspaces of φ.

4. a) Show that the only projection operators in $C(\varphi)$ are ι and 0 if and only if E is irreducible with respect to φ.

b) Show that the set of projection operators in $C(\varphi)$ is a subset of $C^2(\varphi)$ if and only if E is cyclic with respect to φ.

5. a) Define $C^3(\varphi)$ to be the set of all linear transformations in E commuting with every transformation in $C^2(\varphi)$. Prove that

$$C^3(\varphi) = C(\varphi).$$

b) Prove that $C^2(\varphi) = C(\varphi)$ if and only if E is cyclic.

6. Let E be cyclic with respect to φ and S be the set of generators of E. Let G be the set of linear automorphisms in $C(\varphi)$.

a) Prove that G is a group.

b) Prove that for each $\psi \in G$, S is stable under ψ and the restriction ψ_s of ψ to S is a bijection. Show that ψ_s has no fixed points if $\psi \neq \iota$.

c) Let $a \in S$ be a fixed generator and let $\sigma, \tau \in G$ be arbitrary. Prove that $\sigma = \tau$ if and only if

$$\sigma a = \tau a$$

and hence in particular, $\sigma = \tau$ if and only if $\sigma_s = \tau_s$.

d) Prove that G acts transitively on S; i.e. for each $a, b \in S$ there is a $\psi \in G$ such that $\psi a = b$.

e) Conclude that if $a \in S$ is a generator, then the mapping $\Phi: G \to S$ given by $\psi \to \psi(a)$ is a bijection.

7. Let ∂ be a differential operator in E. Consider the set I of transformations $\psi \in C(\partial)$ such that $\psi Z(E) \subset B(E)$ (cf. sec. 6.7).

Show that I is an ideal in $C(\partial)$ and establish an algebra isomorphism

$$C(\partial)/I \xrightarrow{\cong} A(H(E); H(E)).$$

§ 6. Nilpotent and semisimple transformations

13.23. Nilpotent transformations. A linear transformation, φ, is called *nilpotent* if $\varphi^k = 0$ for some integer k or equivalently, if its minimal polynomial has the form

$$\mu = t^m.$$

The exponent m is called the *degree* of φ. It follows from sec. 13.7 that φ is nilpotent if and only if the Fitting null component is the entire space. It is clear that the restriction of a nilpotent transformation to a stable subspace is again nilpotent.

Suppose now that φ and ψ are two commuting nilpotent transformations. Then the transformations $\varphi + \psi$ and $\psi \circ \varphi$ are again nilpotent. In fact, if k and l denote the degrees of φ and ψ, then

$$(\varphi + \psi)^{k+l} = \sum_{j=0}^{k} \binom{k+l}{j} \varphi^j \psi^{k+l-j} + \sum_{j=k+1}^{k+l} \binom{k+l}{j} \varphi^j \psi^{k+l-j} = 0$$

and

$$(\psi \varphi)^k = \psi^k \varphi^k = 0$$

which proves that $\varphi + \psi$ and $\psi \circ \varphi$ are nilpotent.

Assume that E is *irreducible* with respect to the nilpotent transformation φ. Then it follows from sec. 13.14 that the Jordan canonical matrix of φ has the form

$$\begin{pmatrix} 0 & 1 & & 0 \\ & 0 & \ddots & \\ & & \ddots & 1 \\ 0 & & & 0 \end{pmatrix}. \qquad (13.54)$$

Hence, the Jordan canonical matrix of any nilpotent transformation consists of matrices of the form (13.54) following each other along the main diagonal.

Suppose now that φ is any linear transformation, and that μ has the decomposition

$$\mu = f_1^{k_1} \dots f_r^{k_r}.$$

Let f be the polynomial

$$f = f_1 \dots f_r.$$

Then if h is any polynomial, $h(\varphi)$ is nilpotent if and only if $f|h$, as follows at once from sec. 12.12.

13.24. Semisimple transformations. A linear transformation φ is called *semisimple* if every stable subspace $E_1 \subset E$ has a complementary stable subspace.

Example I: Let E be a Euclidean space and φ be a rotation of E. Since the orthogonal complement of every stable subspace is stable (cf. sec. 8.19) it follows that φ is semisimple.

Example II: In a Euclidean space every selfadjoint and every skew transformation is semisimple, as follows from a similar argument.

Example III: Let E be a unitary space. Then every unitary and every selfadjoint transformation is semisimple.

Let φ be a semisimple transformation and suppose that E_1 is a stable subspace. Then the restriction φ_1 of φ to E_1 is semisimple. In fact, suppose $F_1 \subset E_1$ is stable under φ_1. Then F_1 is stable under φ, and hence there exists a complementary stable subspace, F_2, in E,

$$E = F_1 \oplus F_2.$$

Intersection with E_1 yields

$$E_1 = F_1 \oplus (F_2 \cap E_1).$$

Since $F_2 \cap E_1$ is (clearly) stable under φ_1 it follows that φ_1 is semisimple.

Proposition I: Suppose φ is semisimple, and let f be any polynomial. Then $f(\varphi)$ is nilpotent if and only if $f(\varphi)=0$.

Proof: The if part is trivial. Suppose now that $f(\varphi)$ is nilpotent of degree k. Then $K(f^{k-1})$ is stable under φ and so we can write

$$E = K(f^{k-1}) \oplus F$$

where F is stable under φ, and hence under $f(\varphi)$. On the other hand, it is clear that $f(\varphi)F \subset K(f^{k-1})$ whence

$$f(\varphi)F \subset K(f^{k-1}) \cap F = 0$$

i.e.,

$$F \subset K(f).$$

It follows that $E=K(f^l)$ where $l=\max(k-1,1)$. Since k is the degree of nilpotency of $f(\varphi)$ we have

$$l \geq k$$

whence $k=l=1$. Hence $f(\varphi)=0$.

Corollary I: If φ is simultaneously nilpotent and semisimple, then $\varphi=0$.

The major result on semisimple transformations obtained in this section is the following criterion:

Theorem I: A linear transformation is semisimple if and only if its minimum polynomial is the product of relatively prime irreducible polynomials (or equivalently, if the polynomials μ and μ' are relatively prime).

Remark: Theorem I shows that a linear transformation φ is semisimple if and only if it is a semisimple element of the algebra $\Gamma(\varphi)$ (cf. sec. 12.16).

Proof: Suppose φ is semisimple. Consider the decompositions

$$\mu = f_1^{k_1} \dots f_r^{k_r} \qquad f_i \text{ irreducible and relatively prime}$$

of the minimum polynomial, and set

$$f = f_1 \cdots f_r.$$

Then $f(\varphi)$ is nilpotent, and hence by Proposition I of this section, $f(\varphi)=0$. It follows that $\mu | f$. Since $f | \mu$ by definition, we have

$$\mu = f_1 \cdots f_r.$$

This proves the only if part of the theorem.

To prove the other part of the theorem we consider first the special case that the minimum polynomial, μ, of φ is irreducible. To show that φ is semisimple consider the subalgebra, $\Gamma(\varphi)$, of $A(E;E)$ generated by φ and ι. Since μ is irreducible, $\Gamma(\varphi)$ is a field (cf. sec. 12.13). $\Gamma(\varphi)$ contains Γ and hence it is an extension field of Γ, and E may be considered as a vector space over $\Gamma(\varphi)$ (cf. § 3, Chapt. V). Since a subspace of the $\Gamma(\varphi)$-vector space E is stable under φ if and only if it is stable under every transformation of $\Gamma(\varphi)$, it follows that the stable subspaces of E are precisely the $\Gamma(\varphi)$-subspaces of the $\Gamma(\varphi)$-vector space E. Since every subspace of a vector space has a complementary subspace it follows that φ is semisimple.

Now consider the general case

$$\mu = f_1 \dots f_r \qquad f_i \text{ irreducible and relatively prime}.$$

Then we have the decomposition

$$E = E_1 \oplus \dots \oplus E_r$$

of E into generalized eigenspaces. Since the minimum polynomial of the induced transformation $\varphi_i: E_i \to E_i$ is precisely f_i (cf. sec. 13.4) it follows from the above result that φ_i is semisimple. Now let $F \subset E$ be a stable subspace. Then we have, in view of sec. 13.6,

$$F = F \cap E_1 \oplus \dots \oplus F \cap E_r.$$

Clearly $F \cap E_i$ is a stable subspace of E_i and hence there exists a stable complementary subspace H_i,

$$E_i = (F \cap E_i) \oplus H_i.$$

These equations yield

$$E = \sum_i (F \cap E_i) \oplus \sum_i H_i = F \oplus H, \qquad H = \sum_i H_i.$$

Since H is a stable subspace of E it follows that φ is semisimple.

Corollary I: Let φ be any linear transformation and assume that

$$E = F_1 \oplus \dots \oplus F_k$$

is a decomposition of E into stable subspaces such that the induced transformations $\varphi_i: F_i \to F_i$ are semisimple. Then φ is semisimple.

Proof: Let μ_i be the minimum polynomial of the induced transformation $\varphi_i: F_i \to F_i$. Since φ_i is semisimple each μ_i is a product of relatively prime irreducible polynomials. Hence, the least common multiple, f, of the μ_i is again a product of such polynomials. But $f(\varphi)$ annihilates E and hence the minimum polynomial, μ, of φ divides f. It follows that μ is a product of relatively prime irreducible polynomials. Now Theorem I implies that φ is semisimple.

Proposition II: Let $\Delta \subset \Gamma$ be a subfield, and assume that E, considered as a Δ-vector space has finite dimension. Then every (Γ-linear) transformation, φ, of E which is semisimple as a Δ-linear transformation is semisimple considered as Γ-linear transformation.

Proof: Let μ_Δ be the minimum polynomial of φ considered as a Δ-linear transformation. It follows from Theorem I that μ_Δ and μ'_Δ are relatively prime. Hence there are polynomials $g, r \in \Delta[t]$ such that

$$q\,\mu_\Delta + r\,\mu'_\Delta = 1\,. \tag{13.55}$$

On the other hand, every polynomial over $\Delta[t]$ may be considered as a polynomial in $\Gamma[t]$. Since $\mu_\Delta(\varphi) = 0$ we have

$$\mu_\Gamma \mid \mu_\Delta$$

where μ_Γ denotes the minimum polynomial of the (Γ-linear) transformation φ. Hence we may write

$$\mu_\Delta = \mu_\Gamma\, h \quad \text{some} \quad h \in \Gamma[t]$$

and so

$$\mu'_\Delta = \mu'_\Gamma\, h + \mu_\Gamma\, h'\,. \tag{13.56}$$

Combining (13.55) and (13.56) we obtain

$$q\,h\,\mu_\Gamma + h'\,r\,\mu_\Gamma + h\,r\,\mu'_\Gamma = 1$$

whence

$$(q\,h + h'\,r)\,\mu_\Gamma + (h\,r)\,\mu'_\Gamma = 1\,.$$

This relation shows that the polynomials μ_Γ and μ'_Γ are relatively prime. Now Theorem I implies that the Γ-linear transformation φ is semisimple.

Theorem II: Every linear transformation φ can be written in the form

$$\varphi = \varphi_S + \varphi_N$$

where φ_S is semisimple and φ_N is nilpotent. φ_S and φ_N are polynomials in φ and their minimum polynomials are given by

$$\mu_S = f_1 \cdots f_r$$

and

$$\mu_N = t^k, \qquad k = \max(k_1, \ldots, k_r).$$

Moreover, if

$$\varphi = \psi_S + \psi_N$$

is any decomposition of φ into a semisimple and nilpotent transformation such that $\psi_S \circ \psi_N = \psi_N \circ \psi_S$, then

$$\psi_S = \varphi_S \quad \text{and} \quad \psi_N = \psi_N\,.$$

Proof: For the existence apply Theorem I and Theorem IV (sec. 12.16) with $A = \Gamma(\varphi)$.

To prove the uniqueness let $\varphi=\psi_S+\psi_N$ be any decomposition of φ into a semisimple and a nilpotent transformation such that ψ_S commutes with ψ_N. Then the subalgebra of $A(E;E)$ generated by ι, ψ_S and ψ_N is commutative and contains φ. Now apply the uniqueness part of Theorem IV, sec. 12.16.

13.25. The Jordan normal form of a semisimple transformation. Suppose that E is irreducible with respect to a semisimple transformation φ. Then it follows from sec. 13.13 and Theorem I sec. 13.24 that the minimum polynomial of φ has the form

$$\mu = f$$

where f is irreducible. Hence the Jordan canonical matrix of φ has the form (cf. see. 13.14)

$$\begin{pmatrix} 0 & 1 & & & 0 \\ & 0 & 1 & & \\ & & \ddots & \ddots & \\ 0 & & & & 1 \\ -\alpha_0 & -\alpha_1 & \cdots & & -\alpha_{p-1} \end{pmatrix} \tag{13.57}$$

where

$$\mu = \sum_{\nu=0}^{p} \alpha_\nu t^\nu \quad \alpha_p = 1 \quad p = \deg \mu .$$

It follows that the Jordan canonical matrix of an arbitrary semisimple transformation consists of submatrices of the form (13.57) following each other along the main diagonal.

Now consider the special case that E is irreducible with respect to a semisimple transformation whose minimum polynomial is completely reducible. Then we have that $p=1$ and hence E has dimension 1. It follows that if φ is a semisimple transformation with completely reducible minimum polynomial, then E is the direct sum of stable subspaces of dimension 1; i.e., E has a basis of eigenvectors. The matrix of φ with respect to this basis is of the form

$$\begin{pmatrix} \lambda_1 & & & 0 \\ & \lambda_2 & & \\ & & \ddots & \\ 0 & & & \lambda_n \end{pmatrix} \tag{13.58}$$

where the λ_i are the (not necessarily distinct) eigenvalues of φ. A linear transformation with a matrix of the form (13.58) is called *diagonalizable*. Thus semisimple linear transformations with completely reducible minimum polynomial are diagonalizable.

Finally let φ be a semisimple transformation of a *real* vector space E. Then a similar argument shows that E is the direct sum of irreducible subspaces of dimension 1 or 2.

13.26.* The commutant of a semisimple transformation.

Theorem III: The commutant $C(\varphi)$ of a semisimple transformation φ is a direct sum of ideals (in the algebra $C(\varphi)$) each of which is isomorphic to the full algebra of transformations of a vector space over an extension field of Γ.

Proof: Let

$$E = E_1 \oplus \cdots \oplus E_r \tag{13.59}$$

be the decomposition of E into the generalized eigenspaces. It follows from sec. 13.21 that the eigenspaces E_i are stable under every transformation $\psi \in C(\varphi)$. Now let $I_j \subset C(\varphi)$ be the subspace consisting of all transformations ψ such that

$$\psi : E_k \to 0 \qquad k \neq j.$$

Since E_i is stable under each element of $C(\varphi)$ it follows that I_j is an ideal in the algebra $C(\varphi)$. As an immediate consequence of the definition, we have

$$I_j \cap \sum_{k \neq j} I_k = 0 \qquad j = 1, \ldots, r. \tag{13.60}$$

Now let $\psi \in C(\varphi)$ be arbitrary and consider the projection operators $\pi_i : E \to E$ associated with the decomposition (13.59).
Then

$$\psi = \sum_i \pi_i \psi = \sum_i \pi_i^2 \psi = \sum_i \pi_i \psi \pi_i = \sum_i \psi_i \tag{13.61}$$

where

$$\psi_i = \pi_i \psi \pi_i . \tag{13.62}$$

It follows from (13.62) that $\psi_i \in I_i$. Hence formulae (13.60) and (13.61) imply that

$$C(\varphi) = \sum_i I_i .$$

It is clear that

$$I_i \cong C(\varphi_i) \qquad \varphi_i \text{ is the restriction of } \varphi \text{ to } E_i$$

where the isomorphism is obtained by restricting a transformation $\psi \in I_i$ to E_i.

Now consider the transformations $\varphi_i: E \to E$ induced by φ. Since the minimum polynomial of φ_i is irreducible it follows that $\Gamma(\varphi_i)$ is a field. Considering E as a vector space over $\Gamma(\varphi_i)$ we obtain from chap. V, § 3 that

$$C(\varphi_i) = A_{\Gamma(\varphi_i)}(E; E).$$

13.27. Semisimple sets of linear transformations. A set $\{\varphi_\alpha\}$ of linear transformations of E will be called *semisimple* if to every subspace $F_1 \subset E$ which is stable under each φ_α there exists a complementary subspace F_2 which is stable under each φ_α.

Suppose now that $\{\varphi_\alpha\}$ is any set of linear transformations and let $A \subset A(E; E)$ be the subalgebra generated by the φ_α. Then clearly, a subspace $F \subset E$ is stable under each φ_α if and only if it is stable under each $\psi \in A$. In particular, the set $\{\varphi_\alpha\}$ is semisimple if and only if the algebra A is semisimple.

Theorem IV: Let $\{\varphi_\alpha\}$ be a set of commuting semisimple transformations. Then $\{\varphi_\alpha\}$ is a semisimple set.

Proof: We first consider the case of a finite set of transformations $\varphi_1, \ldots, \varphi_s$ and proceed by induction on s. If $s=1$ the theorem is trivial. Suppose now it holds for $s-1$ and assume for the moment that the minimum polynomial of φ_1 is irreducible. Then E may be considered as a $\Gamma(\varphi_1)$-vector space. Since the $\varphi_i (i=2, \ldots, s)$ commute with φ_1 they may be considered as $\Gamma(\varphi_1)$-linear transformations (cf. Chap. V, § 3). Moreover, Proposition II, sec. 13.24 implies that the φ_i, considered as $\Gamma(\varphi_1)$-linear transformations, are again semisimple.

Now let $F_1 \subset E$ be any subspace stable under the $\varphi_i (i=1, \ldots, s)$. Then since F_1 is stable under φ_1, it is a $\Gamma(\varphi_1)$-subspace of E. Hence, by the induction hypothesis, there exists a $\Gamma(\varphi_1)$-subspace of E, F_2, which is stable under $\varphi_2, \ldots, \varphi_s$ and such that

$$E = F_1 \oplus F_2.$$

Since F_2 is a $\Gamma(\varphi_1)$-subspace, it is also stable under φ_1 and so it is a stable subspace complementary to F_1.

Let the minimum polynomial μ_1 of φ_1 be arbitrary. Since φ_1 is semisimple, we have

$$\mu_1 = f_1 \cdots f_r \qquad f_j \text{ irreducible and relatively prime}.$$

Let

$$E = E_1 \oplus \cdots \oplus E_r$$

be the corresponding decomposition of E into the generalized eigenspaces of φ_1.

Now assume that $F_1 \subset E$ is a subspace stable under each φ_i $(i=1, \ldots, s)$.

According to sec. 13.21 each E_j is stable under each φ_i. It follows that the subspaces $F_1 \cap E_j$ are also stable under every φ_i. Moreover the restrictions of the φ_i to each E_j are again semisimple (cf. sec. 13.24) and in particular, the restriction of φ_1 to E_j has as minimum polynomial the irreducible polynomial f_j. Thus it follows that the restrictions of the φ_i to E_j form a semisimple set, and hence there exist subspaces $F^j \subset E_j$ which are stable under each φ_i and which satisfy

$$E_j = (F_1 \cap E_j) \oplus F^j \qquad j = 1, \ldots, r.$$

Setting

$$F_2 = \sum_j F^j$$

we have that F_2 is stable under each φ_i, and that

$$E = F_1 \oplus F_2.$$

This closes the induction, and completes the proof for the case that the $\{\varphi_\alpha\}$ are a finite set.

If the set $\{\varphi_\alpha\}$ is infinite consider the subalgebra $A \subset A(E; E)$ generated by the φ_α. Then A is a commutative algebra and hence every subset of A consists of commuting transformations. In view of the discussion in the beginning of this section it is sufficient to construct a semisimple system of generators for A. But A is finite dimensional and so has a finite system of generators. Hence the theorem is reduced to the case of a finite set.

Theorem IV has the following converse:

Theorem V: Suppose $A \subset A(E; E)$ is a commutative semisimple set. Then for each $\varphi \in A$, φ is a semisimple transformation.

Proof: Let $\varphi \in A$ be arbitrary and consider the decomposition

$$\mu = f_1^{k_1} \ldots f_r^{k_r}$$

of its minimum polynomial μ. Define a polynomial, g, by

$$g = f_1 \ldots f_r.$$

Since the set A is commutative, $K(g)$ is stable under every $\psi \in A$. Hence there exists a subspace $E_1 \subset E$ which is stable under every $\psi \in A$ such that

$$E = K(g) \oplus E_1 . \tag{13.63}$$

Now let

$$h = f_1^{k_1-1} \dots f_r^{k_r-1} .$$

Then we have

$$h(\varphi) E \subset K(g) .$$

On the other hand, since E_1 is stable under $h(\varphi)$,

$$h(\varphi) E_1 \subset E_1$$

whence

$$h(\varphi) E_1 \subset K(g) \cap E_1 = 0 .$$

It follows that $E_1 \subset K(h)$.

Now consider the polynomial

$$p = f_1^{l_1} \dots f_r^{l_r} \quad \text{where} \quad l_i = \max(k_i - 1, 1) .$$

Since $l_i \geq k_i - 1$ it follows that

$$p(\varphi) x = 0 \qquad x \in E_1 \tag{13.64}$$

and from $l_i \geq 1$ we obtain

$$p(\varphi) x = 0 \qquad x \in K(g) . \tag{13.65}$$

In view of (13.63), (13.64) and (13.65) imply that $p(\varphi)=0$ and so $\mu|p$. Now it follows that

$$\max(k_i - 1, 1) \geq k_i \quad i = 1, \dots, r$$

whence

$$k_i = 1 \qquad i = 1, \dots, r .$$

Hence φ is semisimple.

Corollary: If φ and ψ are two commuting semisimple transformations, then $\varphi + \psi$ and $\psi\varphi$ are again semisimple.

Proof: Consider the subalgebra $A \subset A(E;E)$ generated by φ and ψ Then Theorem IV implies that A is a semisimple set. Hence it follows from Theorem V that $\varphi + \psi$ and $\psi\varphi$ are semisimple.

Problems

1. Let φ be nilpotent, and let N_λ be the number of subspaces of dimension λ in a decomposition of E into irreducible subspaces. Prove that

$$\dim \ker \varphi = \sum_\lambda N_\lambda .$$

2. Let φ be nilpotent of degree k in a 6-dimensional vector space E. For each $k\,(1 \le k \le 6)$ determine the possible ranks of φ and show that k and $r(\varphi)$ determine the numbers N_λ (cf. problem 1) explicitly. Conclude that two nilpotent transformations φ and ψ are conjugate if and only if

$$r(\varphi) = r(\psi) \quad \text{and} \quad \deg\varphi = \deg\psi\,.$$

3. Suppose φ is nilpotent and let $\varphi^*: E^* \leftarrow E^*$ be the dual mapping. Assume that E is cyclic with respect to φ and that φ is of degree k. Let a be a generator of E. Prove that E^* is cyclic with respect to φ^* and that φ^* is of degree k. Let $a^* \in E^*$ be any vector. Show that

$$\langle a^*, \varphi^{k-1} a \rangle \neq 0$$

if and only if a^* is a generator of E^*.

4. Prove that a linear transformation φ with minimum polynomial μ is diagonalizable if and only if

i) μ is completely reducible

ii) φ is semisimple

Show that i) and ii) are equivalent to

$$\mu = (t - \lambda_1) \ldots (t - \lambda_r)$$

where the λ_i are distinct scalars.

5. a) Prove that two commuting diagonalizable transformations are simultaneously diagonalizable; i.e., there exists a basis of E with respect to which both matrices are diagonal.

b) Use a) to prove that if φ and ψ are commuting semisimple transformations of a complex space, then $\varphi + \psi$ and $\varphi\psi$ are again semisimple.

6. Suppose φ is a linear transformation of a complex space E. Let $E = \sum_i E_i$ be the decomposition of E into generalized eigenspaces, and let π_i be the corresponding projection operators. Assume that the minimum polynomial of the induced transformation $\varphi_i: E_i \to E_i$ is $(t - \lambda_i)^{k_i}$. Prove that the semisimple part of φ is given by

$$\varphi_S = \sum_i \lambda_i \pi_i\,.$$

7. Let E be a complex vector space and φ be a linear transformation with eigenvalues $\lambda_\nu\,(\nu = 1, \ldots, n)$, not necessarily distinct. Given an arbitrary polynomial f prove directly that the linear transformation $f(\varphi)$ has the eigenvalues $f(\lambda_\nu)$ $(\nu = 1 \ldots n)$ $(n = \dim E)$.

8. Give an example of a semisimple set of linear transformations which contains transformations that are not semisimple.

9. Let A be an algebra of commuting linear transformations in a complex space E.

a) Construct a decomposition $E=E_1\oplus\cdots\oplus E_r$ such that for any $\varphi\in A$, E_i is stable under φ and the minimum polynomial of the induced transformation $\varphi_i: E_i\to E_i$ is of the form

$$(t-\lambda_i(\varphi))^{k_i(\varphi)}.$$

b) Show that the mapping $A\to\mathbb{C}$ given by

$$\varphi\to\lambda_i(\varphi)\qquad \varphi\in A$$

is a linear function in A. Prove that λ_i preserves products and so it is a homomorphism.

c) Show that the nilpotent transformations in A form an ideal which is precisely rad A (cf. chap. V, § 2). Consider the subspace T of $L(A)$ generated by the λ_i. Prove that

$$\operatorname{rad} A = T^{\perp}.$$

d) Prove that the semisimple transformations in A form a subalgebra, A_s. Consider the linear functions λ_i^s in A_s obtained by restricting λ_i to A_s. Show that they generate (linearly) the dual space $L(A_s)$. Prove that the mapping $\lambda_i\to\lambda_i^s$ is a linear isomorphism. $T\xrightarrow{\cong} L(A_s)$.

10. Assume that E is a complex vector space. Prove that every commutative algebra of semisimple transformations is contained in an n-dimensional commutative algebra of semisimple transformations ($n=\dim E$).

11. Calculate the semisimple and nilpotent parts of the linear transformations of problem 7, § 1.

§ 7. Applications to inner product spaces

In this concluding paragraph we shall apply our general decomposition theorems to inner product spaces. Decompositions of an inner product space into irreducible subspaces with respect to selfadjoint mappings, skew mappings and isometries have already been constructed in chap. VIII.

Generalizing these results we shall now construct a decomposition for a *normal* transformation. Since a complex linear space is fully reducible with respect to a normal endomorphism (cf. sec. 11.10) we can restrict ourselves to real inner product spaces.

13.28. Normal transformations. Let E be an inner product space and $\varphi: E \to E$ be a normal transformation (cf. sec. 8.5). It is clear that every polynomial in φ is again normal. Moreover since the rank of φ^k ($k=2, 3 \dots$) is equal to the rank of φ, it follows that φ is nilpotent only if $\varphi = 0$.

Now consider the decomposition of the minimum polynomial into its prime factors,

$$\mu = f_1^{k_1} \dots f_r^{k_r} \tag{13.66}$$

and the corresponding decomposition of E into the generalized eigenspaces,

$$E = E_1 \oplus \dots \oplus E_r. \tag{13.67}$$

Since the projection operators π_i associated with the decomposition (13.67) are polynomials in φ they are normal. On the other hand $\pi_i^2 = \pi_i$ and so it follows from sec. 8.11 that the π_i are selfadjoint. Now let $x_i \in E_i$ and $x_j \in E_j$ be arbitrary. Then

$$(x_i, x_j) = (x_i, \pi_j x_j) = (\pi_j x_i, x_j) = 0 \qquad i \neq j;$$

i.e., the decomposition (13.67) is orthogonal.

Now consider the induced transformations $\varphi_i: E_i \to E_i$. It follows from sec. 8.5 that the φ_i are again normal and hence so are the transformations $f_i(\varphi_i)$. On the other hand, $f_i(\varphi_i)$ is nilpotent. It follows that $f_i(\varphi_i) = 0$ and hence all the exponents in (13.66) are equal to 1. Now Theorem I of sec. 13.24 implies that a normal transformation is semisimple.

Theorem I: Let E be an inner product space. Then a linear transformation φ is normal if and only if

i) the generalized eigenspaces are mutually orthogonal.

ii) The restrictions $\varphi_i: E_i \to E_i$ are homothetic (cf. sec. 8.19).

Proof: Let φ be a normal transformation. It has been shown already that the spaces E_i are mutually orthogonal. Now consider the minimum polynomial, f_i, of the induced transformation φ_i. Since f_i is irreducible over $\mathbb{R}$ it follows that either

$$f_i = t - \lambda_i \qquad \lambda_i \in \mathbb{R} \tag{13.68}$$

or

$$f_i = t^2 + \alpha_i t + \beta_i \quad \alpha_i^2 - 4\beta_i < 0 \qquad \alpha_i, \beta_i \in \mathbb{R}. \tag{13.69}$$

In the first case we have that $\varphi_i = \lambda_i \iota$ and so φ_i is homothetic. Now consider the case (13.69). Then φ_i satisfies the relation

$$\varphi_i^2 + \alpha_i \varphi_i + \beta_i \iota = 0$$

and hence the proof is reduced to showing that a normal transformation $\varphi: E \to E$ which satisfies

$$\varphi^2 + \alpha\varphi + \beta\iota = 0 \quad \alpha, \beta \in \mathbb{R}, \quad \alpha^2 - 4\beta < 0 \tag{13.70}$$

is homothetic. (Thus, if φ is a rotation, it must be proper.)
We prove first that $\tilde{\varphi} - \varphi$ is regular. In fact, let K be the kernel of $\tilde{\varphi} - \varphi$. If $z \in K$ is an arbitrary vector, we have $\tilde{\varphi} z = \varphi z$ whence

$$\tilde{\varphi}(\varphi z) = (\tilde{\varphi}\varphi) z = (\varphi\tilde{\varphi}) z = \varphi(\tilde{\varphi} z) = \varphi(\varphi z).$$

It follows that K is stable under φ and hence stable under $\tilde{\varphi}$. Clearly the restriction of φ to K is selfadjoint. Hence, if $K \neq 0$, φ has an eigenvector in K which contradicts the hypothesis $\alpha^2 - 4\beta < 0$. Consequently, $K = 0$.

Equation (13.70) implies that

$$\tilde{\varphi}^2 + \alpha\tilde{\varphi} + \beta\iota = 0. \tag{13.71}$$

Multiplying (13.70) and (13.71) respectively by $\tilde{\varphi}$ and φ and subtracting we find that

$$(\tilde{\varphi}\varphi - \beta\iota)(\tilde{\varphi} - \varphi) = 0$$

whence, in view of the regularity of $\tilde{\varphi} - \varphi$,

$$\tilde{\varphi}\varphi = \beta\iota. \tag{13.72}$$

Define a transformation, τ, by

$$\tau = \frac{1}{\sqrt{\beta}}\varphi$$

(notice that $\alpha^2 - 4\beta < 0$ implies that $\beta > 0$). Then (13.72) yields $\tilde{\tau}\tau = \iota$ and so τ is a rotation. This proves that every normal mapping satisfies i) and ii). The converse follows immediately from sec. 8.5.

Corollary I: If φ is a normal transformation then the orthogonal complement of a stable subspace is stable.

Proof: Let F be a stable subspace. In view of sec. 13.6 we have

$$F = (E_1 \cap F) \oplus \cdots \oplus (E_r \cap F).$$

Clearly the subspace $E_i \cap F$ is stable under the restriction, φ_i, of φ to E_i. Since φ_i is homothetic it follows that the orthogonal complement, H_i, of $E_i \cap F$ in E_i is again stable under φ_i. Hence, the space $H = \sum_i H_i$ is stable under φ. On the other hand the equations

$$E_i = (E_i \cap F) \oplus H_i$$

yield

$$E = F \oplus H \qquad H = F^{\perp}.$$

Hence $F^{\perp}$ is a stable subspace.

As an immediate consequence of Theorem I we obtain

Theorem II: Let E be an inner product space. Then a linear transformation φ is normal if and only if E can be written as the sum of mutually orthogonal irreducible subspaces such that the restriction of φ to every subspace is homothetic.

13.29. Semisimple transformations of a real vector space. In sec. 13.28 it has been shown that every normal transformation of an inner product space is semisimple. Conversely, let $\varphi: E \to E$ be a semisimple transformation of a real vector space. Then a positive definite inner product can be introduced in E such that φ becomes a normal mapping. To prove this let

$$E = \sum_j F_j$$

be a decomposition of E into irreducible subspaces. In view of Theorem II it is sufficient to define a positive inner product in each F_j such that the restrictions, φ_j, of φ to F_j are homothetic. In fact, we simply extend these inner products to an inner product in E such that the F_j are mutually orthogonal.

Now let F be one of the irreducible subspaces. Since φ is semisimple, F has dimension 1 or 2. If $\dim F = 1$ we choose the inner product in F arbitrarily. If $\dim F = 2$ there exists a basis a, b in F such that

$$\varphi a = b, \quad \varphi b = -\beta a - \alpha b, \quad \alpha^2 - 4\beta < 0$$

(cf. sec. 13.16). Define the inner product by

$$(a, a) = 1, \quad (a, b) = -\frac{\alpha}{2}, \quad (b, b) = \beta.$$

Then we have for every vector

$$x = \xi a + \eta b$$

of F

$$(x, x) = \xi^2 - \alpha \xi \eta + \beta \eta^2.$$

Since $\alpha^2 - 4\beta < 0$ it follows that $(x, x) \geqq 0$ and equality holds only for $x = 0$. Moreover, since

$$(\varphi a, \varphi a) = \beta = \beta(a, a), \quad (\varphi a, \varphi b) = -\frac{\alpha\beta}{2} = \beta(a, b),$$

$$\text{and} \quad (\varphi b, \varphi b) = \beta^2 = \beta(b, b)$$

it follows that

$$|\varphi x|^2 = \beta |x|^2 \qquad x \in F.$$

This equation shows that φ is homothetic and so the proof is complete.

13.30. Lorentz-transformations. As a second example we shall construct a decomposition of the Minkowski-space into irreducible subspaces with respect to a Lorentz-transformation φ (cf. sec. 9.27). For the sake of simplicity we assume that the Lorentz-transformation is proper orthochronous. The condition $\tilde{\varphi} = \varphi^{-1}$ implies that the inverse of every eigenvalue is again an eigenvalue. Since there exists at least one eigenvalue (cf. sec. 9.27) the minimum polynomial, μ, of φ has at least one real root. Now we distinguish three cases:

I. The minimum polynomial μ contains a prime factor

$$t^2 + \alpha t + \beta \qquad \alpha^2 - 4\beta < 0$$

of second degree. Then consider the mapping

$$\tau = \varphi^2 + \alpha \varphi + \beta \iota .$$

The kernel of τ is a stable subspace F of even dimension and containing no eigenvectors. Since φ has an eigenvector in E, $E \neq F$. Thus F has necessarily dimension 2 and hence it is a plane. The intersection of the plane F and the light-cone consists of two straight lines, one straight line, or the point 0 only. The two first cases are impossible because the plane F does not contain eigenvectors (cf. sec. 9.26). Thus the inner product must be positive definite in F and the induced transformation φ_1 is a proper Euclidean rotation. (An improper rotation of F would have eigenvectors.) Now consider the orthogonal complement $F^\perp$. The restriction of the inner product to $F^\perp$ has index 1. Hence $F^\perp$ is a pseudo-Euclidean plane. Denote by φ_2 the induced transformation of $F^\perp$. The equation

$$\det \varphi = \det \varphi_1 \det \varphi_2$$

implies that $\det \varphi_2 = +1$, showing that φ_2 is a proper pseudo-Euclidean rotation. Choosing orthonormal bases in F and in $F^\perp$ we obtain an orthonormal basis of E in which the matrix of φ has the form

$$\begin{pmatrix} \cos\omega & \sin\omega & & 0 \\ -\sin\omega & \cos\omega & & \\ & & \cosh\theta & \sinh\theta \\ 0 & & \sinh\theta & \cosh\theta \end{pmatrix} \qquad \omega \neq 0, \pi.$$

II. The minimum polynomial is completely reducible, and not all its roots are equal to 1. Then φ has eigenvalues $\lambda \neq 1$ and $\frac{1}{\lambda} \neq 1$. Let e and e' be corresponding eigenvectors

$$\varphi e = \lambda e \quad \varphi e' = \frac{1}{\lambda} e'.$$

The condition $\lambda \neq 1$ implies that e and e' are light-vectors. These vectors can be chosen to be linearly independent (even if $\lambda = -1$), whence $(e, e') \neq 0$ (cf. sec. 9.21). Let F be the plane generated by e and e' and let

$$z = \xi e + \eta e'$$

be any vector of F. Then

$$(z, z) = 2(e, e')\xi\eta.$$

This equation shows that the induced inner product has index 1. The orthogonal complement $F^{\perp}$ is therefore a Euclidean plane and the induced mapping is a Euclidean rotation. The angle of this rotation must be 0 or π, because otherwise the minimum polynomial of φ would contain an irreducible factor of second degree. Select orthonormal bases in F and $F^{\perp}$. These two bases form an orthonormal basis of E in which the matrix of φ has the form

$$\begin{pmatrix} \cosh\theta & \sinh\theta & & 0 \\ \sinh\theta & \cosh\theta & & \\ & & \varepsilon & 0 \\ 0 & & 0 & \varepsilon \end{pmatrix} \qquad \begin{matrix} \theta \neq 0 \\ \varepsilon = \pm 1 \end{matrix}$$

III. The minimum polynomial of φ has the form

$$\mu = (t-1)^k \qquad (1 \le k \le 4).$$

If $k=1$, φ reduces to the identity map. Next, it will be shown that the case $k=2$ is impossible. If $k=2$, applying φ^{-1} to the equation $(\varphi - \iota)^2 = 0$ yields

$$\varphi + \tilde{\varphi} = 2\iota$$

whence

$$(x, \varphi x) = (x, x) \qquad x \in E.$$

Inserting for x a light-vector, l, we see that $(l, \varphi l) = 0$. But two light-vectors can be orthogonal only if they are linearly dependent. We thus obtain $\varphi l = \lambda l$. Since φ does not have eigenvalues $\lambda \neq 1$, it follows that $\varphi l = l$ for all light-vectors l. But this implies that φ is the identity. Hence the minimal polynomial is $t-1$ in contradiction to our assumption $k=2$.

Now consider the case $k \geq 3$. As has been shown in sec. 9.27 there exists an eigenvector e on the light-cone. The orthogonal complement E_1 of e is a 3-dimensional subspace of E which contains the light-vector e. The induced inner product has rank and index 2 (cf. sec. 9.21). Let F be a 2-dimensional subspace of E in which the inner product is positive definite. Selecting an orthonormal basis e_1, e_2 in F we can write

$$\begin{aligned} \varphi e_1 &= e_1 \cos\omega + e_2 \sin\omega + \alpha_1 e \\ \varphi e_2 &= -e_1 \sin\omega + e_2 \cos\omega + \alpha_2 e \\ \varphi e &= e\,. \end{aligned} \tag{13.73}$$

The coefficients α_1 and α_2 are not both zero. In fact, if $\alpha_1 = 0$ and $\alpha_2 = 0$ the plane F is invariant under φ and we have the direct decomposition $E = F \oplus F^{\perp}$ of E into two 2-dimensional invariant subspaces. This would imply that $k \leq 2$.

Now consider the characteristic polynomial, χ_1, of the induced mapping $\varphi_1 : E_1 \to E_1$. Computing the characteristic polynomial from the matrix (13.73) we find that

$$\chi_1 = (t^2 - 2t\cos\omega + 1)(1 - t)\,. \tag{13.74}$$

At the same time we know that

$$\chi_1 = (1 - t)^3\,. \tag{13.75}$$

Comparing the polynomials (13.74) and (13.75) we find that $\omega = 0$. Hence, equations (13.73) reduce to

$$\begin{aligned} \varphi e_1 &= e_1 + \alpha_1 e \\ \varphi e_2 &= e_2 + \alpha_2 e \\ \varphi e &= e\,. \end{aligned}$$

Now consider the vector

$$y = \alpha_1 e_2 - \alpha_2 e_1\,.$$

Then

$$(y, y) = \alpha_1^2 + \alpha_2^2 > 0$$

and

$$\varphi y = \alpha_1 \varphi e_2 - \alpha_2 \varphi e_1 = \alpha_1 (e_2 + \alpha_2 e) - \alpha_2 (e_1 + \alpha_1 e) = y\,.$$

In other words, y is a space-like eigenvector of φ. Denote by Y the 1-dimensional subspace generated by y. Then we have the orthogonal decomposition

$$E = Y \oplus Y^{\perp}$$

into two invariant subspaces. The orthogonal complement $Y^{\perp}$ is a 3-dimensional pseudo-Euclidean space with index 2.

The subspace $Y^{\perp}$ is irreducible with respect to φ. This follows from our hypothesis that the degree of the minimal polynomial μ is ≥ 3. On the other hand, μ can not have degree 4 because then the space E would be irreducible.

Combining our results we see that the decomposition of a Minkowski-space with respect to a proper orthochronous Lorentz-transformation φ has one of the following forms:

I. E is completely reducible. Then φ is the identity.

II. E is the direct sum of an invariant Euclidean plane and an invariant pseudo-Euclidean plane. These planes are irreducible except for the case where the induced mappings are $\pm\iota$ (Euclidean plane) or ι (pseudo-Eudidean plane)

III. E is the direct sum of a space-like 1-dimensional stable subspace (eigenvalue 1) and an irreducible subspace of dimension 3 and index 2.

Problems

1. Suppose E is an n-dimensional vector space over Γ. Assume that a symmetric bilinear function $E\times E\to\Gamma$ is defined such that $\langle x, x\rangle \neq 0$ whenever $x\neq 0$.

a) Prove that $\langle\,,\rangle$ is a scalar product and thus E is self dual.

b) If $F\subset E$ is a subspace show that

$$E = F\oplus F^{\perp}.$$

c) Suppose $\varphi: E\to E$ is a linear transformation such that $\varphi\varphi^* = \varphi^*\varphi$. Prove that φ is semisimple.

2. Let φ be a linear transformation of a unitary space. Prove that φ is normal if and only if, for some polynomial f,

$$\tilde{\varphi} = f(\varphi).$$

3. Suppose φ is a linear transformation of a complex vector space such that $\varphi^k = \iota$ for some integer k. Show that E can be made into a unitary space such that φ becomes a unitary mapping.

4. Let E be a real linear space and let φ be a linear transformation of E. Prove that a positive definite inner product can be introduced in E such that φ becomes a normal mapping if and only if the following conditions are satisfied:

a) The space E can be decomposed into invariant subspaces of dimension 1 and 2.

b) If τ is the induced mapping in an irreducible subspace of dimension 2, then

$$\tfrac{1}{4}(\operatorname{tr}\tau)^2 - \det\tau < 0 .$$

5. Consider a 3-dimensional pseudo-Euclidean space E with the index 2. Let l_i $(i=1, 2, 3)$ be three light-vectors such that

$$(l_i, l_j) = 1 \qquad i \neq j .$$

Define a linear transformation, φ, by the equations

$$\begin{aligned}
\varphi l_1 &= l_1 \\
\varphi l_2 &= \alpha(\alpha-1)l_1 + \alpha l_2 + (1-\alpha)l_3 \qquad \alpha \neq 1 \\
\varphi l_3 &= (\alpha-2)(\alpha-1)l_1 + (\alpha-1)l_2 + (2-\alpha)l_3 .
\end{aligned}$$

Prove that φ is a rotation and that E is irreducible with respect φ.

6. Show that a real 3-dimensional vector space cannot be irreducible with respect to a semisimple linear transformation. Conclude that the pseudo-Euclidean rotation of problem 5 is not semisimple. Use this to show that a linear transformation in a self dual space which satisfies $\varphi\varphi^* = \varphi^*\varphi$ is not necessarily semisimple (cf. problem 1).

7. Let a Lorentz-transformation φ be defined by the matrix

$$\begin{pmatrix}
\frac{1}{2} & \frac{2}{3} & -1 & -\frac{5}{6} \\
\frac{2}{3} & \frac{1}{9} & \frac{4}{3} & \frac{10}{9} \\
1 & -\frac{4}{3} & 1 & \frac{5}{3} \\
\frac{5}{6} & -\frac{10}{9} & \frac{5}{3} & \frac{43}{18}
\end{pmatrix}$$

Construct a decomposition of E into irreducible subspaces.

8. Consider the group G of Lorentz transformations.

Let e be a time-like unit vector and F be the orthogonal complement of e. Consider the subgroup $H \subset G$ consisting of all Lorentz transformations φ such that $\varphi F = F$.

a) Prove that H is a compact subgroup.

b) Prove that H is not properly contained in a compact subgroup of G.

Hint: Show first that if K is a compact subgroup of G and $\varphi \in K$, then every real eigenvalue of φ is ± 1.

Bibliography

[1] BAER, R. Linear Algebra and Projective Geometry, New York, Academic Press Inc. 1952.

[2] BELLMAN, R. Introduction to Matrix Analysis. Mac Graw-Hill Publ. Comp. Inc. 1960.

[3] BIRKHOFF, G., and S. MAC LANE. A Survey of Modern Algebra. New York, The MacMillan Co. 1944.

[4] BOURBAKI, N. Elements de mathematique, Première Partie, Livre II.

[5] CHEVALLEY, C. Fundamental Concepts of Algebra, Academic Press Inc., New York, 1956.

[6] CULLEN, C. Matrices and Linear Transformations, Addison-Wesley Publ. Comp. 1966.

[7] GANTMACHER, F. R. Matrizenrechnung. Berlin, Deutscher Verlag der Wissenschaften, 1958.

[8] GEL'FAND, I. M. Lectures on Linear Algebra, Interscience Tracts in Pure and Applied Mathematics No. 9, New York, 1961.

[9] GRÖBNER, W. Matrizenrechnung, München, R. Oldenbourg, 1956.

[10] HALMOS, P. Finite Dimensional Vector-spaces. Princeton, D. van Nostrand Co., 1958. Springer-Verlag, 1974.

[11] HALMOS, P. Naive Set Theory, D. van Nostrand Co., 1960. Springer-Verlag, 1974.

[12] HOFFMAN, K., and R. KUNZE. Linear Algebra. Prentice-Hall Inc., 1961.

[13] JACOBSON, N. Lectures in Abstract Algebra, II. Princeton, D. van Nostrand Co. 1953.

[14] JACOBSON, N. The Theory of Rings, American Math. Soc. 1943.

[15] KELLER, O. H. Analytische Geometrie und lineare Algebra. Berlin, Deutscher Verlag der Wissenschaften, 1957.

[16] KELLEY, J. General Topology, D. van Nostrand Co., 1955.

[17] KOWALSKI, H. Lineare Algebra, Walter de Gruyter Co., Berlin, 1963.

[18] KUIPER, N. Linear Algebra and Geometry. North-Holland Publ. Comp. Amsterdam, 1961.

[19] LICHNEROWICZ, A. Lineare Algebra und Lineare Analysis, Berlin, Deutscher Verlag der Wissenschaften, 1956.

[20] MAL'CEV, A. Foundations of Linear algebra, W. H. Freeman, San Francisco and London.

[21] NEVANLINNA, R. u. F. Absolute Analysis, Grundlehren, Math. Wiss. 102, Berlin, Göttingen-Heidelberg, Springer, 1958.

[22] NOMIZU, K. Fundamentals of Linear Algebra, MacGraw-Hill Book, Publ. Comp. Inc.

[23] PICKERT, G. Analytische Geometrie, Leipzig, Akademische Verlagsgesellschaft, 1953.

[24] REICHARDT, H. Vorlesungen über Vector- und Tensorrechnung, Berlin, Deutscher Verlag der Wissenschaften, 1957.

[25] SCHREIER, O. and E. SPERNER, Modern Algebra and Matrix Theory, New York, Chelsea Publ. Co., 1955.

[26] SHILOV, G. Theory of Linear Spaces, Prentice Hall Inc., 1961.

[27] Smirnow, W. I. Lehrgang der höheren Mathematik Teil III, 1, Berlin, Deutscher Verlag der Wiss., 1958.
[28] Stoll, R. Linear Algebra and Matrix Theory, London, New York-Toronto, McGraw-Hill Book Publ. Comp. Inc., 1952.
[29] Van der Waerden, B. L. Algebra, Grundlehren Math. Wiss. Berlin-Göttingen-Heidelberg, Springer, 1955.

Additions to the Bibliography

[7] Dickson, L. Linear Algebras: Cambridge University Press, 1914.
[8] Dieudonné, J. Algèbre linéaire et géométrie elémentaire: Hermann & Cie, Paris.
[15] Husemoller, D. Fibre Bundles: McGraw-Hill Book Company.
[22] Lang, S. Algebra: Addison Wesley 1965.
[23] Lang, S. Linear Algebra: Addison Wesley 1970.

Subject Index

Graduate Texts in Mathematics

continued from page ii

61 WHITEHEAD. Elements of Homotopy Theory.
62 KARGAPOLOV/MERLZJAKOV. Fundamentals of the Theory of Groups.
63 BOLLOBAS. Graph Theory.
64 EDWARDS. Fourier Series. Vol. I. 2nd ed.
65 WELLS. Differential Analysis on Complex Manifolds. 2nd ed.
66 WATERHOUSE. Introduction to Affine Group Schemes.
67 SERRE. Local Fields.
68 WEIDMANN. Linear Operators in Hilbert Spaces.
69 LANG. Cyclotomic Fields II.
70 MASSEY. Singular Homology Theory.
71 FARKAS/KRA. Riemann Surfaces. 2nd ed.
72 STILLWELL. Classical Topology and Combinatorial Group Theory. 2nd ed.
73 HUNGERFORD. Algebra.
74 DAVENPORT. Multiplicative Number Theory. 2nd ed.
75 HOCHSCHILD. Basic Theory of Algebraic Groups and Lie Algebras.
76 IITAKA. Algebraic Geometry.
77 HECKE. Lectures on the Theory of Algebraic Numbers.
78 BURRIS/SANKAPPANAVAR. A Course in Universal Algebra.
79 WALTERS. An Introduction to Ergodic Theory.
80 ROBINSON. A Course in the Theory of Groups. 2nd ed.
81 FORSTER. Lectures on Riemann Surfaces.
82 BOTT/TU. Differential Forms in Algebraic Topology.
83 WASHINGTON. Introduction to Cyclotomic Fields.
84 IRELAND/ROSEN. A Classical Introduction to Modern Number Theory. 2nd ed.
85 EDWARDS. Fourier Series. Vol. II. 2nd ed.
86 VAN LINT. Introduction to Coding Theory. 2nd ed.
87 BROWN. Cohomology of Groups.
88 PIERCE. Associative Algebras.
89 LANG. Introduction to Algebraic and Abelian Functions. 2nd ed.
90 BRØNDSTED. An Introduction to Convex Polytopes.
91 BEARDON. On the Geometry of Discrete Groups.
92 DIESTEL. Sequences and Series in Banach Spaces.
93 DUBROVIN/FOMENKO/NOVIKOV. Modern Geometry—Methods and Applications. Part I. 2nd ed.
94 WARNER. Foundations of Differentiable Manifolds and Lie Groups.
95 SHIRYAEV. Probability. 2nd ed.
96 CONWAY. A Course in Functional Analysis. 2nd ed.
97 KOBLITZ. Introduction to Elliptic Curves and Modular Forms. 2nd ed.
98 BRÖCKER/TOM DIECK. Representations of Compact Lie Groups.
99 GROVE/BENSON. Finite Reflection Groups. 2nd ed.
100 BERG/CHRISTENSEN/RESSEL. Harmonic Analysis on Semigroups: Theory of Positive Definite and Related Functions.
101 EDWARDS. Galois Theory.
102 VARADARAJAN. Lie Groups, Lie Algebras and Their Representations.
103 LANG. Complex Analysis. 3rd ed.
104 DUBROVIN/FOMENKO/NOVIKOV. Modern Geometry—Methods and Applications. Part II.
105 LANG. $SL_2(\mathbf{R})$.
106 SILVERMAN. The Arithmetic of Elliptic Curves.
107 OLVER. Applications of Lie Groups to Differential Equations. 2nd ed.
108 RANGE. Holomorphic Functions and Integral Representations in Several Complex Variables.
109 LEHTO. Univalent Functions and Teichmüller Spaces.
110 LANG. Algebraic Number Theory.
111 HUSEMÖLLER. Elliptic Curves.
112 LANG. Elliptic Functions.
113 KARATZAS/SHREVE. Brownian Motion and Stochastic Calculus. 2nd ed.
114 KOBLITZ. A Course in Number Theory and Cryptography. 2nd ed.
115 BERGER/GOSTIAUX. Differential Geometry: Manifolds, Curves, and Surfaces.
116 KELLEY/SRINIVASAN. Measure and Integral. Vol. I.
117 SERRE. Algebraic Groups and Class Fields.
118 PEDERSEN. Analysis Now.

119 ROTMAN. An Introduction to Algebraic Topology.
120 ZIEMER. Weakly Differentiable Functions: Sobolev Spaces and Functions of Bounded Variation.
121 LANG. Cyclotomic Fields I and II. Combined 2nd ed.
122 REMMERT. Theory of Complex Functions. *Readings in Mathematics*
123 EBBINGHAUS/HERMES et al. Numbers. *Readings in Mathematics*
124 DUBROVIN/FOMENKO/NOVIKOV. Modern Geometry—Methods and Applications. Part III.
125 BERENSTEIN/GAY. Complex Variables: An Introduction.
126 BOREL. Linear Algebraic Groups.
127 MASSEY. A Basic Course in Algebraic Topology.
128 RAUCH. Partial Differential Equations.
129 FULTON/HARRIS. Representation Theory: A First Course. *Readings in Mathematics*
130 DODSON/POSTON. Tensor Geometry.
131 LAM. A First Course in Noncommutative Rings.
132 BEARDON. Iteration of Rational Functions.
133 HARRIS. Algebraic Geometry: A First Course.
134 ROMAN. Coding and Information Theory.
135 ROMAN. Advanced Linear Algebra.
136 ADKINS/WEINTRAUB. Algebra: An Approach via Module Theory.
137 AXLER/BOURDON/RAMEY. Harmonic Function Theory.
138 COHEN. A Course in Computational Algebraic Number Theory.
139 BREDON. Topology and Geometry.
140 AUBIN. Optima and Equilibria. An Introduction to Nonlinear Analysis.
141 BECKER/WEISPFENNING/KREDEL. Gröbner Bases. A Computational Approach to Commutative Algebra.
142 LANG. Real and Functional Analysis. 3rd ed.
143 DOOB. Measure Theory.
144 DENNIS/FARB. Noncommutative Algebra.
145 VICK. Homology Theory. An Introduction to Algebraic Topology. 2nd ed.
146 BRIDGES. Computability: A Mathematical Sketchbook.
147 ROSENBERG. Algebraic K-Theory and Its Applications.
148 ROTMAN. An Introduction to the Theory of Groups. 4th ed.
149 RATCLIFFE. Foundations of Hyperbolic Manifolds.
150 EISENBUD. Commutative Algebra with a View Toward Algebraic Geometry.
151 SILVERMAN. Advanced Topics in the Arithmetic of Elliptic Curves.
152 ZIEGLER. Lectures on Polytopes.
153 FULTON. Algebraic Topology: A First Course.
154 BROWN/PEARCY. An Introduction to Analysis.
155 KASSEL. Quantum Groups.
156 KECHRIS. Classical Descriptive Set Theory.
157 MALLIAVIN. Integration and Probability.
158 ROMAN. Field Theory.
159 CONWAY. Functions of One Complex Variable II.
160 LANG. Differential and Riemannian Manifolds.
161 BORWEIN/ERDÉLYI. Polynomials and Polynomial Inequalities.
162 ALPERIN/BELL. Groups and Representations.

9 780387 901107